STO

EXPLORING
PHYSICAL
SCIENCE

ROBERT E. KILBURN
Science Curriculum Coordinator—Newton, Massachusetts

PETER S. HOWELL
Junior High Science Teacher—Needham, Massachusetts

ALLYN AND BACON, INC.
Boston   Rockleigh, N.J.   Atlanta   Dallas   Belmont, Calif.

ROBERT E. KILBURN has taught junior high science for nine years full-time, and for 16 years part-time in his present position as Science Curriculum Coordinator for the Newton, Massachusetts public schools. Prior to teaching, Dr. Kilburn spent four years as a science researcher, the last two as research biophysicist at the General Electric Research Laboratory.

PETER S. HOWELL has taught junior high science for 16 years. Besides teaching, Mr. Howell has been a contributing author and editor of textbooks in science, and has written science materials for educational television.

Special thanks go to William R. Radomski, Collaborator for the Energy unit. Presently Science Consultant for the Newton, Massachusetts public schools, Mr. Radomski has taught junior high science for nine years in New Haven, Connecticut and Newton.

Cover photo: H. Armstrong Roberts

Title page photo: Jerry Irwin

Editor: Gene Moulton
Designer: Debby Welling
Photo Researcher: Janice Thalin
Preparation Buyer: Patricia Hart

ISBN: 0-205-06738-7

Library of Congress Catalog Card Number 79-54812

1   2   3   4   5   6   7   8   9   88   87   86   85   84   83   82   81   80

PREFACE

The recent changes in science teaching are a response to the mushrooming growth of scientific information. Teachers can no longer depend upon easy generalizations as in the past; ideas become obsolete almost as soon as they are presented. Therefore, modern science programs emphasize the processes by which information is obtained and made meaningful. Such programs help young people "uncover" science rather than "cover" it.

Modern approaches to science teaching demand modern textbooks. This revision of EXPLORING PHYSICAL SCIENCE has the same two goals which made the previous editions successful: (1) to bring young people into contact with their environment in such a way as to stimulate a desire to investigate, and (2) to provide them with an understanding of the methods and philosophies of science so that they can carry out investigations with a minimum of direct guidance from others.

This major revision has incorporated the suggestions of hundreds of teachers who used the previous editions. Their suggestions included organizing the book into fewer topics. This change allows classes more time to focus on each major theme. Each major theme, or unit, has been divided into three parts. The first part consists of several chapters of guided investigations, designed to give students an accurate sense of the nature of science and its limitations. The second part of each unit consists of one chapter devoted to having students work independently on a research project of their own design. There can be no better way to understand the strengths and weaknesses of scientific investigation than to experience it first-hand. The final chapters of each unit utilize readings to extend the learnings of earlier chapters, and deal with important topics not possible to study first-hand. This unit structure allows students to (1) learn the skills of scientific investigation and knowledge-building, (2) practice the skills independently, and (3) utilize their understandings in readings on related topics. Critical attitudes and habits developed through independent thinking will stir continued interest in learning long after the years of formal schooling have ended.

The present authors are indebted to Dr. Walter A. Thurber whose vision and efforts led to the original editions of this series. We hope the present editions maintain the high quality associated with all his work.

CONTENTS

CHEMISTRY 1

BALANCED AND UNBALANCED FORCES   97

ELECTRICITY   289

ENERGY   385

CHEMISTRY

Working Safely
with Chemicals

In this unit, you will be working with laboratory equipment. In this way, you will learn about some of the tools chemists use. Good chemists must not only know which tools to use, but they must also learn how to use them safely.

1

In the science laboratory classroom, safety has two important parts. One part is the teacher's responsibility. THE OTHER PART IS YOUR RESPONSIBILITY! Most laboratory classroom accidents would never occur if students followed the rules of laboratory safety described in this chapter. Remember, good laboratory safety depends on you.

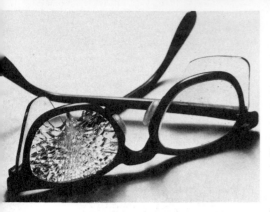

Safety Glasses. Few parts of your body are as important to you as are your eyes. Eyes are easily damaged by many different things. If you're looking into a boiling test tube, or watching someone break a rock apart, your eyes are right in the way if something comes flying at you. Whenever anyone around you is working with chemicals, hot solutions, or breaking rocks apart, protect your eyes with safety glasses.

You need two eyes . . . wear safety glasses!

What's wrong here?

What's wrong here?

What's wrong here?

Proper Clothing for Lab Work. Lab workers with loose-fitting clothes have more accidents than people who are properly dressed. People with long, loose hair are more likely to have their hair catch fire. These two ideas are shown in the drawings below. Before you start your laboratory work, think about these two ideas. If your hair is long, tie it back. Take off loose jackets and fix other loose-fitting clothes.

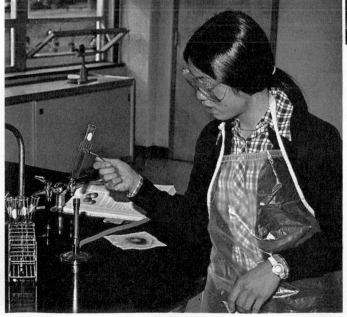

What's wrong here?

Alcohol Burners. Alcohol is a liquid which burns. In an alcohol burner, alcohol rises up the wick. At the top of the wick, heat from a match vaporizes the alcohol and ignites these vapors. The burning vapors are called a *flame*. Liquid alcohol evaporates quickly and ignites easily. For this reason, **never bring a container of alcohol near a flame. Never try to refill your alcohol burner.** This is a job for your teacher to do in a room away from all flames. The top of the burner should fit tightly so that the alcohol inside the burner cannot catch fire.

The use of alcohol burners causes more serious accidents in school than any other single cause of accidents. Some people say that alcohol burners cause more serious accidents than all other causes put together. Always work with the burner under the safest conditions you can set up. Do not handle or attempt to move a lighted burner. The little bit of extra effort thinking about being safe is worthwhile.

What's wrong here?

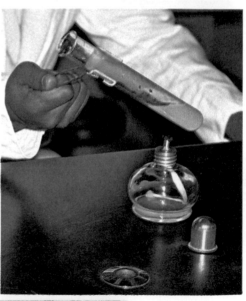

What's wrong here?

Time saved being careless is not worth being burned by an alcohol fire.

Heating Safely. Heating test tubes and beakers is an important part of science. But there is a right way to do it. The right way is easy. The wrong way can sometimes lead to serious accidents.

When heating liquids in a test tube, make sure the test tube is not cracked. Use a clamp or wire test tube holder. Do not squeeze the holder so that the test tube falls out! Do not fill the tube more than half full. Heat the test tube slowly at first, near the top of the liquid. If the bottom is heated, the contents may bubble up and out of the tube. **Never** heat the test tube with the end pointing at another person or at yourself.

Before heating anything, check with a teacher to find out if the heating container can withstand high temperatures. Never heat a container with a tightly sealed lid . . . it will explode.

What's wrong here?

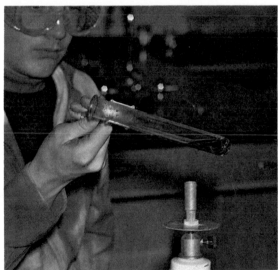

What's wrong here?

7

What's wrong here?

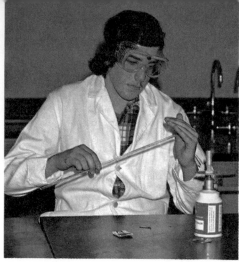

Working Safely with Hot Glass. Burns are a common accident in laboratories. One frequent source of burns is hot glassware. Glass is a very poor conductor of heat. If a piece of glass is heated, the heat does not spread out quickly. As a result, the heated place on the glass remains very hot for a long time. If you have been heating glass or other equipment, treat every piece as though it might be hot. What is wrong in the left drawing above?

Labeling Containers. Many chemicals and solutions look the same, but some are harmless and others are very dangerous. For this reason, it is very important that all chemicals and solutions be in containers with labels. Look at the label below. What useful information is given in each line of this label?

Only take chemicals from bottles which are labeled. Read the labels before pouring. Store chemicals and solutions in labeled containers. What is wrong in the drawing at the left?

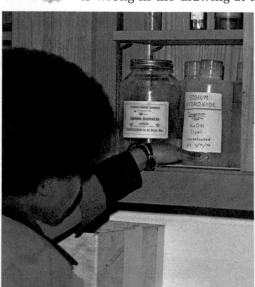

What's wrong here?

8

What's wrong here?

Spills and Breakage. When you are ready to work with chemicals, your teacher believes you are careful enough not to hurt yourself. Your teacher also believes that you are careful and considerate enough not to hurt others. Sometimes one student carelessly hurts another person by being a poor housekeeper. Two examples are (1) carelessly disposing of broken glass, and (2) carelessly disposing of used chemicals and hot equipment.

Most laboratories have special containers for broken glass so the school custodian or others won't get cut accidentally. Put used chemicals in special waste containers so that plumbing won't be damaged. Allow hot equipment to cool before putting it where others might get burned. If you break or spill something, quickly tell your teacher so that you can clean it up correctly.

What's wrong here?

What's wrong here?

Laboratory Techniques. Scientists and other laboratory workers learn hundreds of special skills or laboratory techniques while working. Some of these skills involve the safe handling of materials. Other skills conserve equipment and supplies. Still other skills are important in order to get accurate results. Only a few such skills are shown in this chapter. You will learn many other skills from your teacher in the later chapters of this unit.

Three examples of laboratory techniques are shown on this page. Study the drawing at the left. Why is this method of shaking unsafe? Why might this method of shaking produce a different result than expected?

The other drawings show examples which waste supplies and equipment, which means that money and natural resources are wasted. What is wrong in each case? What is being wasted?

What's wrong here?

What's wrong here?

Concentration vs. Horseplay. One cause of "accidents" in laboratories is shown in the drawing at the right. Why is this method of sitting dangerous to the student? Why is it dangerous to others sitting nearby? Why is unnecessary talking in a laboratory dangerous?

If accidents happened only to careless people, such accidents would be bad enough. Unfortunately, accidents sometimes happen to careful people because careless people near them cause accidents to others. Such accidents cannot be tolerated. If you act so that you endanger other people, you may expect to be isolated from others. In this way, accidents are less likely to happen to you, and others who are careful are also less likely to get hurt.

First Aid Resources. When accidents happen, quick action can often reduce the damage. Quick action occurs best when students and teachers know what resources are available and how to use

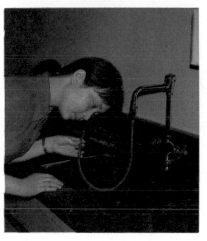

When eyes are splashed by chemicals, flood eyes with water for 15 minutes and call the doctor. Report all accidents.

Safety equipment such as safety blankets, fire extinguishers, and first-aid kits should be located in work areas close to where accidents might happen.

them. Look around your science room. What first aid resources are available? Ask your teacher for instructions in the use of any safety equipment you don't know how to use.

Some laboratory accidents include (1) cuts, (2) burns, (3) chemicals splattered in the eyes or on the skin or clothing, (4) spilled chemicals, (5) chemical or clothing fires. Discuss the first aid treatment for each type of accident.

REVIEW QUESTIONS

1. How many examples of unsafe practices can you find in the drawing above?
2. How many examples of poor techniques described in this chapter can you find in the drawing above?
3. Some examples of poor techniques above were not described in the chapter. Identify some examples. Why are these techniques poor?
4. What single piece of laboratory equipment is involved in the most serious laboratory accidents? Suggest reasons why this piece of equipment is so dangerous.
5. Why are test tubes filled with liquid heated near the top of the liquid?
6. Why can some glass containers be heated safely while others cannot?

THOUGHT QUESTIONS

1. Why are teachers more strict in laboratory classes than in classes without laboratory materials?
2. Why does some glass crack when heated while other glass does not?

Matter and Its States

Chemistry is the study of matter and how it changes. To help explain matter, chemists invented the idea that matter is composed of small particles called molecules. Chemists say that the differences in matter result from the differences in the molecules which make up matter.

For example, steel is a strong metal. Therefore, particles which make up steel must be strongly attracted to each other. Lead is a heavy metal. Therefore, the particles which make up lead must be very heavy. Helium and hydrogen are light gases. What would you predict about the molecules which make up these gases?

A molecule of water is composed of three smaller particles called atoms. Two hydrogen atoms and one oxygen atom make up one water molecule.

A THEORY ABOUT SMALL PARTICLES

Chemists always talk about atoms and molecules, even though they never see them. There is a simple reason. The idea that matter consists of these small particles has proven very useful. Chemists can explain many interesting changes they observe by using this idea. They can also accurately predict how substances will act when heated or mixed by using their knowlege of atoms and molecules in these substances. Throughout this unit you will have many opportunities to explore the idea that all matter is composed of small particles.

Observing with your Nose. Close the windows and doors of the classroom, and shut off the ventilating system. The air in the classroom should be still for this activity.

Pour some strong perfume on a cloth at the front of the room. Ask members of the class to raise their hands as soon as they smell the perfume. Note the spreading of the perfume through the room.

Many people have observed that odors travel across a room by spreading out from a source, such as a perfume bottle, cooking pan, or smelly sock. Curious people start asking, "What does this observation mean about the nature of matter?" This question may be answered by inventing the idea of atoms and molecules. In order to explain how odors travel, scientists have agreed that matter, such as air and perfume, behave as though they were made up of tiny particles. These particles are called molecules. If odors travel through air, then molecules of these chemicals which produce odors must be able to move through the air.

Air Molecules Bounce Off Walls. Blow air into a plastic bag. How did the size of the bag change? Why did the size change?

The first question can be easily answered by looking at or *observing* the plastic bag. The second question is more difficult. The idea of molecules is useful to explain why the bag gets larger.

Air is believed to be composed of molecules in motion. The diagram at the left shows molecules of air both inside and outside the plastic bag. As these molecules bounce off the wall of the bag, they exert a push or *pressure* on the bag's walls. Because air molecules push against both sides of a plastic bag with the same pressure, the bag does not change size.

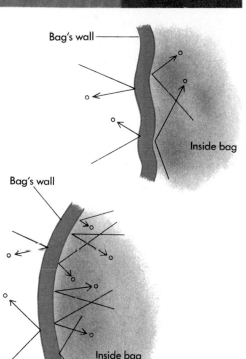

Bag's wall

Inside bag

Bag's wall

Inside bag

The second diagram shows a view of part of the bag as you blow into it. What happens to the number of molecules inside the bag? What happens to the number of times air molecules inside the bag bounce off the wall each second? What will happen to the pressure inside the bag? Why does the bag get larger?

The Effect of Heat on Air Molecules. Tie a balloon (or small plastic bag) to a test tube. Heat the test tube, being careful not to burn the balloon. What happens to the balloon?

Develop an explanation for this observation. Use the idea of air molecules to help you with your explanation. Compare your explanation with those made by your classmates. Which explanations seem best? Why?

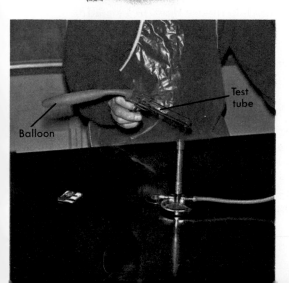

Test tube

Balloon

THEORIES. Observations are perhaps the easiest part of science. However, careful observations are of greatest use when they are used to try to develop an idea of how the world around us behaves. The idea that air is composed of molecules in motion which move faster when heated is called a *theory*.

Theories help us to explain what we observe. However, useful theories should enable us to predict how matter will behave when we test it. For example, how do you think the balloon in the last activity will change if the test tube is cooled? Try it. Does your theory help you to predict how and why the balloon will act when the test tube is cooled? Is the theory useful?

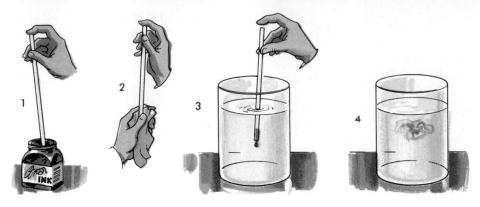

Diffusion in Liquids. Dip a glass tube or medicine dropper into a bottle of ink and lift out some of the ink as shown above. Wipe off the outside of the tube. Then lower the tube carefully into a glass of water and let one drop of ink flow into the water. Watch what happens to the drop of ink. Use your theory of molecules of water and ink to explain your observations.

Most scientists accept the idea that molecules in liquids and gases are in motion. This spreading out of molecules of gases and liquids is called *diffusion.*

The Effect of Temperature on Diffusion. Fill a glass with very hot water. Fill another with very cold water. Let both glasses stand until the water has stopped swirling about. Then carefully add a drop of ink to each glass as you did in the previous activity.

Watch the ink diffuse through the water. What are some differences in the way the ink diffuses? What conclusions can you draw?

The Effect of Heat on Molecules. The molecular theory helps explain the different rates of diffusion in hot and cold water. The theory does so by stating that molecules move faster when heated. Why would ink diffuse faster when molecules are speeded up?

At Start of Experiment

Hot water   Cold water

SUMMARY QUESTIONS

1. What evidence have you seen which supports the idea that matter is composed of small particles?
2. What evidence do you have that supports the idea that heat speeds up atoms and molecules?
3. Why do scientists invent theories?

CHANGING STATE

Water, like most matter, can be found in three different states. These states are solid, liquid, and gas. The special names for water in these three states are ice, water, and water vapor.

The terms freezing and melting describe matter (such as water) changing from the liquid state to the solid state, and changing from the solid state to the liquid state. The terms boiling and condensing describe matter changing from the liquid state to the gas state, and from the gas state to the liquid state.

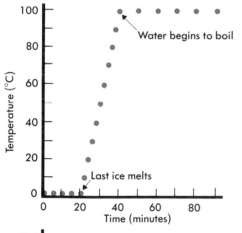

Adding Heat to Ice. Add about 1/4 cup of ice to a beaker. Hang a thermometer in the beaker as shown here. Record the temperature of the ice. Use a burner to slowly heat the beaker and its contents while stirring. Record the temperature every 30 seconds. Also record the times when the last piece of ice melts and when the water begins to boil. Continue recording temperatures until the water boils for five minutes.

Plot your observations on a graph like the one at the left. Mark the time on the graph when the last piece of ice melted, and when the water began to boil.

Studying the Graphs. The graphs at the left show a similar experiment which was performed under "ideal" conditions. In this experiment, heat was added very slowly. As a result, the water temperature did not increase until all the ice had melted. Note the second graph. Did the amount of water change when the solid water (ice) changed to liquid water? What happened to the amount of water in the beaker after the water boiled? Why?

Note the middle part of the upper graph. In this region, the heat which was added caused the temperature of the water to increase. There are, however, two other parts of this graph. In these regions, heat was added to the beaker but the temperature did not go up. What other effect does heat have on substances in addition to changing their temperature?

Compare the lengths of time required in the lower graph to (1) melt the ice, (2) change the temperature of water from 0°C to 100°C, and (3) change the water from a liquid to a gas. Which required the most heat? The least heat? Indicate on your graph the time during which water was melting. Indicate when water was boiling.

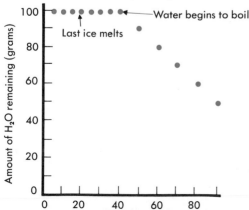

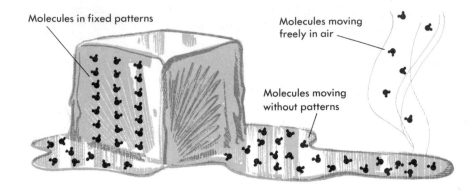

Molecules in fixed patterns

Molecules moving freely in air

Molecules moving without patterns

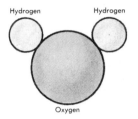

Hydrogen   Hydrogen

Oxygen

A Model of Water Molecules. The diagram at the left shows how chemists picture a molecule of water. This molecule consists of one atom of oxygen and two atoms of hydrogen. The diagram above shows one model of how water molecules are pictured in the solid, liquid, and gaseous states. Note the molecules in ice. The molecules are arranged in patterns. The patterns indicate that the molecules are attracted to each other. Each molecule is moving slightly. It vibrates. If more heat is added to the molecule, it vibrates faster. When enough heat is added to the molecules, their individual motion is stronger than the attracting force between the molecules. What happens to the ice now?

Although most liquids contract when they freeze, water expands slightly. What does this observation suggest about the distance between molecules of water in the liquid and solid states?

Note the distance between the water molecules in the gas state. What observations have you made in this chapter which support this idea?

The Changing Mass. Tie two sponges to the ends of a meterstick as shown at the left. Dip one of the sponges in water. Then squeeze it gently so that water does not drip from it. Now, use a third piece of string to balance the meterstick as nearly level as possible as shown in the drawing.

Observe the meterstick after 5 or 10 minutes. Which end of the meterstick is higher? How might a change in mass of the dry sponge have caused the stick to change? How might a change in mass of the wet sponge have caused the stick to change? What other possible changes in mass might affect the stick this way? Which explanation seems most likely to you? Why? Propose a different experiment which could be done to test your hypothesis.

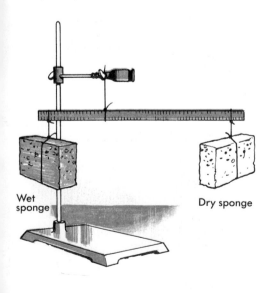

Wet sponge

Dry sponge

HYPOTHESIS—A proposed explanation.

Experimenting with Evaporation. Most observers agree that the changes just described are caused by the mass of the wet sponge decreasing. Wet sponges lose mass slowly. After several hours, the sponge is dry and its mass no longer changes. The disappearance of water as things dry out is called *evaporation*.

Evaporation is a process which can be readily investigated. The experiments pictured on this page are planned to learn more about evaporation. Study each picture. What condition or *variable* is changing in each picture?

The variable which the experimenter changes in an experiment is called the *independent variable*. The condition which changes because of changes in the independent variable is called the *dependent variable*. What is the dependent variable in each experiment?

Most experimenters plan their experiment so that it has only one independent variable. Which experiment pictured here has more than one independent variable? Why would results of such experiments be difficult to interpret?

Experimenters plan most of their experiments so that a change in the dependent variable can be said to be caused by changes in the independent variable. To do this, experimenters must make sure that all other possible variables act the same throughout the experiment. For example, in the upper right experiment on this page, the humidity, wind, and surface area are kept the same on both sponges. What variables should be kept the same in the other experiments?

Experiments where other variables are kept the same are called *controlled experiments*. Set up a controlled experiment to study the effect of one variable on the evaporation of water from a sponge. Share your findings with your class. What variables affect the evaporation of water from sponges?

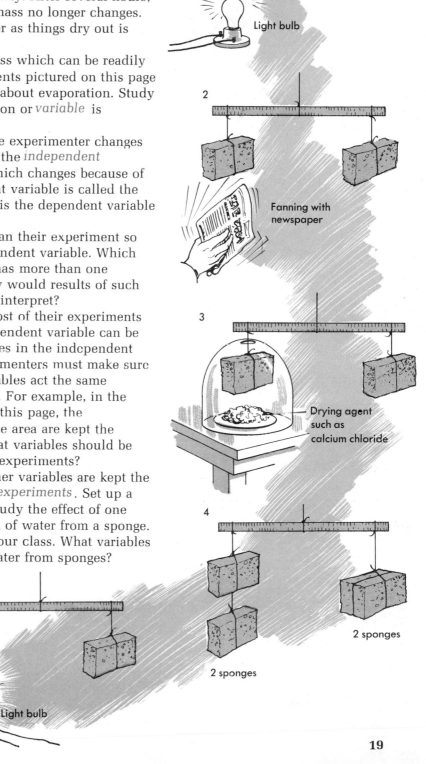

1 Wet sponges / Light bulb

2 Fanning with newspaper

3 Drying agent such as calcium chloride

4 2 sponges / 2 sponges

5 Fanning / Light bulb

19

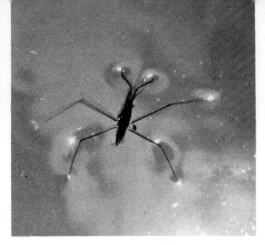

The Molecular Theory and Evaporation

Observations. According to the molecular theory, molecules of water are in motion, constantly bumping into each other. The average speed of the molecules determines the water's temperature. As the average speed of the molecules increases, the temperature also increases. At any moment, some molecules are going slower and some faster than the average rate. In the diagram at the left, the speed of each molecule is shown by the length of its arrow.

The molecules at the surface of the water behave differently. These molecules are attracted strongly to each other. As a result, water acts as though its surface is composed of a thin "skin." You have probably seen this surface support small insects or objects such as razor blades and needles. Note the photographs above.

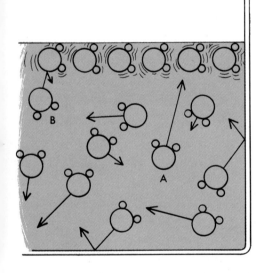

The diagram at the left shows a molecule, labeled *A*, moving toward the surface of the water. This molecule is moving very fast, as shown by the length of the arrow. What will happen to this molecule if it is going fast enough to break through the surface of the water? Suggest why the molecule at *B* did not break through the surface.

Use this model to explain your observations about the variables which affect evaporation.

Evaporation Cools. During evaporation, molecules traveling at speeds greater than the average speed escape from the liquid. What happens to the average speed of the molecules remaining behind in the liquid? What happens to the temperature of the remaining liquid during evaporation?

Wet finger

The student at the left has moistened a finger and is using it to tell which way the wind is blowing. Use the molecular theory to explain why the finger feels colder on the side facing the wind. Make a list of other observations you have seen which show that evaporation cools.

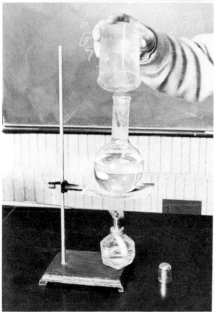

Condensed water vapor

Water vapor

Boiling Point and Molecular Theory. According to the molecular theory, when a liquid is heated, the molecules move faster. As a result, evaporation increases. At the boiling point, many molecules in the liquid are moving fast enough to change to a gas, all at the same time.

Boiling water usually receives its heat from the bottom of the container. Thus, liquid water at the bottom of a boiling pan changes into water vapor. These bubbles of water vapor quickly float to the top where the water vapor mixes with air.

Changing Water Vapor to Liquid Water. Boil some water in a flask with a narrow neck similar to the one at the left. When the water is boiling, place a beaker of cold water about 3 cm above the mouth of the flask. Notice what collects on the bottom of the beaker. Where does this substance come from? Use the molecular theory to explain why the substance forms there. This process of changing a gas to a liquid is called *condensation.*

Water Vapor, Condensed Water Vapor, and Steam. Water vapor is a colorless gas. The gas which forms as water evaporates, or the invisible gas in the bubbles of boiling water, are examples of water vapor. Note the region of invisible water vapor directly above the spout of the tea kettle.

Sometimes a mass of water vapor cools, slowing down some of the molecules so that they change to tiny droplets of liquid water. Clouds and fog are composed of tiny droplets of water which condensed from water vapor.

Small whitish clouds can often be seen above the invisible water vapor coming from a tea kettle. Other condensed water vapor clouds can be seen as you breathe out on a cold day. Usually such small clouds quickly disappear as the tiny droplets of water evaporate in the surrounding air.

Condensed water vapor droplets can be hot, as shown in the photograph, or cold as you breathe out on a cold day. Many people refer to any visible condensed water vapor as *steam.* Other people use the term, steam, to refer to water vapor above the boiling point of liquid water. Others refer to condensed hot water vapor as steam.

Often the gases coming from a smokestack are invisible until they are several meters above the top of the stack. What causes this and what is one probable substance in the smoke?

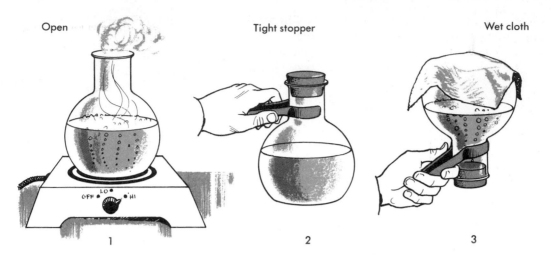

Open	Tight stopper	Wet cloth
1	2	3

Boiling by Cooling. The drawings above show how water can be made to boil by cooling! Boil some water in an open, thick-walled boiling flask until steam comes rapidly from its mouth. Quickly remove the flask and plug it with a tightly fitting stopper. Use a wet cloth to cool the flask or place it under a faucet. What happens to the water in the flask?

Remove the stopper from the flask. What does this show you about the pressure in the flask?

Boiling water fills the space above the water with water vapor. This water vapor forces most of the air molecules out of the flask. When the stopper is put in the flask and the flask is cooled, the water vapor condenses. What happens to the pressure inside the flask? How is boiling temperature affected by this change in pressure?

Repeat the experiment with a one-hole stopper in the flask and insert a thermometer into the water. Take the temperature of the boiling water. What is the lowest temperature at which the water boils?

Boiling Point and Elevation. The boiling point of water is 100°C at the pressure often found at sea level. However, at higher altitudes the pressure is less and the boiling point is lower than 100°C. Make a graph of the data above. Use this graph to estimate the boiling point of water for your elevation, if you live away from sea level.

The Effect of Increased Pressure. People who live where the pressure is lower (at high altitudes) must boil foods longer when cooking because the water boils at a lower temperature. The idea that pressure affects the boiling point can be used to

WEAR SAFETY GLASSES

Elevation	Boiling Point
0 m	100°C
1 000 m	96.6°C
2 000 m	93.4°C
3 000 m	90.1°C

Place	Elevation (m)
Atlanta, Ga	320
Boise, Id	825
Chicago, Il	183
Denver, Co	1 600
Mt. McKinley, Al	6 200
Mt. Mitchel, NC	2 040
Mt. Whitney, Ca	4 440
Mt. Washington, NH	1 920
Pike's Peak, Co	4 300
Phoenix, Ar	340
Salt Lake City, Ut	1 300

cook foods faster by increasing the pressure. Special cooking pots called pressure cookers do this. Note the dial of a pressure cooker at the right. What is the boiling point of water when the pressure inside the cooker is increased by 750 mm of mercury pressure above normal air pressure? This is about twice the normal pressure of the atmosphere.

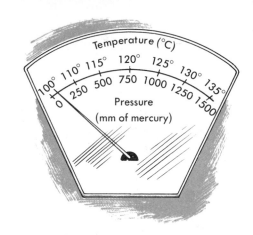

Distilling. Set up the apparatus shown below, using a clean glass tube and test tube. Put 125 ml of water and a few drops of ink in the flask. Heat the flask to boiling. Continue heating until several milliliters of condensed water collect in the test tube.

Examine the vaporized and condensed water. Does it contain ink? Discuss these observations and use the molecular theory to propose an explanation.

This process of separating or purifying liquids is called *distillation*. The water which is produced is called *distilled water*. Chemists often use distilled water in preparing their chemicals.

Mix some perfume or rubbing alcohol with water in the distillation flask. Boil the mixture. Smell the liquid which condenses. Explain what has happened.

Rain and Snow. Rain and melted snow can be used in many experiments for which tap water is not suitable. They are forms of distilled water. Describe the process by which rain and snow are formed, and compare this process with the distillation experiment just performed. Can rain or snow near most large cities be used for distilled water? Why?

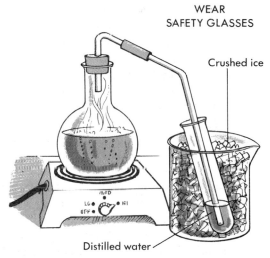

WEAR SAFETY GLASSES

Crushed ice

Distilled water

CAUTION: Do not pour alcohol near a flame.

Melting snow spilling over a plugged gutter.

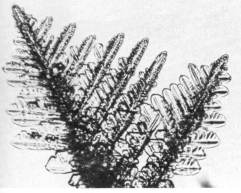

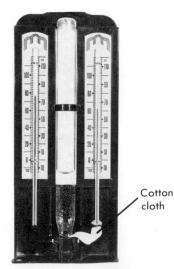

Cotton
cloth

Freezing and Sublimation. The photographs at the left show two examples of water in the solid state. The upper photograph shows a pond which lost a great amount of heat during the night. Some of the energy of motion of the molecules became less than the attractive force between the molecules. As a result, some of the water molecules became a part of the molecules forming solid water, or ice. This is called *freezing*.

The next photograph shows an example of a solid which formed when molecules of water changed from the gaseous state to the solid state. This is called *sublimation*. The same name is used to describe changes from the solid to the gas state, such as when dry ice changes to a gas.

REVIEW QUESTIONS

1. What is happening in the apparatus below at the left?
2. What evidence do you have that molecules of gases and liquids spread out?
3. Solid iodine changes directly to a gas. What is the name for this observation?
4. Use the idea of molecules to explain why a bicycle tire increases in size when you put air in it.
5. How does heat affect a gas? How does the theory of molecules and atoms help you to explain this observation?
6. How does heat affect diffusion? How does the theory of molecules and atoms help you to explain this observation?
7. What is the name for the process in which liquids change to gases?
8. What are some factors affecting the rate of evaporation?

THOUGHT QUESTIONS

1. Why are more bacteria likely to be killed when food is cooked in a pressure cooker than when cooked in an open pan?
2. Why do the thermometers at the left give different temperatures?
3. Why do boiled potatoes take longer to cook on a mountain top than at sea level?

How many mixtures are shown here?

Mixtures

Blood, sea water, milk, and ginger ale are four examples of mixtures. These mixtures are each composed mostly of water; however, each is quite different from the other. The reason for this difference is because each mixture has other different chemicals added to the water.

Generally, mixtures have the combined properties of (or resemble) the chemicals making up the mixtures. For example, each mixture above is a liquid because liquid water is an important part of them. Salt has a strong taste. How does sea water taste? Why? If blood is mostly water, how can you explain its red color? How many mixtures can you find in the photograph?

There are very few examples of pure substances in our daily life. Most substances are mixtures; that is, they are two or more chemicals mixed together displaying the properties of each chemical. Some common mixtures shown above are milk, soup, coffee, bread, brass, and jelly. Name one or two different chemicals in each of these mixtures.

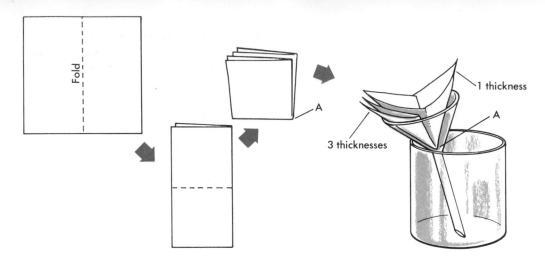

Fold

1 thickness

A

3 thicknesses

A

SEPARATING MIXTURES

Many tasks in chemistry involve separating mixtures. For example, crude oil comes from the ground as a mixture of thousands of different chemicals. Crude oil, by itself, has very few uses. However, our world would be very different if we did not have the hundreds of different substances which are separated from crude oil.

Filtering. Add some fine dirt to a half cup of water. The dirt particles should be small enough so that they do not quickly settle out.

Fold a square piece of paper towel as shown above. Place the folded paper in a funnel so that three thicknesses of paper are on one side of the funnel. One thickness is on the other part, as shown. Pour the dirty water into the funnel. Describe the appearance of the water which comes through the paper filter.

Two or more different chemicals mixed together are called a *mixture*. In this activity, the dirty water is a mixture. You separated the mixture. The differences which can be used to separate chemicals in mixtures are called *physical properties*. What difference in physical properties did you use to separate the mixture?

The photograph here shows a series of sand filters which are used in a city water purification plant. Water from a nearby river is pumped into the top of the sand filters. Filtered water is removed from the bottom of the filters. Then chlorine is added to the filtered water to remove any harmful organisms which passed through the filters. Describe some other filters which are a part of your daily life.

Filtering and Evaporating. Mix together 5 grams of salt and 10 grams of sand. Now, how would you separate this mixture and recover all or almost all of the sand and salt?

There are many different ways which could be used to separate this mixture. Do the following. Add the mixture to a half cup of water. Stir the mixture until the salt dissolves in the water. Then filter the mixture. Where is the salt after you filter? Where is the sand?

Dry the sand on the filter paper. How much sand did you recover?

Devise a way to separate the salt from the water. How much salt did you recover? What difference in physical properties did you use to separate the salt and sand? What difference in physical properties did you use to separate the salt and water?

Separating Other Mixtures. Separate several other mixtures such as:

1. Sulfur and sugar
2. Iron and sulfur
3. Sawdust and sand
4. Sand and sulfur
5. Iron, salt, and sawdust
6. Rubbing alcohol and water
7. Salt and sugar

Keep records of your success in separating these mixtures. What difference in physical properties did you use in each case to separate them? In which cases did you use physical changes to separate the mixtures?

Substance	Boiling Point	Dissolves in Water	Magnetic	Density
Water	100°C	---		---
Rubbing alcohol	82°C	✓		Low
Iron	3000°C	---	Yes	High
Sawdust	breaks down	---		Less than water
Sand	2200°C	---		Medium
Sulfur	445°C	---		Low
Sugar	breaks down	✓		Low
Salt	1450°C	✓		Low

SOLUTIONS

Many mixtures consist of one or more chemicals dissolved in a liquid such as water. This physical property of being able to dissolve is called solubility. *A substance which does not dissolve in a liquid is said to be* insoluble.

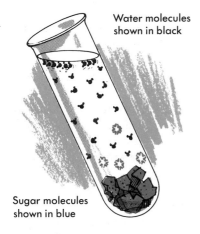

Water molecules shown in black

Sugar molecules shown in blue

Making a Solution. Put a few grains of sugar into a test tube of water. Use a hand lens to observe what happens to them. When they disappear, have they been destroyed or are they still in the test tube? How can you prove your statement?

The process that went on in the test tube is called *dissolving*. This liquid mixture is called a *solution*.

Why Things Dissolve. According to the molecular theory, both the water and the sugar are made up of molecules in constant motion. The sugar, however, is in the solid state; its molecules are held together by strong forces and cannot move far.

Every now and then a rapidly moving water molecule strikes a sugar molecule. The sugar molecule is knocked out of the crystal with enough energy to move about in the water. In time, all the sugar molecules become mixed with the water.

Diffusion from Crystals. Put a large piece of copper sulfate into a jar of water. Cover the jar and set it on a window sill where it can be watched without being disturbed. Notice the spread of blue color outward from the crystals. What is happening? Does the copper sulfate diffuse throughout the entire jar of water?

Making Things Dissolve Faster. The experiments shown below and on the top of the next page test three conditions that affect the speed with which sugar dissolves.

Put a tablespoonful of sugar into each of two glasses of water. Stir one but do not stir the other. Which dissolves faster? Test table salt, baking soda, and other chemicals in the same way.

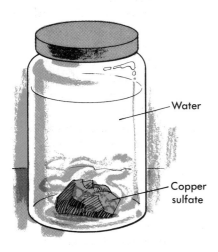

Water

Copper sulfate

A

Without stirring   Stirring

B

Crushed sugar cube

Sugar cube

C

Cold water

Hot water

For a second experiment, use two sugar cubes. Crush one cube into a powder and put it into a glass of water. Put the other cube into a second glass of water. Stir each the same number of times. Which dissolves faster?

For the third experiment, put a tablespoonful of sugar into each of two glasses. Add a cupful of cold water to one glass. Add a cupful of hot water to the other glass. Stir each mixture the same number of times. Which dissolves faster?

Explain the results of these three experiments in terms of the molecular theory.

How Temperature Affects Dissolving. Put 100 ml of water into a glass. Add a teaspoonful of sugar and stir the mixture until the sugar has dissolved. Record the temperature of the water and the time needed for the sugar to go into solution.

Repeat these measurements with water at different temperatures. A change of temperature during the experiments can be reduced. To do this, set the glass in a pan of water which has the same temperature as the water in the glass.

Make a graph of the results of the experiment.

Test table salt and other chemicals in the same way and make graphs of the results.

How Dissolving Affects Temperature. Have you observed that some solutions get warmer as substances dissolve while other solutions seem to get colder? Set up an experiment to find out how dissolving chemicals affect the temperature of the solution. (**CAUTION:** Treat all chemicals as harmful, unless told otherwise.) In order to compare which chemicals raise or lower temperatures most, use the same amount of chemical each time. How much water should you use each time? What should the water temperature be at the start of each experiment?

Keep a record of the solutions you test, such as the record shown at the left. Which chemicals warm the solutions? Which ones cool them? Which ones warm or cool the best?

Effect of Temperature on Time
Needed to Dissolve Sugar

Time (minutes)

Temperature (°C)

Testing which chemicals warm
or cool solution when dissolved.

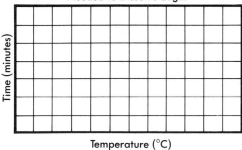

Chemical	Cools (°C)	Warms (°C)
Hypo		
Baking soda		
Sugar		
Salt		

Saturated Solutions. Add sugar slowly to 100 ml of water, stirring the mixture constantly. Notice that after a certain amount of sugar has been added, no more can be dissolved. The solution is *saturated.*

A sugar molecule may be attracted back into a crystal if one passes close by. This does not happen often when the solution contains only a few sugar molecules. A saturated solution, however, contains so many molecules that some of them are attracted back into the crystals as fast as others leave. The amount of undissolved solid remains unchanged.

Comparing Saturation. Weigh some sugar in a paper muffin cup. Add this sugar a little at a time to 100 ml of water, stirring it constantly. When no more dissolves, weigh the sugar remaining in the paper cup. Calculate the amount that dissolved.

Repeat the measurements with table salt, copper sulfate, and other chemicals. What variables must be kept controlled if accurate conclusions are to be drawn from the experiment?

Effect of Temperature on Saturation. Set up an experiment to show the effect of temperature upon the amount of a chemical that can be dissolved in water. Try to keep all factors constant except the temperature, which will be changed in steps of five or ten degrees.

One problem you may have is keeping the water from cooling off while the chemical is dissolving. Put the glass in a pan of water having the same temperature. (You can fill the pan first and then dip a glassful of water from it.)

Compare your results with those of other students who tested other substances, such as sugar, table salt, and copper sulfate. How does temperature affect the amount of each substance that dissolves? Make graphs of the results.

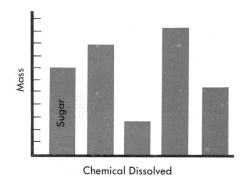

Chemical Dissolved

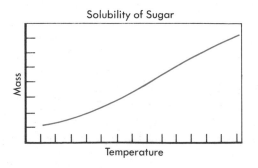

Solubility of Sugar

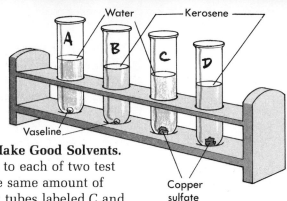

Water — Kerosene —

A B C D

Vaseline

Copper
sulfate

Predicting What Liquids Make Good Solvents.
Add a pea-sized bit of vaseline to each of two test
tubes labeled A and B. Add the same amount of
copper sulfate to two other test tubes labeled C and
D. Half fill test tubes A and C with water. Half fill
test tubes B and D with kerosene. Stir each test tube
for several minutes. Which test tubes show that a
solution forms? Which do not?

Chemists who study solutions have observed
that "like dissolves like." For example, kerosene
and vaseline are alike. Both are produced when a
naturally-occurring mixture (crude oil) is separated
by distillation. Chemists who study molecules say
that vaseline molecules are similar in shape to
kerosene molecules, but the vaseline molecules are
larger. Solids usually dissolve in liquids composed
of molecules similar to the shapes of the molecules
of the solid. The shapes of kerosene and vaseline
molecules are described as *nonpolar*.

Water molecules are described as *polar* in
shape. The table at the right lists the shapes of
molecules of a number of substances. Following the
rule that "like dissolves like," which substances in
the table will dissolve in others?

Soap: A Solution for a Dirty Problem. Fat is a
nonpolar substance. It does not dissolve in water.
Soapy water, however, can dissolve fat. The
diagrams at the right show one explanation. Note
the soap molecules in diagram A. Soap is believed
to be composed of molecules which are polar at one
end and nonpolar at the other end.

Diagram B shows how soap molecules behave
on the surface of water. Which part of the molecule
"dissolves" in the polar water? Which part of the
soap molecule would come in contact with
nonpolar fat molecules?

Diagram C shows tiny particles of grease
dissolved in soapy water. Now the grease can be
washed away. Suggest how you might use this
information to develop experiments and theories to
understand how stains can be removed from fabrics.

Substance	Polar	Nonpolar
Copper sulfate	+	
Cooking oil		+
Salt	+	
Vaseline		+
Wax		+
Baking soda	+	

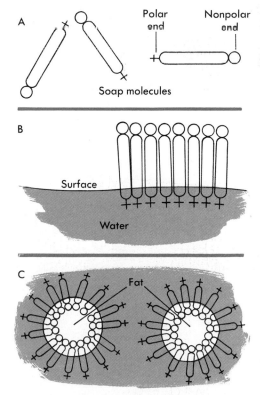

A

Polar
end

Nonpolar
end

Soap molecules

B

Surface

Water

C

Fat

Which of these quartz crystals grew in a supersaturated solution that was cooled slowly?

SEPARATING SOLUTIONS

Solutions consist of two types of substances. The substance which is dissolved is called the solute. The substance the solute dissolves in is called the solvent. In the previous section, you learned how solutes dissolved in solvents. In this section you will study the processes by which solutes and solvents can be separated.

Making Crystals. Dissolve as much sugar as possible in 50 ml of water. (Note: About 125 grams of table sugar dissolves in this amount of cool water.)

Stir the solution until all the sugar dissolves. Place the saturated solution in an open dish and place it on a windowsill. Observe what happens to the saturated solution in a day or two.

Use the theory of molecules to explain your observations.

The Theory of Supersaturation. The graph at the left compares the amount of table sugar which dissolves in different amounts of water at 20°C. How does the amount of water affect the amount of sugar which can dissolve?

Imagine that a student added 600 grams of sugar to 300 ml of water. Would the solution be saturated or unsaturated? Explain your answer.

Now imagine that the student left the solution in a beaker on a windowsill where the water in the solution slowly evaporated. The dashed line on the graph describes the conditions in the beaker as the water evaporated. At what water volume did the solution become saturated?

As the saturated solution continues to evaporate, the solution is said to become *supersaturated.* Supersaturated solutions are unstable. To become stable, some sugar must leave

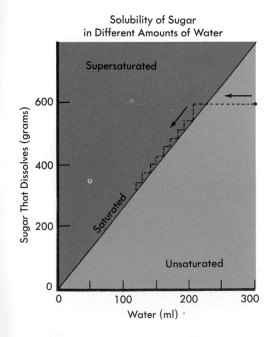

Solubility of Sugar
in Different Amounts of Water

Supersaturated

Sugar That Dissolves (grams)

600

400

Saturated

200

Unsaturated

0

0   100   200   300

Water (ml)

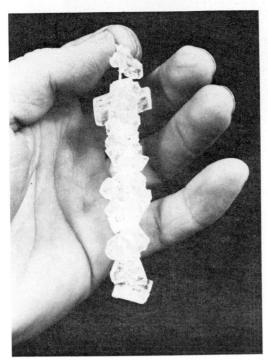

the solution. When this happens, sugar molecules attract each other, forming a solid. If the solid has regular sides, the solid is called a crystal. As more solution evaporates, more sugar leaves the solution, forming either more or larger crystals.

Note the photograph of rock (sugar) candy at the left. This type of candy was a favorite in colonial times. Propose how the early settlers might have made rock candy.

Crystals from Hot Solutions. Using your past studies of solutions and the graph below, prepare 25 ml of hot, saturated potassium nitrate (KNO_3) solution.

Hot water can dissolve much more potassium nitrate than cooler water. Can all the potassium nitrate remain dissolved if the solution cools? Find out. Explain your observations.

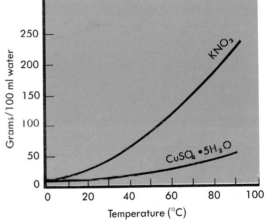

Rapid and Slow Cooling. The graph above also describes the solubility of copper sulfate. Use the graph to prepare a nearly saturated, hot solution of copper sulfate. When all the copper sulfate has dissolved, split the solution into two parts. Place half the solution in an insulated cup. Cover the cup with a good heat insulator. Leave the other half of the solution in a beaker. Allow both containers to cool. Which solution will cool more quickly?

When both containers have cooled, pour out the remaining solutions. Which container has the larger crystals? What condition was different in the containers?

A Theory to Explain Differences in Crystal Size. A good theory about molecules must be able to explain observations of how solutions behave.

Hot saturated copper sulfate in covered insulated cup.

Hot saturated copper sulfate in beaker.

33

Thus, the molecular theory should be able to explain differences in crystal size.

According to this theory, particles of copper sulfate are moving about when in solution. As the temperature drops, the solution becomes supersaturated. Soon, copper sulfate crystals will begin to develop.

Molecules in solids are strongly attracted to each other, giving solids their strength. In supersaturated solutions, molecules in crystals must attract the same type of molecules in solution. If the solution is cooling slowly, the slower molecules still in solution travel a greater distance before being attracted to similar atoms already in crystals. Under these conditions, the slower molecules are more likely to stop being attracted to the ends of growing crystals. As a result, the crystal slowly grows larger.

In a rapidly cooling supersaturated solution, many molecules of dissolved matter are moving slowly. The chances of two or more slow molecules in solution being attracted to each other is great. As a result, many new, small copper sulfate crystals develop.

Study the crystals of quartz in the photograph on page 32. Propose a difference in the way these two crystals developed.

Gases Leaving Solution. Open a bottle of soda pop. Pour some of the contents into a drinking glass. What do you see rising in the glass? Was this happening in the bottle before you opened it? How did removing the bottle top change the conditions inside the bottle?

Soft drinks are bottled under pressure. There is more pressure than normal inside the bottle. By increasing the pressure on the liquid, the solubility of the gas also increases. This gas, called carbon dioxide, is used in all carbonated beverages.

The bottle top is tightly sealed so the gas inside will not leak out and lower the pressure. When the bottle cap is removed, the pressure on the liquid is reduced. As a result, the solubility of carbon dioxide is also decreased. The drink now has more carbon dioxide dissolved in it than it can hold. What is the name for this condition?

The soda pop in a freshly opened bottle is an unstable solution. The solution is highly supersaturated with carbon dioxide gas. There are

gas. What are some ways you have previously discovered on your own? Set up experiments to test any of the ways which you have not already experienced: temperature, shaking, stirring, adding particles such as sand, sugar, or salt.

How Much Gas in Soda Pop? Plug a bottle of ice-cold soda pop with a one-hole stopper. Connect a rubber tube to the bottle and lead the tube into a jug of water as shown at the right. Set the bottle in a pan of hot water to drive the gas from the soda pop. Shake the bottle as needed to speed up the process.

Hot water

After bubbles have stopped forming, turn the jug right side up. Empty the soda pop bottle and use it to pour water into the jug. How many bottles of water are needed to fill the jug again? How much gas came from one bottle of soda pop?

Use your observations to describe the distance between carbon dioxide molecules as a free gas and when the gas is dissolved in soda pop.

Driving Gases from Tap Water. Fill a flask or other glass container with cold tap water. Heat the water on a hot plate. Watch for the formation of tiny bubbles on the side of the container. At what temperature do they begin to collect? Where do they collect first?

Try water from other sources such as a hot water faucet, an aquarium, and a stream. Do the bubbles always begin to form at the same temperature? How do the numbers of bubbles in each flask compare?

The graph shows the solubility of oxygen in water. This gas, like most gases, is less soluble as the water temperature increases. Is this similar to the way solids behave when dissolved in water?

Suppose you heated 100 ml of water with 2.8 cm^3 of oxygen dissolved in it. At what temperature would the solution become saturated? Above what temperature would bubbles begin to collect on the sides of the container? Use this procedure to estimate the amount of dissolved oxygen in water that you heat.

Gases in Tap Water. Water for an aquarium is usually taken directly from a faucet. Few people give any thought as to whether or not the water contains gases needed by fish and plants. What gas do green plants need? What gas do fish need?

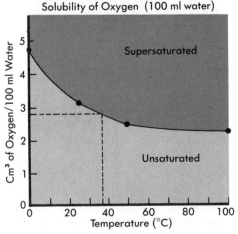

Solubility of Oxygen (100 ml water)

Cm^3 of Oxygen/100 ml Water

Supersaturated

Unsaturated

Temperature (°C)

What purpose do the bubbles serve in this fish tank?

80°C ——

1. What are some examples of physical properties?
2. What is a solute? A solvent?
3. What is a mixture?
4. How are physical properties useful in separating mixtures?
5. What are some conditions which speed up dissolving?
6. What does saturated mean?
7. What does supersaturated mean? What are two different ways that saturated solutions can become supersaturated?
8. How is the molecular theory expanded to show that some liquids will dissolve one chemical group and not a second group, while other liquids will dissolve the second chemical group but not the first?
9. How does the theory in the last question help to explain why soap can be used in water to dissolve chemicals in both groups?

THOUGHT QUESTIONS

1. A student once drank only one glass of pop, but claimed to have burped up three glasses worth. How is this possible?
2. Some chemicals cool solutions when they dissolve; others warm the solution. Which temperature change is easier to explain using the theory of molecular motion described in this chapter?
3. Why is more stirring needed when sugar is dissolved in iced tea than in hot tea?
4. How can more copper sulfate be dissolved in a saturated solution without adding more water?
5. The thermometer in the middle photograph reads 80°C. What do the bubbles of air on the side of the glass show you about the solubility of air in water as the temperature increases?
6. The lower photograph was taken 30 minutes after the glass was filled from a faucet. The water in the faucet is under high pressure. What do the air bubbles indicate about how pressure affects the solubility of air in water?
7. Why do fish die in a lake during hot summers?
8. Why do fish in late summer spend more of their time at the bottom of the lake?

4

Chemical Properties
and Chemical Changes

 Chemists have developed the idea of chemical properties to help them predict and understand what happens when two or more chemicals are mixed. Sometimes a new substance forms. When this happens, a chemical reaction is said to have occurred. During a chemical reaction, the original substances are changed into new substances with new physical and chemical properties.

 Many different chemicals behave the same way when added to another substance. These chemicals are said to have the same chemical properties. Names such as acids, bases, salts, sugars, and alcohols describe groups of many different chemicals with similar chemical properties.

 In the previous chapters you have seen only a few chemical reactions. Now you will see many more. The chemical reactions in this chapter are grouped together. The reason for this is to assist you in using the idea of similar chemical properties in predicting and understanding chemical changes.

NATURALLY-OCCURRING ELEMENTS[1]

Atomic Number	Element	Symbol	Atomic Number	Element	Symbol	Atomic Number	Element	Symbol
1	Hydrogen	H	32	Germanium	Ge	63	Europium	Eu
2	Helium	He	33	Arsenic	As	64	Gadolinium	Gd
3	Lithium	Li	34	Selenium	Se	65	Terbium	Tb
4	Beryllium	Be	35	Bromine	Br	66	Dysprosium	Dy
5	Boron	B	36	Krypton	Kr	67	Holmium	Ho
6	Carbon	C	37	Rubidium	Rb	68	Erbium	Er
7	Nitrogen	N	38	Strontium	Sr	69	Thulium	Tm
8	Oxygen	O	39	Yttrium	Y	70	Ytterbium	Yb
9	Fluorine	F	40	Zirconium	Zr	71	Lutetium	Lu
10	Neon	Ne	41	Niobium	Nb	72	Hafnium	Hf
11	Sodium	Na	42	Molybdenum	Mo	73	Tantalum	Ta
12	Magnesium	Mg	43	Technetium	Tc	74	Tungsten	W
13	Aluminum	Al	44	Ruthenium	Ru	75	Rhenium	Re
14	Silicon	Si	45	Rhodium	Rh	76	Osmium	Os
15	Phosphorus	P	46	Palladium	Pd	77	Iridium	Ir
16	Sulfur	S	47	Silver	Ag	78	Platinum	Pt
17	Chlorine	Cl	48	Cadmium	Cd	79	Gold	Au
18	Argon	Ar	49	Indium	In	80	Mercury	Hg
19	Potassium	K	50	Tin	Sn	81	Thallium	Tl
20	Calcium	Ca	51	Antimony	Sb	82	Lead	Pb
21	Scandium	Sc	52	Tellurium	Te	83	Bismuth	Bi
22	Titanium	Ti	53	Iodine	I	84	Polonium	Po
23	Vanadium	V	54	Xenon	Xe	85	Astatine	At
24	Chromium	Cr	55	Cesium	Cs	86	Radon	Rn
25	Manganese	Mn	56	Barium	Ba	87	Francium	Fr
26	Iron	Fe	57	Lanthanum	La	88	Radium	Ra
27	Cobalt	Co	58	Cerium	Ce	89	Actinium	Ac
28	Nickel	Ni	59	Praseodymium	Pr	90	Thorium	Th
29	Copper	Cu	60	Neodymium	Nd	91	Protactinium	Pa
30	Zinc	Zn	61	Promethium	Pm	92	Uranium	U
31	Gallium	Ga	62	Samarium	Sm			

LABORATORY-MADE ELEMENTS[2]

Atomic Number	Element	Symbol	Atomic Number	Element	Symbol	Atomic Number	Element	Symbol
93	Neptunium	Np	97	Berkelium	Bk	101	Mendelevium	Md
94	Plutonium	Pu	98	Californium	Cf	102	Nobelium	No
95	Americium	Am	99	Einsteinium	Es	103	Lawrencium	Lr
96	Curium	Cm	100	Fermium	Fm			

[1] Some of these naturally-occurring elements have been known for centuries. A few of them were in common use before the dawn of history, such as tin, copper, lead, gold, silver, iron, and sulfur.

[2] These laboratory-made elements do not exist in nature. They have been produced by bombarding known elements with atomic particles. Scientists have reported the discovery of elements 104 and 105, but these elements have not yet been named.

CHEMICAL ELEMENTS AND THEIR PROPERTIES

The table on page 38 lists the simplest substances, called elements. The atoms of an element are all the same. Elements are found in nature either free or combined with other elements. Oxygen in air and copper in pennies are two examples of free elements. Water (H_2O) is an example of two elements combined.

In this section you will combine elements, break down substances into elements, and use the properties of elements to identify them. You will also learn to predict which elements will combine.

The Chemical Elements. Make a collection of as many elements as you can find. Use steel wool to remove any coatings that may hide the metallic elements. Make a table of the elements in your collection. Include on the table: (1) the name of the element, (2) the uses of the element, and (3) the properties of the element which make it useful.

Classifying Elements. There are so many elements that chemists would have a difficult time remembering the different properties of all of them. Fortunately, many elements have similar properties. Thus, elements can be put into groups with similar properties. One method of grouping considers the elements as falling into several groups. Two important groups are (1) metals, and (2) nonmetals.

Study the collection of elements your class has made. Which elements in the collection can be classified as metals? List some physical properties that all the metallic elements have in common.

Elements	Use	Useful Properties
Helium	In balloons	Lighter than air
Silver	Jewels	Malleable

Burning Metallic Elements. Clean a 3-centimeter piece of magnesium ribbon with steel wool. Observe and record the properties, such as color, shine, strength, and so forth. Hold the magnesium with forceps (tweezers) over an asbestos pad. Light the end of the magnesium. (**CAUTION:** Do not look directly at the burning metal.)

Observe the magnesium after burning. Record the properties that have changed as a result of this experiment.

Test small pieces of other common metals provided by your teacher to find out the ones that burn. Use thin pieces of each metal. Compare the properties of each metal with the metal after heating it in a flame. Test metals such as zinc, iron (steel wool), copper, and tin. Record your observations in a table.

Wear safety glasses when working with fire.

DATA — Burning Metals

Element	Properties of Pure Metal	Properties of Metal Oxide	Activity
Magnesium	Silvery, bends	White ash, brittle	High — gives off heat, light
Iron	Silvery		
Gold			
Tin			
Lead			

Chemists have observed that metals can burn. Some metals burn very rapidly. Chemists say these metals have high chemical activity. Other metals burn slowly or not at all. Such metals are not chemically active. Use another column on your data table to describe the activity of each metal you test.

Rusting. Chemists believe that the burning of metal and the rusting of metal are very similar. Burning happens more quickly, releasing large amounts of heat and light. As a result, burning may seem more interesting. Rusting, however, has the advantage of occurring more slowly. The changes which occur are easier to observe.

Set up the apparatus shown at the right. In one bottle put new steel wool such as that sold in hardware stores. Wet the steel wool so that it will rust. Then insert the stopper and dip the end of the glass tube in colored water.

The second bottle is set up in the same way as the first except that there is no steel wool in it. What is the purpose of this second bottle?

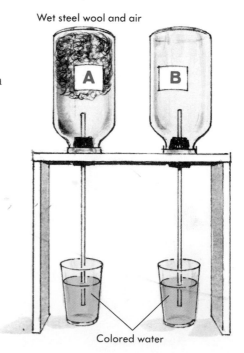

Wet steel wool and air

A   B

Colored water

Check the experiment an hour and a day later. What happens? Name some possible explanations.

Testing the Remaining Gases. Remove the two rubber stoppers from the bottles in the previous experiments. Then put a burning wood splint into each bottle at the same time as shown here. Do the splints burn the same in each bottle?

Chemists agree that flames occur only when oxygen gas is present. Oxygen is a nonmetallic element found in air. Which bottle contained less oxygen after the rusting experiment? How does the change in the water level also provide evidence that part of the air was removed during rusting?

Chemistry students have repeated the experiment on rusting which you have done. In their experiment, the steel wool was weighed before and after rusting with a very sensitive balance. The steel wool weighed more after rusting. Propose an explanation to fit this observation as well as your other observations during this experiment.

Chemical Changes. Burning and rusting are two examples of *chemical changes*. During chemical changes, new substances with different properties form, in place of other substances which no longer can be detected. When iron rusts, it changes, and a new substance called iron oxide forms.

Although there are only 92 different naturally occurring elements, there are many thousands of different chemical substances, each with its own properties. Compare this idea with the English alphabet. The 26 letters can be used alone in our speech, but the number of words would be very small if our language had only 26 "words." Most of the English language consists of words with two or more letters combined together. Similarly, most of the chemical world consists of compounds. Compounds are chemical substances consisting of two or more elements combined chemically.

Chemical Properties. Many elements, such as magnesium and iron, combine with oxygen to form oxides and release heat. This ability to combine is called a *chemical property*. Many metals have this ability; therefore, they have a similar chemical property. An important part of chemistry is learning the chemical properties of many groups of chemicals such as metals, nonmetals, acids, salts, and so on. If you had an active metal and heated it, what would you predict would happen to the metal?

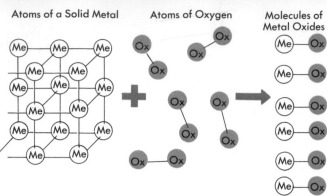

Atoms of a Solid Metal   Atoms of Oxygen   Molecules of Metal Oxides

A Model to Explain Chemical Changes. The diagram above shows one idea of atoms combining to form new substances with new properties. When a metal is heated, the metal atoms vibrate fast enough to knock apart the pairs of oxygen atoms in the surrounding air. The now separated oxygen atoms each attach to a metal atom, forming a molecule of metal oxide. As more and more metal oxide molecules form, the amount of metal and oxygen decreases. The reaction ends when the supply of metal atoms or oxygen atoms is used up.

Preparing the Element Oxygen. Oxygen can be prepared by breaking apart molecules which have oxygen in them. Hydrogen peroxide is a compound whose molecules contain oxygen. Hydrogen peroxide molecules can be easily broken apart. When this happens, two new substances form, water and oxygen.

Mix in a jar two tablespoonsful of hydrogen peroxide with a pinch or two of manganese dioxide. Use 3% hydrogen peroxide, which is sold in drugstores as an antiseptic. The black powder from dry cells may be used to provide manganese dioxide.

Note the bubbles in the mixture. The bubbles show that a gas is being given off. Cover the jar with a flat piece of glass or metal.

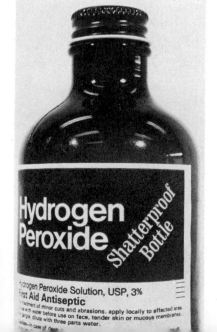

Hydrogen Peroxide Solution, USP, 3%
First Aid Antiseptic

Glass or metal cover

Manganese dioxide and carbon

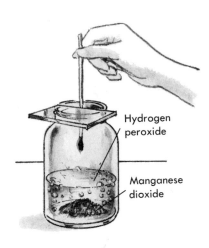

Hydrogen
peroxide

Manganese
dioxide

Testing the Gas. Thrust a lighted splint into the jar. What happens? What conclusion can you draw?

Light a splint, blow out the flame, and thrust it into the jar while the splint is still glowing. What happens? Compare the result with what happens when a glowing splint is thrust into a jar that contains only air. What is a chemical property of oxygen? What are some of its physical properties, such as color, odor, and density?

Chemical Tests. For many years, people have been using the burning splint test as proof that oxygen is present. This process is called a chemical test for oxygen. Chemical tests can be used to identify all the elements and thousands of compounds. Chemical tests are used to analyze the minerals in soil, the nutrients in foods, the composition of meteorites, and hundreds of other substances.

The Basic Assumptions. When the test for oxygen is used, two assumptions must be made:

1. The gas that keeps wood burning is oxygen.

2. There is no other gas that keeps wood burning.

Chemists tell us that we are safe in making these assumptions. They have found that fire without oxygen is uncommon.

Limitations of the Test. There are, however, at least three things that the test for oxygen does not tell us:

1. Oxygen may be present but in combination with other substances.

2. There may be other gases besides oxygen.

3. If the flame goes out, there may still be oxygen but not enough oxygen present to keep the wood burning.

> RECOGNIZING THE LIMITATIONS OF CHEMICAL TESTS. *In this unit you will perform many different chemical tests. Very few tests tell you all that you might want to know. Usually each test gives only a little information. You should be careful when drawing conclusions from the results of any one test.*

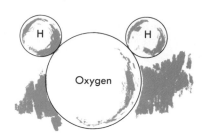

A water (H_2O) molecule

Where Oxygen is Found. Oxygen is, by far, the most common element in our lives. Every molecule of water consists of two atoms of hydrogen and one atom of oxygen. Thus, water is called H_2O by chemists. However, the oxygen atom in H_2O is much heavier than the two hydrogen atoms. Almost 90% of the mass of the water molecule is caused by the mass of the oxygen atom in the molecule.

Air is about 20% oxygen. What observation have you made which supports this idea? The oxygen in the air is used by animals, including yourself. Oxygen in air readily combines with many metals, forming rust.

About 50% of the mass of rocks and minerals in the earth's crust is oxygen. Identify the elements in the following minerals:

Calcite (in limestone and marble)—$CaCO_3$
Quartz (in granite, sand, sandstone)—SiO_2
Feldspar (in granite, clay, shale)—$KAlSi_3O_8$

Pure oxygen is used in many places including hospitals and steel mills. Most pure oxygen is made by cooling air to a liquid. Then the gases are separated by distillation.

Summary Questions

1. What evidence do you have that air is about 20% oxygen?
2. What is one chemical property of metals?
3. What is the test for oxygen?
4. What are three different substances that contain oxygen?

Wood chips

FREEING ELEMENTS FROM COMPOUNDS

Many important elements, such as iron, magnesium, and aluminum, are found in nature only as parts of compounds. Elements in compounds are held together by attractive forces between atoms. To separate the elements, energy must be added to overcome these attracting forces. In this section, many different examples are studied which add energy to compounds to separate or free the elements.

Using Heat to Break Down Wood Molecules.
Wood is a complex mixture of hundreds of different molecules. Most of the molecules are composed of many atoms of carbon, hydrogen, and oxygen. Add some small pieces of wood to a test tube as shown above. Heat the test tube with a gas burner. How does the wood change when heated? What forms after the wood has been heated for several minutes? What element makes up this substance?

Using Heat to Break Down Pyrite. Use a mortar and pestle like the one at the right to crush a piece of iron pyrite into a powder. Test the powder with a magnet. Record your observations.

Add ½ tspn of the powdered pyrite to a test tube. Heat it over a gas flame. What forms in the top of the test tube?

Allow the contents of the test tube to cool. Then pour out the remaining powder and test it with a magnet. How has this physical property changed? What element has this physical property?

Chemists tell us that iron pyrite molecules contain one iron (Fe) atom and two sulfur (S) atoms. Thus, the chemical formula for iron pyrite is FeS_2. What evidence do you have that heating iron pyrite produces free iron? What evidence do you have that heating iron pyrite produces free sulfur?

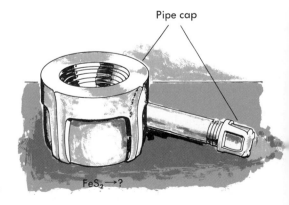
Pipe cap

$FeS_2 \longrightarrow ?$

Using Light to Break Apart a Silver Compound.
Add 5 milliliters of saturated table salt (sodium chloride—NaCl) solution to a test tube. Use a medicine dropper to add 30 drops of silver nitrate ($AgNO_3$) solution.

When silver nitrate is added to sodium chloride, a cloudy, white substance forms. Chemists have analyzed this substance. They say it is composed of silver atoms and chlorine atoms. They call it silver chloride. What atoms are probably still dissolved in the solution?

Set up a filter to separate the silver chloride from the other substances. Remove the filter paper from the funnel. Smear the silver chloride around on the paper. Then cover half the silver chloride with a piece of cardboard and place it near a bright light. After 5–10 minutes, compare the silver chloride in the light with the same material in the shadow. How has it changed?

Photographic film makes use of an interesting property of many silver compounds. These compounds break apart when light strikes them. Tiny specks of free silver form. These tiny pieces are black. The black parts of a photographic film negative are composed of tiny bits of silver.

Using a Catalyst to Speed Up Natural Decomposition. The chemical hydrogen peroxide (H_2O_2) is another example of a chemical which is not stable. It slowly breaks down or *decomposes* into two simpler substances, water and oxygen. Perhaps you bought a bottle of hydrogen peroxide a year ago to use as an antiseptic. Today, you probably have in your medicine chest only a bottle of water! Chemists have observed that this chemical change occurs more quickly in the presence of light. For this reason, hydrogen peroxide is sold in dark bottles.

Some chemicals speed up chemical changes without themselves changing. Such chemicals are called *catalysts*. What was the name of the catalyst you used on page 42 to speed up the natural decomposition of hydrogen peroxide?

Set up three test tubes, each containing 1−2 teaspoonsful of hydrogen peroxide. Add a tiny piece of animal fat or chicken skin to one test tube. Add the same amount of liver or red meat to the second tube. Add the same amount of bone or hair to the third tube. Which part of an animal contains

SILVER NITRATE SOLUTION: Add 17 grams of $AgNO_3$ to 100 ml of distilled water. Store in an amber bottle away from light.

$AgNO_3 + NaCl \rightarrow AgCl + ?$

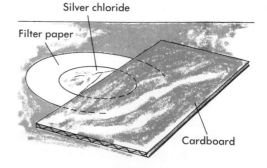

Silver chloride

Filter paper

Cardboard

a catalyst which increases the rate at which hydrogen peroxide decomposes? What gas forms?

Using Electricity to Decompose H₂O. Cut two strips from a tin can. Wrap plastic electrician's tape around the strips, as shown below. Bend the tin strips to fit over the edge of a beaker.

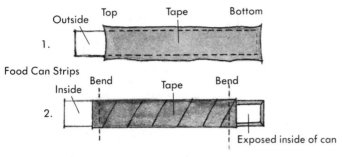

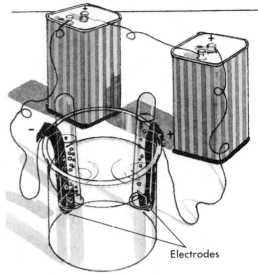

Electrodes

Fill the beaker three-quarters full of water to which a teaspoonful of sodium sulfate (Na₂SO₄) has been added. Put a test tube filled with water over the lower end of each of the strips. Connect the strips to two dry cells in series. At which strip is a gas produced faster?

Compare the amount of gases in the two test tubes at any one time. What does this show about the rates at which the gases are produced?

When one test tube is filled, lift it from the water without tipping it over. Bring a burning match near the opening. If an explosive pop is heard, the gas is hydrogen.

Test the gas in the other test tube with a glowing splint. What is the gas?

The separation of a compound by the process just described is called *electrolysis*. Many other useful elements are separated from other compounds the same way.

Chemists agree that many chemicals in solution consist of atoms which have electrical charges. These charged atoms are called *ions*.

The diagram at the right shows a chemist's idea about the way ions behave when electricity passes through a solution. Note the positively and negatively charged ions. Note the positively and negatively charged metal strips that are connected to the batteries. Chemists tell us that opposite charges attract. Toward which metal strip will the positively charged hydrogen ions (H⁺) move? Toward which metal strip will the negatively charged oxygen ions (O⁻⁻) move?

Using Heat to Break Down Copper Sulfate. Put a piece of copper sulfate in a test tube. Put a one-hole stopper in the test tube. Use a piece of bent glass tubing, a second test tube, and a beaker filled with cold water to set up the apparatus as shown at the right. Then heat the copper sulfate.

What collects in the empty test tube? How does the copper sulfate change?

Continue heating until all the copper sulfate has changed. Allow the heated test tube to cool and pour out the contents. What different physical properties does the changed copper sulfate have? Save the clear liquid to study in the next section.

The Test for Water. Dip a piece of cobalt chloride paper into water. How does the paper change? Chemists tell us that water is the only clear liquid which affects cobalt chloride in this way. As a result, this paper is used to find out if a clear liquid is water. Note the cobalt chloride paper on the next page. Test the clear liquid which formed in the last section. What is it?

The chemical test for water does not tell you all you might want to know about the liquid. Refer to the assumptions and limitations of the test for oxygen on page 43. What doesn't this test tell you? Can you use the cobalt chloride test to tell you if water is safe to drink? Explain your answer.

Qualitative Analysis and Quantitative Analysis. When chemists analyze a substance, they can determine two things. If they find out what chemicals are in the substance, it is called *qualitative analysis*. If they find out how much of a chemical is present, it is called *quantitative analysis*.

In the previous activity you determined that heating copper sulfate drives out water. What kind of analysis was this?

In the next analysis, you will repeat the previous experiment to determine how much water is in copper sulfate. What kind of analysis is this?

Quantitative Analysis of Water in Copper Sulfate. Weigh a clean, dry test tube. Crush some copper sulfate to a powder and add it to the test tube. Weigh the tube and contents. Calculate the mass of copper sulfate in the tube.

Heat the test tube and contents to drive off the water. Continue heating until all the copper sulfate has changed color. Reweigh the test tube and contents. How much mass was lost?

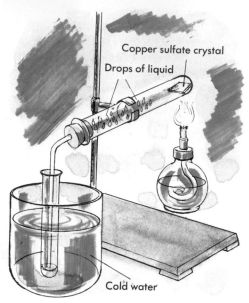

Copper sulfate crystal

Drops of liquid

Cold water

Quantitative Analysis of Water in Copper Sulfate	
A. Weight of test tube	g
B. Weight of test tube and blue copper sulfate	g
C. Weight of blue copper sulfate (B-A)	g
D. Weight of test tube plus heated copper sulfate	g
E. Weight of heated copper sulfate (D-A)	g
F. Water lost (C-E)	g
G. Fraction water (F/C)	

Dry cobalt
chloride paper

Wet cobalt
chloride paper

Turns pink

If blue cobalt chloride paper
turns pink when dipped in a
liquid, the liquid contains water.

Express as a fraction the mass of water in
copper sulfate compared to the total mass of the
blue copper sulfate. What percent of blue copper
sulfate is water? Review this experiment. What
errors may affect your answer? For example,
suppose you did not drive off all the water. How
would this error affect your answer? Consider each
error as to whether it would make your answer
smaller or larger than other students might get.

Comparing Your Results with the Theory.
Chemists have determined the relative mass of the
atoms of each element. For example, an oxygen
atom has 16 times more mass than a hydrogen
atom. A sulfur atom has 32 times more mass than a
hydrogen atom. The atomic masses of some other
familiar elements are given at the left.

The chemical formula of blue copper sulfate is
$CuSO_4 \cdot 5H_2O$. The mass of a molecule of blue
copper sulfate can be calculated as shown at the
left. How many times heavier is such a molecule
compared with a hydrogen atom?

The amount of blue copper sulfate which is
due to water is also shown at the left. According to
this theory, what percent of the blue copper sulfate
molecule is due to water. How does this value
compare with the value you determined
experimentally? Suggest some possible reasons for
any differences between your value and the
theoretical value.

Atomic Masses of Some Elements

Element	At. Mass
hydrogen (H)	1
helium (He)	4
carbon (C)	12
oxygen (O)	16
sulfur (S)	32
chlorine (Cl)	35.5
copper (Cu)	63.5
lead (Pb)	207

Calculating the Mass of $CuSO_4 \cdot 5H_2O$

$$1 \text{ Cu} = 63.5$$
$$1 \text{ S} = 32$$
$$4 \text{ O } (4 \times 16) = 64$$
$$5 \text{ (H}_2\text{O)} =$$
$$5\begin{cases}2 \text{ H} = 2 \\ 1 \text{ O} = 16\end{cases}$$
$$\overline{18 \times 5 = 90}$$

Weight of $CuSO_4 \cdot 5 H_2O = 249.5$
Weight of $5 H_2O = 90$

Percent of water in
blue copper sulfate

$$\frac{90}{249.5} \times 100 =$$

Summary Questions

1. In this section, different forms of energy have
   been used to break apart molecules. Name
   several of these energy forms.
2. What is the difference between qualitative
   analysis and quantitative analysis?
3. You can buy copper sulfate in the hydrated or
   anhydrous (an—without) forms. Explain what
   these forms are. (Hint: What does dehydrate
   mean?)

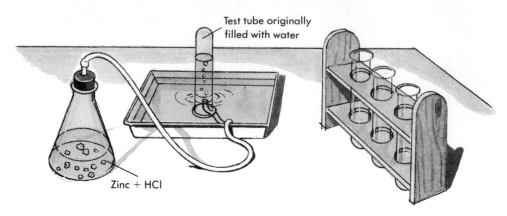

Test tube originally filled with water

Zinc + HCl

REPLACING ATOMS IN COMPOUNDS

You have already studied two types of chemical reactions in this chapter. The first type of reaction combined two elements, a metal and oxygen. The second type of reaction used energy to break apart compounds into simpler substances, including elements. In this section you will exchange one element in a compound with a different element. This type of reaction is called a single replacement reaction.

Replacing Hydrogen Ions with Metal Ions. Compounds which dissolve in water form positively charged and negatively charged atoms. You learned in the last section that charged atoms are called ions. Ions of elements which are metals are positively charged. Ions of nonmetallic elements are negatively charged. Hydrochloric acid (HCl) consist of H^+ and Cl^- ions.

$$Zn + H^+Cl^- \rightarrow$$

The symbols in the margin describe zinc atoms placed in hydrochloric acid. Set up such a condition. Add a small piece of zinc (the casing around a dry cell is zinc) to a flask containing weak hydrochloric acid (prepared by your teacher). Collect some of the gas using the method shown above.

Test the gas for hydrogen. What are some chemical and physical properties of hydrogen gas?

A Theory of Atoms and Ions. The diagrams on page 51 show a chemist's idea of a hydrogen atom and a hydrogen ion. Atoms similar to the hydrogen atom consist of a positively charged nucleus at the center and negatively charged electrons orbiting around the nucleus.

Compare the number of positive and negative charges in the hydrogen atom. Chemists say that

THE TEST FOR HYDROGEN: Bring a test tube of gas near a flame. If the gas burns quickly with a "barking" sound, the gas is probably hydrogen.

Hydrogen atom (H)   Hydrogen ion (H+)

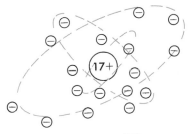

Chlorine atom (Cl)

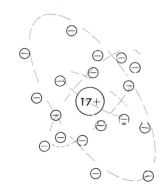

Chlorine ion (Cl⁻)

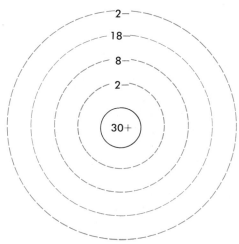

Zinc atom (simplified)

atoms are electrically neutral. Explain why.

Part of the diagram shows a hydrogen ion. What is the charge of this ion?

How many positive and negative charges are in a neutral chlorine atom? In a negatively charged chlorine ion?

The lower diagram shows a model of a zinc atom. Which negatively charged electrons are least attracted to the positively charged nucleus? Which electrons are most easily lost? What does it mean when chemists write Zn^{++} for the zinc ion?

Using the Theory of Ions. At the start of the experiment, the flask contained water, zinc atoms, hydrogen ions (H^+), and chlorine ions (Cl^-). At the end of the experiment, the hydrogen ions had become neutral hydrogen atoms (hydrogen gas) and the zinc atoms had disappeared.

Chemists believe that zinc atoms lose electrons to the hydrogen ions in an acid. Neutral hydrogen atoms and positively charged zinc ions (Zn^{++}) are produced. This can be written:

$$Zn + 2H^+Cl^- \rightarrow H_2 + Zn^{++}Cl_2^-$$

This statement of what happens is called a *chemical equation*.

Testing the Theory of Ions. Whenever you observe a chemical reaction, ask some good questions. What other similar chemicals probably react the same way? In the last reaction, a metal (zinc) reacted with an acid (HCl). Will other metals do the same thing? Will other acids do the same thing?

Make a study to find out if other metals become ions in the presence of acid, releasing neutral hydrogen atoms. Add a teaspoonful of weak hydrochloric acid to each of several test tubes. Add a different metal, such as copper, iron, magnesium ribbon, tin, or aluminum to each test tube.

Which metals react rapidly with acid and become ions? Which react slowly? Which do not react?

Make a table of the metals you have tested with acid. Put at the top of the table the metal which reacted quickest with acid. Put those metals which did not react on the bottom of the table. How does this ranking of chemical activity compare with a similar table you made earlier in the chapter when you prepared metal oxides?

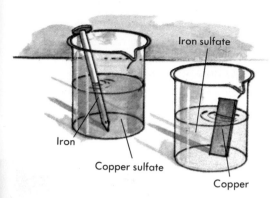

Iron sulfate

Iron

Copper sulfate

Copper

Reacting Metals with Metal Ions. Prepare solutions of copper sulfate and iron sulfate. Place an iron nail in the copper sulfate solution. Place a piece of copper in the iron sulfate solution.

Observe what happens after one day. How have the solutions changed? Examine the metals carefully. How have the metals changed? Does copper or iron have the greater tendency to lose electrons and exist as ions? Compare your answer to your previous experiments where you determined an activity scale for metals.

The Activity Scale for Metals. Many elements have a tendency to lose electrons, forming positive ions. This group of elements is called metals. Some metals are more active than others; that is, they have a greater tendency to lose electrons and exist as ions. Sodium and calcium are two examples of active metals. Why are these metals never found in nature as pure metals?

Gold is rarely found combined with other ions in nature. Describe its chemical activity. Compare your answer with the value in the table below at the left. What value is this low activity to jewelry makers?

Study the table. Which element in the table have you not considered as a metal until now? What evidence have you observed in this chapter which supports the idea that this element behaves like a metal? What properties of this element are different from the properties of metals?

Copper Sulfate Solution:

Dissolve 10 grams of $CuSO_4 \cdot 5H_2O$ in 100 ml of water.

Iron Sulfate Solution:

Dissolve 10 grams of $FeSO_4 \cdot 7H_2O$ in 100 ml of water.

Activity Scale of Metals

Calcium	(Ca)
Sodium	(Na)
Magnesium	(Mg)
Aluminum	(Al)
Zinc	(Zn)
Tin	(Sn)
Lead	(Pb)
Iron	(Fe)
Hydrogen	(H)
Copper	(Cu)
Silver	(Ag)
Platinum	(Pt)
Gold	(Au)

Summary Questions

1. Use your knowledge from this section, plus the activity scale at the left, to predict which mixtures shown below will react.

2. For each reaction, show the chemicals which form.

   a. $Zn + H_2SO_4$ (sulfuric acid) →

   b. $Cu + HNO_3$ (nitric acid) →

   c. $Zn + CuSO_4$ (copper sulfate) →

   d. $Cu + FeSO_4$ (iron sulfate) →

   e. $Au + NaNO_3$ (sodium nitrate) →

   f. $Zn + NaCl$ (sodium chloride) →

   g. $Cu + ZnSO_4$ (zinc sulfate) →

   h. $H_2 + AgNO_3$ (silver nitrate) →

WHEN TWO COMPOUNDS REACT

When sodium chloride (NaCl) dissolves in water, chemists describe the solution as containing positive sodium ions and negative chlorine ions. As a beginning chemist, you need to know what will happen when two solutions of different ions are mixed. Will the chemicals react, forming a new substance, or will nothing happen? In this section you will learn to predict what will happen when solutions are mixed.

Reactions Occur When a Precipitate is Formed. Add 10 ml of NaI (sodium iodide) solution to one test tube and 10 ml of $Pb(NO_3)_2$ (lead nitrate) to a second test tube. Using both test tubes, add both clear liquids to a beaker at the same time as shown here. What evidence do you see that a chemical reaction has occurred, forming a new substance?

Analyzing Ions in a Reaction. Study the table below. In this table, S means that a compound is soluble. The compound dissolves in water, forming ions. An I means that the compound is insoluble. Positive and negative ions do not separate. As a result, the compound does not dissolve in water.

The blue shaded part of the table describes how sodium iodide behaves in water. Does sodium iodide dissolve in water? Does lead nitrate dissolve in water?

The two compounds listed above readily break down in water into ions. However, when the two solutions are mixed together, four types of ions are mixed. When lead and iodide ions meet, the ions are so strongly attracted that an insoluble compound, lead iodide, forms. What is the yellow substance which formed in the beaker in the above experiment? What atoms probably still remain dissolved in the water as ions?

The insoluble, cloudy chemical which forms in some reactions is called a *precipitate*. When two clear liquids are mixed and a precipitate forms, this observation indicates that a new substance has formed. A chemical change has taken place. How can tables like the one at the right be used to predict when a chemical reaction will occur between two or more solutions?

Solubility Table

Negative ion

Positive ion	I^-	NO_3^-
Na^+	S	S
Pb^{++}	I	S

Ions and Radicals. Compounds form when metal ions and nonmetal ions combine. The name of the new compound is very similar to the name of the two elements from which it forms. For example, sodium and chlorine ions form sodium chlor*ide*. Magnesium and oxygen ions form magnesium ox*ide*. What would be the name of the compound of zinc and sulfur ions? What is the general rule for naming such compounds?

Many compounds consist of more than two elements. Frequently part of this compound, consisting of several different atoms, reacts as though it were a metal or nonmetal ion. Such groups of charged atoms are called *radicals*. Identify the radicals in the table below.

Solubilities of Compounds in Water
(S, soluble in water   I, insoluble in water)

	Acetate $C_2H_3O_2^-$	Carbonate CO_3^{--}	Chlorate ClO_3^-	Chloride Cl^-	Chromate CrO_4^{--}	Hydroxide OH^-	Iodide I^-	Nitrate NO_3^-	Oxide O^{--}	Phosphate PO_4^{---}	Sulfate SO_4^{--}	Sulfide S^{--}
Aluminum Al^{+++}	S		S	S		I	S	S	I	I	S	
Ammonium NH_4^+	S	S	S	S	S		S	S		S	S	S
Calcium Ca^{++}	S	I	S	S	S	S	S	S	I	I	I	I
Copper Cu^{++}	S		S	S		I		S	I	I	S	I
Iron Fe^{+++}	S		S	S	I	I	S	S	I	I	I	
Lead Pb^{++}	S	I	S	S	I	I	I	S	I	I	I	I
Magnesium Mg^{++}	S	I	S	S	S	I	S	S	I	I	S	
Manganese Mn^{++}	S	I	S	S		I		S	I	I	S	I
Mercury Hg^{++}	S		S	S	I	I	I	S	I	I		I
Potassium K^+	S	S	S	S	S	S	S	S	S	S	S	S
Silver Ag^+	I	I	S	I	I		I	S	I	I	I	I
Sodium Na^+	S	S	S	S	S	S	S	S	S	S	S	S
Zinc Zn^{++}	S	I	S	S	I	I	S	S	I	I	S	I

Predicting Precipitates. The table on the previous page lists the solubility of nearly 150 different compounds. These compounds are only a tiny fraction of the thousands of compounds whose solubilities have been studied and catalogued. The table can be used to help you predict which chemicals will react when solutions are mixed. Teach yourself to predict when two chemicals will react to form precipitates. Use the two examples below. Then test your knowledge on the examples at the bottom of the page.

Example 1:
  Will these two chemicals react?
  silver nitrate + sodium chloride → ?
  $Ag^+NO_3^- + Na^+Cl^- →$?

  Answer: Silver nitrate and sodium chloride are both soluble, so they will both dissolve. However, when mixed together, sodium nitrate and silver chloride compounds are able to form. According to the table, silver chloride is insoluble. As a result, a precipitate will form. (This reaction was observed in the previous chapter.)

Example 2:
  Will these two chemicals react?
  copper chloride + sodium sulfate → ?
  $Cu^{++}Cl_2^- + Na_2^+SO_4^{--} →$?

  Answer: Copper chloride and sodium sulfate are both soluble in water. When mixed together, the other two possible arrangements of ions, copper sulfate and sodium chloride, are also soluble. No precipitate will form; no chemical reaction occurs.

Now test your skill. Which solutions of these mixtures will react?

1. ammonium carbonate and silver chloride → ?

2. calcium nitrate and magnesium sulfate → ?

3. ammonium phosphate and sodium chloride →?

4. potassium nitrate and sodium chlorate → ?

5. sodium acetate and potassium chloride → ?

6. calcium chloride and zinc sulfate → ?

Some Common Gases

H_2 hydrogen

H_2S hydrogen sulfide
(rotten egg smell)

$H_2CO_3 \rightarrow CO_2 + H_2O$
(spontaneously)

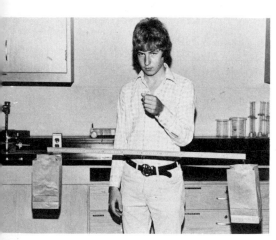

CO₂ gas

Meterstick

Paper bags

Reactions Occur When Gas is Formed. Put a piece of calcium carbonate ($CaCO_3$) such as chalk, egg shell, sea shell, marble or limestone into a bottle. Slowly add enough weak hydrochloric acid (HCl) to cover the calcium carbonate. Put a metal or glass plate loosely over the mouth of the bottle.

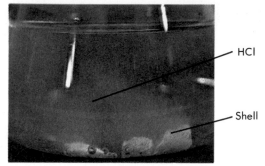

HCl

Shell

What is happening when the calcium carbonate and hydrochloric acid are in contact with each other?

After a few minutes, test the gas in the bottle with a glowing splint. Does the gas which is forming behave like oxygen?

Put a burning splint into the bottle. Does the gas behave like hydrogen would? What do such test results tell you about the gas?

The chemical reaction above is written:

$$Ca^{++}CO_3^{--} + 2H^+Cl^- \rightarrow H_2^+CO_3^{--} + Ca^{++}Cl_2^-$$

The chemical, H_2CO_3, however, is not stable. It breaks down quickly, as shown below.

$$H_2^+CO_3^{--} \rightarrow H_2O + CO_2 \uparrow$$

The vertical arrow is written to show that carbon dioxide is a gas. What properties of this gas did you observe while testing for hydrogen and oxygen? What other interesting property of carbon dioxide gas is shown at the left?

More Predicting. Chemists try to learn which mixtures will react. When two solutions are mixed and bubbles of gas appear, a reaction has occurred. Note the chemical formula of three common gases. Use the formulas below to predict which mixture will react:

1. $H^+Cl^- + Fe^{++}S^{--} \rightarrow$

2. $H^+(NO_3)^- + Pb^{++}S^{--} \rightarrow$

3. $Zn + H^+Cl^- \rightarrow$

4. $Cu^{++}(SO_4)^{--} + Na^+Cl^- \rightarrow$

5. $Na^+(HCO_3)^- + H^+(C_2H_3O_2)^- \rightarrow$
   (vinegar)

6. $Zn^{++}Cl_2^- + Cu^{++}Cl_2^- \rightarrow$

ACIDS

H^+Cl^- hydrochloric acid

$H^+NO_3^-$ nitric acid

$H_2^+SO_4^{--}$ sulfuric acid

$H^+C_2H_3O_2^-$ acetic acid (vinegar)

$H_2^+CO_3^{--}$ carbonic acid
(unstable)

BASES

Na^+OH^- sodium hydroxide

$Ca^{++}(OH)^-$ calcium hydroxide

$NH_4^+OH^-$ ammonium hydroxide

K^+OH^- potassium hydroxide

$Mg^{++}(OH)_2^-$ magnesium
hydroxide

IDENTIFY ACIDS AND BASES

$NaCl$	$HgCl_2$
$MgSO_4$	$BaSO_4$
H_2SO_4	HF
$Cu(NO_3)_2$	$Al(OH)_3$

Acids and Bases. Chemists do not try to memorize how thousands of different chemicals behave. Instead, they think of a chemical as belonging to one of several different groups. Chemicals which belong to the same group have similar chemical properties. Thus, chemicals which belong to the same group will often behave the same.

Two important groups of chemicals are called *acids* and *bases*. The first table at the left lists several important acids. The second table lists several important bases. Now look at the third table. Which chemicals are acids? Which are bases? Which are neither acids nor bases?

Testing for Acids and Bases. Place red and blue litmus strips into small samples of weak acids and bases which your teacher has prepared. (**CAUTION:** Do not spill acids or bases on your skin or clothes.) How do acids and bases react with litmus paper? What do these tests fail to tell you about each solution you tested?

Reactions Occur When Water is Formed. So far, you have learned in this section that two chemicals will react if their ions produce (1) an insoluble product or (2) a gas. There is a third rule that chemists have observed: Two chemicals will react if their ions can combine to form water. Two chemicals which react to form water are an acid and a base. Study the reaction shown below:

$$K^+OH^- + H^+Cl^- \rightarrow$$
$$\text{(base)} + \text{(acid)}$$

What are the products in this reaction?

Slowly add weak HCl to weak KOH, testing the mixture with litmus paper. Stop adding the acid when neither red nor blue litmus will change color. Is there any acid or base left in the mixture at this time? How do you know? What chemicals are now in the mixture?

When an acid and base combine, two products are formed. One product is water, the other product is called a *salt*. Thousands of different chemicals are called salts. The salt you use on foods is called table salt (NaCl). What base and acid would you react to produce water and table salt?

Which chemicals in the third table at the left are salts? What base and acid would you combine to form each salt?

REVIEW QUESTIONS

1. Which of the following are elements and which
    are compounds:
    a. Fe, Al, H_2O, CO_2, HCl, S, C, MgO, H.
    b. Gold, sodium chloride, sodium bicarbonate,
        helium, copper sulfate, potassium
        hydroxide.
    c. Table salt, baking soda, quicksilver.
2. Which part of question 1 above is easiest — a, b,
    or c? Which is hardest? Why?
3. Name some physical properties of oxygen,
    hydrogen, and carbon dioxide that you have
    observed in this chapter.

Activity
Scale of
Metals

Ca
Na
Mg
Al
Zn
Sn
Pb
Fe
H
Cu
Ag
Pt
Au

4. Refer to the activity scale at the left. Explain
    why pure sodium is never found in nature.
    Explain why gold is usually found free (not
    combined with other elements).
5. Heat sometimes causes molecules to break
    down into simpler substances. Heat sometimes
    causes atoms or molecules to combine, forming
    larger compounds. What examples of each type
    of reaction have you observed?
6. Which of the following mixtures will react?
    Why?
    a. Fe + HCl
    b. Ag + H_2SO_4
    c. Zn + O_2
    d. NaOH + HCl
    e. $NaHCO_3$ + HNO_3
    f. Cu + $FeSO_4$

THOUGHT QUESTIONS

1. Thinking in terms of atoms, why does a
    molecule of carbon dioxide weigh more than a
    molecule of oxygen (O_2)?
2. How do chemists know that copper sulfate has
    $5H_2O$ and not $4H_2O$ or $6H_2O$?
3. How can computers be helpful to chemists?
4. How can there be thousands of different
    substances when there are only 92 naturally
    occurring elements?

Investigating on Your Own

The study of chemicals and their changes offers many opportunities for doing science projects. Most of the projects suggested in this section require that you plan, carry out, and think about your results in the same way that research scientists and technicians do. In this way, you can consider whether a career in some area of science has interest for you.

Many of the activities suggested in this section require more time than one class period, or special materials not found in some schools. For these reasons, you should identify one or more activities from these pages to carry out on your own, outside of class. Ask your teacher to help you plan your work during class time. Then you will have all the materials you need to carry out your research. Plan to share your findings with others during later class periods.

THE SKILL OF OBSERVING

Scientists use many special skills in their work. One of the most important skills is knowing how to look carefully at something. No one is born a careful observer. You must practice to develop this skill. In this section you will be given practice in observing. Then you can select an investigation to do on your own which requires a high degree of observational skill.

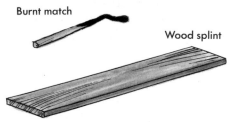
Burnt match

Wood splint

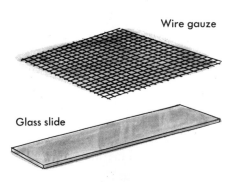
Wire gauze

Glass slide

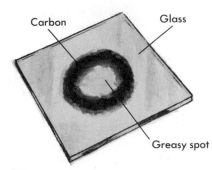
Carbon

Glass

Greasy spot

The Nature of an Observation. When you use one or more senses to describe something, you are observing. For example, "The candle is shaped like a cylinder" is an observation. "The candle is shorter now than it was five minutes ago" is another observation. Both of these observations use the sense of sight. However, "The candle is composed of atoms of carbon and hydrogen" is not an observation. Why? Make another statement about candles which would not be called an observation.

Observing a Burning Candle. Observe a burning candle with one or two classmates. Keep a record of your observations. How many different observations can you make in 15 minutes? Compare your observations with others in your class. How many different observations were made by the class in this length of time?

Spend a minute or two analyzing why you did not make all the observations your classmates made. Then take five more minutes to observe the candle even more carefully than before. What new observations can you record?

Working with the Burning Candle. The drawing at the left shows four simple pieces of equipment to assist you in observing parts of a burning candle. Take a few more minutes to observe (and record) the following:

1. Did you observe the pool of liquid at the top of the candle? Sprinkle a few tiny bits of ash from a burnt match into this pool and observe it.

2. Did you observe the white smoke which rises when the flame is blown out? Light a wooden splint and use this flame to study the white smoke.

3. Use wire gauze and a lighted wooden splint to study the flame.

4. Hold a glass slide in the flame for 2 seconds. Observe the slide.

Compare your observations with others in the class. Did these simple tools aid you in observing? Why?

Organizing Observations. By now, your classmates have dozens of different observations about burning candles. They probably have so many different observations that they lose track of them. Obviously, the observations need to be organized. There are, however, good and bad ways of organizing observations. Good ways of organizing aid the observer, bad ones do not. Would organizing the observations alphabetically be helpful? Is it a good way to organize?

One way to organize candle observations is to group them by candle regions. For example, group the observations about the solid wax cylinder together. What other observations fit together? Arrange the rest of your observations this way. Why might grouping observations this way be useful?

Combining Observations and Past Information. What is an explanation? Some teachers say an explanation is a statement which allows a person to take a new observation and make it "fit" with the person's previous knowledge. You probably have made many unexplained observations. That is, you probably have not thought how to make these candle observations "fit" with your previous knowledge. In this section, many aids will be given to help you think about your observations and explain them.

1. Observation: When a screen is placed across the flame, it appears to be hollow. Only the outer edge is burning.

A. Previous knowledge: Burning occurs only if there is (1) a fuel, (2) oxygen, and (3) enough heat to raise fuel to its kindling temperature. Now, combine observation (1) with your previous knowledge (1A) to produce an explanation for this observation.

2. Observation: The candle wick is white at the bottom, black in the middle, and glowing red at the top.

A. Previous knowledge: In the absence of air, wood turns to charcoal when heated. Charcoal glows red when combining with oxygen (burning). Combine observation (2) with your previous knowledge (2A) to produce an explanation for this observation.

Wire screen

Candlemaking as it was done in the 1700's and 1800's.

Candlemaking as it is done today.

3. Observation: When a screen is placed across the flame, the flame above the screen goes out.

A. Previous knowledge: Flames go out when fuel, oxygen, or heat is removed. Explain this statement.

4. Observation: When the screen in number 3 above is held across the flame for several minutes, the flame reappears above the screen. Explain this.

Now, try to explain your observations.

Some Additional Candle Knowledge. You probably noticed the pool of liquid wax at the top of the candle. Many observers have noted that the sides surrounding the pool of wax are higher, as shown at the left. This is caused by the shape of the flame which melts the solid wax. In drafty rooms, however, the candle flame may blow about, melting the edge of the pool and allowing hot wax to drip down the candle. To prevent dripping, expensive candles are dipped into a wax that has a higher melting point. Such candles are sold as dripless candles. Why would these candles be dripless?

Candle manufacturers make wicks from string consisting of several strands. One strand is pulled slightly, causing the string to kink (curl). As a result, the wick at the top of the candle always bends over. The end of the wick extends to the edge of the flame. What does this do to the end of the wick? Why might a short wick be useful?

The Explanations of Others. Observers should not expect to explain all their observations. Good observers, however, organize their observations so they are easier to remember. Then, later, when possible explanations are proposed, the observers can determine if these explanations fit their earlier observations. The following explanations of candle burning have been proposed by others.

The fuel of a burning candle is wax, in the gas state. Wax consists of a mixture of large molecules, such as $C_{26}H_{54}$. Wax molecules resemble a string of carbon atoms bonded to each other. This string is surrounded by hydrogen atoms. Hydrogen burns (combines with oxygen) producing a blue flame. At this time, the remaining carbon atoms without hydrogen atoms have the property of tiny black pieces of carbon. These pieces are heated to incandescence, giving off the light seen in the flame. White-hot carbon particles combine with oxygen (burn) forming carbon dioxide. If these white-hot glowing carbon particles are cooled below incandescence, black carbon (soot) escapes unburned.

The combining of hydrogen atoms and oxygen atoms releases enough heat to vaporize more liquid wax on the wick. This assures a continuing supply of fuel in the flame. The heat also melts the solid wax, adding to the pool at the base of the wick. Liquid wax rises up the wick by a process called capillary action.

How many observations of yours fit this explanation? What parts of the explanation were you unable to observe?

OBSERVING ON YOUR OWN

This page and the next provide many suggestions for observing chemical changes. Carry out one or more of the activities on your own. Try to develop an explanation for your observations. Report your observations and explanation to the rest of the class.

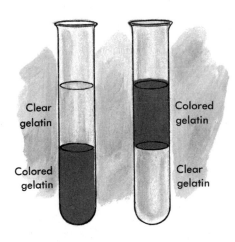

Clear gelatin

Colored gelatin

Colored gelatin

Clear gelatin

Types of fabrics	Berry juice	Types of dyes RIT		
Cotton				
Wool				

1.   Polish strips of different metals—aluminum, copper, zinc, lead, brass, steel with fine steel wool. Wax one side of each strip; leave the other side uncovered. Make a record of the changes in the metals. Explain the changes.

2.   Find out whether water can diffuse through gelatin. Make up gelatin according to the directions on the package. Color half the gelatin with food dye; leave the other half clear. Put some colored gelatin into one test tube and some clear gelatin into another. Chill both to harden the gelatin. Then put additional gelatin into the test tubes as shown here. Chill the gelatin again. Observe the changes in the test tubes for several days.

3.   Fill balloons or thin plastic bags with air, carbon dioxide, and natural gas. Compare the actions of the balloons when floating in the air. Assume each bag has about the same number of gas molecules in it. Which gas has the lightest molecules? The heaviest molecules?

4.   Study the effect of dyes on different fabrics. Test natural and synthetic dyes and fabrics. Glue each dyed cloth onto a chart, as shown at the left.

White card

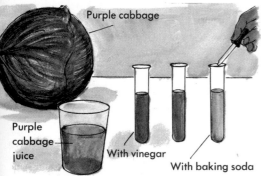

Purple cabbage

Purple cabbage juice

With vinegar

With baking soda

Cold water

Angel food cake pan

Boiling water

Microphotograph of wool fiber

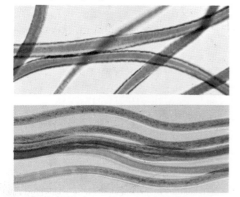

Microphotograph of nylon fiber

5.   Compare the color in the light-protected seams of old clothes with the colors in the rest of the fabric. Which colors withstood fading best? Poorest?

6.   Discover the hottest regions in a candle flame by cautiously holding white paper cards in the flame until they begin to scorch.

7.   Add several leaves of red cabbage to a pan of water. Boil the water until it is purple. Allow the water to cool and add it to a number of test tubes or juice glasses. Add small amounts of household chemicals to each jar, such as aspirin, baking soda, and so on. Keep a record of your observations. Propose an explanation.

8.   Make a device for preparing distilled water as shown at the left. The center pan is an angel food cake pan. The top pan is kept full of cold water (with ice or snow if possible).

9.   Observe a number of different pieces of plastic when heated or when small amounts are held in a flame. Organize the plastics into groups which behave similarly.

10.   Use a microscope to compare the strands in natural and synthetic fibers. Make sketches of each type of fiber studied and point out the difference between them.

11.   Study the solubility of carbon dioxide in water. Collect a bottle of carbon dioxide gas. Add three tablespoonfuls of cold water to the bottle and plug it with a one-hole stopper and nozzle tube as shown below. Shake the tube vigorously while plugging the nozzle. Then open the nozzle tube under water. Explain the results. Test the water in the jar with litmus paper and explain the results.

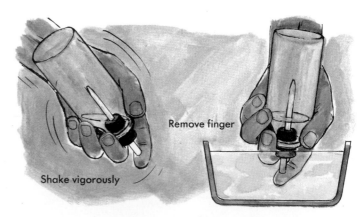

Remove finger

Shake vigorously

ANALYZING SUBSTANCES ON YOUR OWN

One important part of chemistry involves knowing how to identify and analyze chemicals. Analyze some chemicals by doing one of the activities described below. Report your results to the class.

1. Analyze which chemicals release the most heat energy when they burn. Put an alcohol burner, candle, or canned heat on a balance. Burn exactly one gram of fuel. How much can you increase the temperature of 250 grams of water by using this amount of each fuel? Calculate the calories absorbed per gram of fuel burned.

2. Determine the weight of the water of crystallization that is driven off by heating a sample of Epsom salt or borax. Calculate the percentage of the original weight made up of water.

3. Test several different kinds of soap with litmus paper or pH paper. Are soaps generally acidic, neutral, or basic? Is there a difference in the pH of different soaps?

4. Using commercial pH test paper, determine the pH of several common foods. Prepare a chart of your findings.

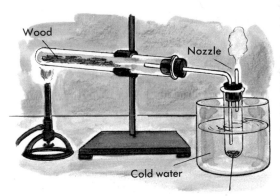

Wood

Nozzle

Cold water

Wood condensate

5. Analyze the water content in different foods. Weigh a sample of food such as green beans. Heat the sample in an oven for 15 minutes at a temperature just above the boiling point. Reweigh the sample. Then repeat the process until the weight remains constant. Calculate the percentage of water in the food.

6. Locate a soil testing kit in school or the directions for testing soil in a library. Analyze some local soil. Keep a record of your findings.

7. Analyze the liquid which condenses when charcoal is prepared. Test this liquid with litmus and cobalt chloride paper. Heat a sample of the liquid, plotting temperature vs. time until all the liquid is boiled away. Analyze the graph and identify some of the chemicals with the aid of your teacher.

8. Prepare coke from coal using the apparatus described in number 7. Analyze the liquid which condenses in the test tube as described above.

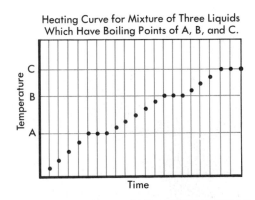

Heating Curve for Mixture of Three Liquids Which Have Boiling Points of A, B, and C.

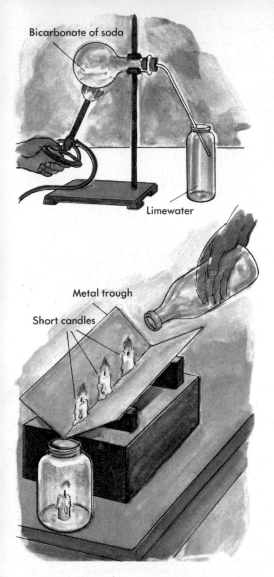

Bicarbonate of soda

Limewater

Metal trough

Short candles

MORE LABORATORY INVESTIGATIONS

1. Find out whether carbon dioxide can be produced from baking soda by heating it as shown.

2. Find out if carbon dioxide can be produced by adding baking soda to water at different temperatures.

3. Extract the juice from different fruits. Mix each with acidic and basic solutions to see which juices change color. Prepare a chart showing the color changes.

4. Chop up a beet, boil it, and filter off the liquid. Determine the range of colors that can be produced by acidic and basic solutions. Compare the colors with those of purple cabbage juice.

5. Put colored flowers or leaves into water that is slightly acidic or basic. Note whether the colors change.

6. Produce carbon dioxide in a large bottle. Pour the gas down a trough containing small candles as shown at the left. Explain the results to your classmates.

7. Compare heat produced by different sizes and kinds of candles. Measure the rise in temperature of water heated by each candle for one minute.

8. Freshly peeled potatoes and apples often turn brown in air. Set up an experiment to find out whether this change is caused by oxygen.

9. Find out whether sulfur dioxide will bleach a slice of apple or potato that has turned brown. If so, does the slice turn brown again as quickly as a fresh slice?

10. Hydrogen peroxide is sold in bottles made of dark glass. Find out if light affects this chemical. Divide the contents of a new bottle in half, putting one part in a clear glass bottle and the remainder in the dark glass bottle. Store both bottles in a lighted place for several days. Measure the oxygen produced from each.

11. Distill natural wintergreen oil from black birch bark or wintergreen leaves.

Joseph Priestley

Hard water

Soft water

OTHER INVESTIGATIONS AND PROJECTS

1. Find out how Joseph Priestley discovered oxygen. Repeat his experiments if possible.

2. Find out how oxygen is produced for use in welding, hospitals, and making steel.

3. Visit a place where acetylene torches are being used. Report on your observations.

4. Make a model of an automatic sprinkler as shown at the left. Punch holes in the bottom of a can and seal the holes with candle wax. Fill the can with water and hang it outdoors. Build a fire under the can.

5. Arrange for a demonstration of different kinds of fire extinguishers.

6. Prepare an exhibit of different kinds of fuels.

7. Choose one of the elements from the table on page 84 of this unit. Find out all you can about the element and make a report to the class.

8. Visit a garage and make a study of the antifreeze mixtures used in radiators. Report on the chemicals used and the percentages of each which provide protection at different temperatures.

9. Visit a bottling plant and find out how water is carbonated for use in soda pop.

10. Make "sympathetic" ink. Write with dilute cobalt chloride solution on white paper, using a steel pen. The writing is pink and almost invisible until it is dried over heat. It then turns blue.

11. Set up a demonstration of diffusion by suspending a tea bag in a large jar of water.

12. Make pH paper by dipping strips of paper towel or filter paper into purple cabbage juice. Dry the strips. Use them as you would litmus paper.

13. Several devices are made for removing dissolved minerals from water that is to be used in steam irons. Collect water from one of these devices, evaporate it, and note the amount of solid materials remaining. Compare the results with similar tests of untreated water and distilled water.

Home-made
Solid alcohol

erno

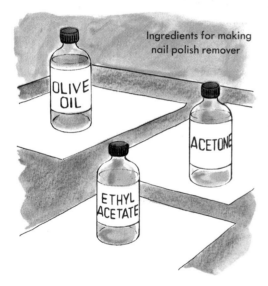

Ingredients for making
nail polish remover

OLIVE
OIL

ACETONE

ETHYL
ACETATE

CAUTION: Acetone is
highly flammable.

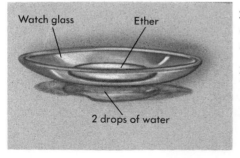

Watch glass   Ether

2 drops of water

14. Demonstrate the preparation of "solid" alcohol. Dissolve as much calcium acetate as possible in 10 milliliters of water. Pour this solution into a beaker containing 60 milliliters of methyl alcohol or ethyl alcohol. After a few seconds, tip the jar upside down, demonstrating the change in the material in the beaker. Place a small amount of this gel in a metal can and light it. Compare this flame with the flame produced by commercial "canned heat."

15. Prepare a report about such manufactured silicone chemicals as silicone oil and silicone rubber.

16. Set up a display of materials made from coal tar.

17. Use an organic chemistry book from a library or high school chemistry teacher to locate the structural formulas for: (1) trinitrotoluene, (2) 2,4-dichlorophenoxyacetic acid, and (3) dichlorodiphenyltrichloroethane. Explain to your class what each name means, its common abbreviation, and the use of each substance.

18. Prepare nail polish remover by mixing one teaspoonful of ethyl acetate ($CH_3COOC_2H_5$), three teaspoonfuls of acetone (CH_3COCH_3), and one teaspoonful of olive oil. Compare this polish remover with commercial brands.

19. Demonstrate the cooling produced by evaporating ether (CH_3OCH_3) as shown below. As the ether evaporates and cools the watch glass, the water beneath the glass freezes. (**CAUTION: DO NOT USE IN A ROOM WITH FLAMES.**)

20. Make artificial lemon juice by mixing a small quantity of citric acid ($H_3C_6H_5O_7$) and table sugar ($C_{12}H_{22}O_{11}$) in water. Adjust the amounts of each ingredient until a lemonade taste is produced.

21. Arrange a visit to the local gas company. Find out the composition of the gas distributed locally and its geographical source.

22. Read about the methods of preparing home-made soap in pioneer days. Follow these procedures to make soap. Compare your results with commercial soaps.

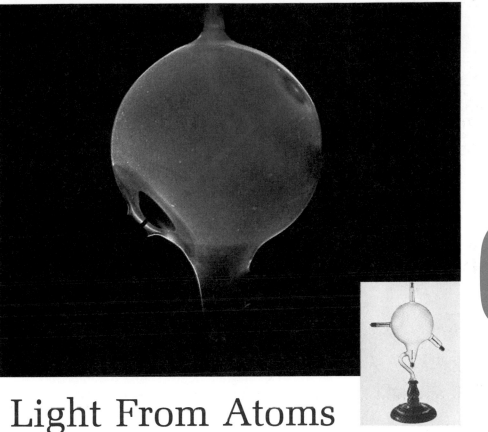

Light From Atoms

In the late nineteenth century, physical scientists discovered that glass containers, filled with gas at low pressure, gave off light if electricity was passed through the gas. The photograph above shows such a tube containing a small amount of oxygen. Note the light in the tube.

In the early twentieth century, a young Danish physical scientist, Niels Bohr, made some observations using tubes filled with hydrogen. Niels Bohr used his observations to develop an idea about what atoms must be like based on the light given off. Much of our present understanding of atoms still resembles his ideas of 70 years ago.

LIGHT, VISIBLE AND INVISIBLE

White light can be separated into many colors. The different colors which form are called a spectrum. *The spectrum which is made from light we can see is called the* visible spectrum. *There is, however, "light" that we cannot see. Some of this light is too red for our eyes to detect. This light is called infrared light. Light beyond the purple end of the visible spectrum is called ultraviolet light. Study the spectrum above. Where would infrared and ultraviolet light fall on this diagram?*

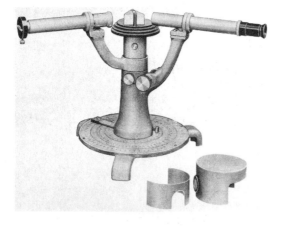

Wavelengths and Energy. Perhaps you have wondered why some light is red and other light is blue. In this chapter you will find it helpful to think of light as packets of energy which travel as waves. Different colors of light are caused by waves of different lengths.

The diagram below shows the longest wavelength of red light and the shortest wavelength of violet light that we can see. The length of a wave is measured in angstroms. One angstrom is 1/10 000 000 of a millimeter. Estimate the wavelength of green light.

The diagram below also shows which waves have the most energy. Which waves have more energy, red light or violet light? Ultraviolet light or infrared light? Light scientists say that the energy in light varies inversely with the wavelength. Discuss what this means in class.

Spectroscopes. Spectra are produced for study by means of spectroscopes. One type of spectroscope shown at the left uses a prism to separate light into its colors.

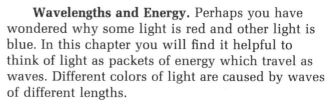

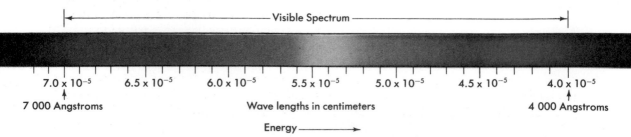

|← Visible Spectrum →|

| 7.0 x 10⁻⁵ | 6.5 x 10⁻⁵ | 6.0 x 10⁻⁵ | 5.5 x 10⁻⁵ | 5.0 x 10⁻⁵ | 4.5 x 10⁻⁵ | 4.0 x 10⁻⁵ |

7.0×10^{-5} 6.5×10^{-5} 6.0×10^{-5} 5.5×10^{-5} 5.0×10^{-5} 4.5×10^{-5} 4.0×10^{-5}

7 000 Angstroms   Wave lengths in centimeters   4 000 Angstroms

Energy ——→

Glass tube

Diagram of an Atom

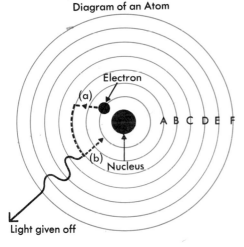

Electron

(a)

A B C D E F

(b) Nucleus

Light given off

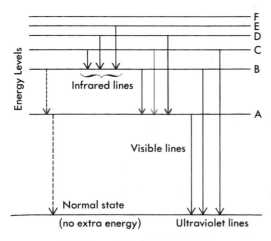

Energy Levels

F
E
D
C
B

Infrared lines

A

Visible lines

Normal state
(no extra energy)

Ultraviolet lines

Hydrogen, the Simple Spectrum. The glass tube at the left contains hydrogen gas. Two wires pass through the glass. The other ends of the wires are connected to a source of high voltage. When electricity flows through this circuit, the hydrogen gas gives off light. The visible spectrum of hydrogen gas is shown above.

If the spectroscope is changed so that light passing through can land on a fluorescent screen, an interesting observation results. Not only can three similarly spaced lines be seen on the fluorescent screen, but similar sets of lines appear in the ultraviolet (higher energy) region and the infrared (lower energy) region. All this information was known by the scientists studying light in the early 1900's. Niels Bohr, however, was the first person to use these observations to propose our present model of an atom.

According to his theory, the single electron of hydrogen usually revolves in an orbit close to the nucleus. If energy is added to the atom, the excited electron is forced to a more distant orbit, as shown by (a) in the diagram of an atom. Upon moving back (b), the extra energy is given off as a packet of light, or *photon*.

The atomic diagram shows the orbits in which the electron may travel. An electron traveling in any of the higher orbits (A–F) would require more energy than in the closest orbit (normal state). For this reason, it is easier to think of energy levels (shown here) instead of orbits. Suppose that an electron absorbs energy, forcing it up to level C. The far right arrow shows the loss in energy as the electron falls back. One high energy photon of ultraviolet light is given off.

Suppose that the electron absorbs less energy and is forced to level B. The electron may return in one jump, giving off a photon of ultraviolet at a longer wavelength (lower energy) than before. Or it may return to level A, producing photons of visible (red) light and then continue to the stable level by giving off a photon of ultraviolet light. The greater the drop, the shorter the wavelength (and greater energy) of the light given off.

71

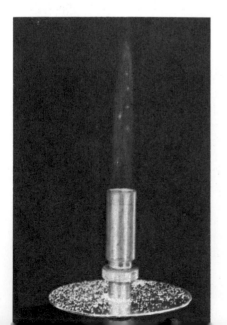

Other Bright Line Spectra. The 6 spectra above show the lines produced by atoms of other elements. These spectra can be produced from gas-filled tubes, such as in a neon sign or by heating a compound in a nearly colorless flame. For example, table salt (sodium chloride) produces yellow light when bits of salt are dropped into a gas flame as shown here. What would this yellow light look like when viewed through a spectroscope? Compare the neon spectrum with the color of the neon sign shown in the photograph on page 73.

When a piece of glass is heated to a very high temperature, the flame around the glass gives off light. What could you do to identify one element in the glass? Why do chemists find bright line spectra useful?

Continuous spectrum of an incandescent lamp

7.0 x 10⁻⁵   6.5 x 10⁻⁵   6.0 x 10⁻⁵   5.5 x 10⁻⁵   5.0 x 10⁻⁵   4.5 x 10⁻⁵   4.0 x 10⁻⁵

Wave length in centimeters

Continuous plus bright line spectra from a fluorescent lamp

Incandescent light bulb

Fluorescent light bulbs

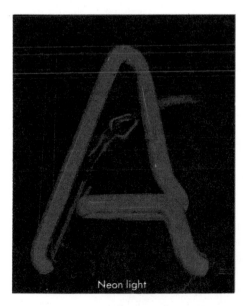

Neon light

Continuous Spectra. Note the spectrum from an incandescent light bulb shown above. Why is this spectrum said to be *continuous*?

When solids are heated to a high temperature, they give off light. If less electricity passed through the incandescent bulb, its color would become redder. How would its spectrum change?

If more electricity passed through the bulb, its color would become bluish-white. How would its spectrum change? Why do chemists believe that red stars are cooler than blue-white stars?

Fluorescent Materials. Certain substances give off visible light when exposed to ultraviolet radiation. This is called *fluorescence*.

According to the theory, electrons may absorb ultraviolet radiation and be driven to higher energy levels. As electrons return to the normal level, they give off photons with the wavelengths of visible light.

Fluorescent Light: A Combination Spectrum. Study the diagram below. Which part of a fluorescent lamp contains fluorescent material?

The two filaments serve first to warm the tube and vaporize the mercury. Then they serve as electrodes for the current to pass through the gas.

Mercury gas gives off both ultraviolet and visible light. The ultraviolet light is absorbed by the fluorescent material, causing the material to fluoresce. This fluoresced light produces a continuous spectrum. Explain the fluorescent lamp spectrum shown above.

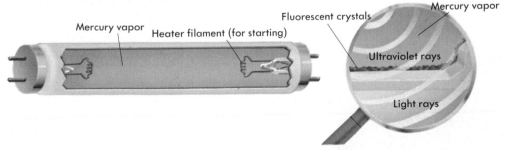

Mercury vapor   Heater filament (for starting)   Fluorescent crystals   Mercury vapor   Ultraviolet rays   Light rays

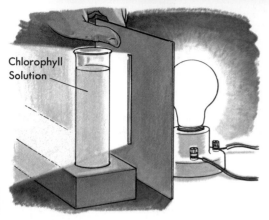

Chlorophyll
Solution

Absorption Band Spectra. The diagram at the left shows another method of studying chemicals with spectroscopes. In the diagram, white light is passed through a green-colored solution, and the light coming through the solution is studied.

Study the graph. Line *A* represents the amount of light of each wavelength which passes through the test tube containing pure water. Study line *B*. Which wavelengths do not pass through the green solution? What happens to this wavelength energy in the solution? Compare the graph with the absorption band spectrum above.

All colored solutions produce absorption patterns which can be diagramed, but most solutions of chemicals are clear. Clear solutions, however, absorb different wavelengths of infrared light. As a result, chemists have developed the infrared absorption "fingerprints" for thousands of chemicals. In this way, they quickly identify unknown chemicals by comparing the way the unknown chemical absorbs infrared light with the absorption pattern of known chemicals.

Absorption Line Spectra. The diagram at the left shows another way of studying chemicals with light. In this case, white light is passed through a region of sodium gas. The spectrum which results is shown below.

Compare this spectrum with the spectrum for hot sodium vapor on page 72. Describe the probable condition of the sodium electrons in the gas after light has been absorbed. How might this idea be used to identify unknown gases?

Spectroscopic Analysis. A spectroscope can be used for analyzing substances quickly and easily. A bit of material is vaporized, usually by means of an electric arc. The spectrum is photographed and compared with known spectra.

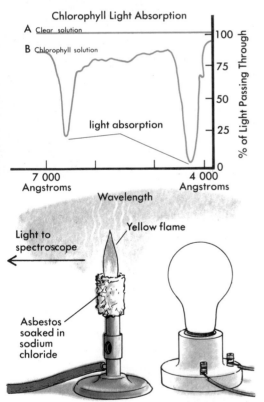

Chlorophyll Light Absorption

A Clear solution

B Chlorophyll solution

light absorption

% of Light Passing Through

100
75
50
25
0

7 000
Angstroms

4 000
Angstroms

Wavelength

Light to
spectroscope

Yellow flame

Asbestos
soaked in
sodium
chloride

Dark line absorption spectrum of sodium

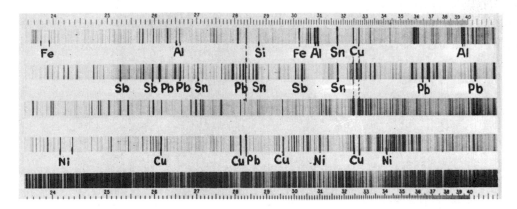

Hold glass tubing in a gas flame until glowing vapors are given off. Analyze the light with a spectroscope and identify one element in the glass.

The spectrum shown above was made by vaporizing a piece of aluminum. Some of the important lines are labeled. What impurities were present in the aluminum?

A spectroscope is sensitive to exceedingly small quantities. It can detect one one-millionth of a gram of some substances. Consequently, it is an important instrument in chemical research, industry, and crime laboratories.

Three spectra are shown below. The top spectrum represents a bit of paint found on a burglar's "jimmy" tool. The middle spectrum represents paint from a door that was "jimmied." The bottom spectrum serves as a reference.

How closely do the top and middle spectra match? If you were on a jury, would you believe that the person who owned the "jimmy" had burglarized the house?

How might a spectroscope be used to find and convict a "hit-and-run" driver? What are some other ways police chemists might use spectroscopes?

Courtesy of Beckman Instruments, Inc.

A scientist using an infrared spectro-photometer to study crystal surfaces under various conditions.

Paint from suspect's tool

Paint from window sill

Reference spectrum

Solar spectrum (continuous spectrum of sunlight)

Bright line spectrum of helium

INVESTIGATING THE UNIVERSE

Nearly all that has been learned about the universe has been learned through the study of visible and nonvisible light from objects in space. Fortunately, our atmosphere is nearly transparent to visible radiations, making direct observations possible by using telescopes and cameras.

The atmosphere is much less transparent to ultraviolet and infrared light. Today, however, satellites carry spectroscopes above the atmosphere so that scientists can learn about the ultraviolet and infrared light from the stars.

The Solar Spectrum. The solar spectrum is shown above. What does the continuous part of the spectrum represent? What do the lines represent?

Astronomers believe that this spectrum consists of two parts. The continuous part suggests that the main part of the sun glows, giving off a continuous spectrum. The dark lines suggest that gases in the sun's atmosphere absorb some light from the sun.

One element was discovered on the sun before being found on earth. A set of lines in the solar spectrum failed to match any known at the time. The element was named *helium* (Helios was the Greek sun god).

The spectrum below is lengthened to permit easier study. For truly detailed study, astronomers have prepared spectra over 13 meters long.

Notice some of the labeled lines. Which elements are present in the sun's atmosphere?

Part of the solar spectrum

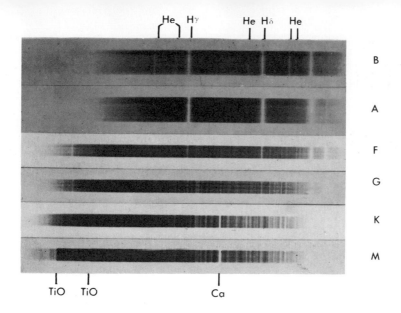

Star Types. A spectrum may reveal more than the elements in a star. The spectrum may indicate that some of the elements are combined into compounds. Since molecules tend to break up at high temperatures, their presence may indicate relatively cool conditions.

A spectrum may also tell whether atoms have lost electrons. The loss of each additional electron represents an enormous increase in energy. Therefore, something about the conditions of a star can be inferred from the number of missing electrons.

Astronomers have divided the stars into classes on the basis of their spectra. The spectra of several stars are shown above. Note the great differences. What are some characteristics of the classes as indicated in the table?

Type	Typical Stars	Temp.	Color	Spectrum
O	Zeta Puppis	33 000°C	Blue-white	Helium lines
B	Rigel, Spica	22 000°C	Blue-white	Helium lines
A	Sirius, Vega	11 000°C	White	Hydrogen lines; metals appear
F	Canopus, Procyon	7 700°C	Yellowish-white	Metals stronger
G	Sun, Capella	6 100°C	Yellow	Many metallic lines
K	Aldebaran, Arcturus	4 400°C	Orange	Calcium lines very strong
M	Betelgeuse, Antares	2 200°C to 3 300°C	Red	Titanium oxide
R	U Cygni			Carbon compounds
N	Y Canum Venaticorum			Carbon compounds
S	R Andromadae			Zirconium oxide

Determining Chemical Distribution. A spectroscope is helpful in determining the distribution of elements and compounds. For example, a photograph made by the light emitted from hydrogen atoms shows where these atoms are located.

To make such a photograph, a spectrum is projected on a screen which contains a narrow slit. The slit may be lined up with one of the lines of the spectrum. In this case, only light of this wavelength falls on photographic film behind the screen.

The photograph of a distant nebula shown above in color shows the nebula as it would appear to the eye, if the eye were sufficiently sensitive. What information does the photograph provide?

The photograph at the left below was made by light from the red line of hydrogen. What information is provided by this photograph?

The photograph at the right below was made by light from one of the oxygen lines. What information is provided?

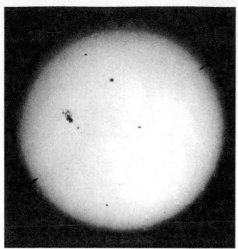

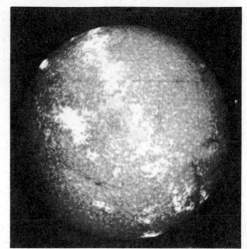

Chemical Distribution on the Sun. The top left photograph shows the sun as viewed by white light. What information about the sun can you obtain from this picture? Thinking in terms of temperature, do the spots probably represent holes in the sun's atmosphere or projections above it?

The top right photograph shows the sun as viewed by light from its calcium atoms. Where are these atoms especially numerous? Note the granular appearance of the sun. It has been suggested that these features are convection cells. If so, where are hot gases ascending?

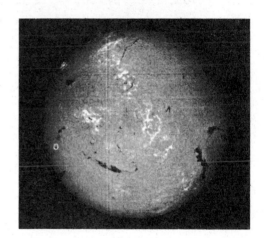

The right photograph shows the sun as viewed by the light of hydrogen. It is believed that this view shows a higher level of the atmosphere than the second photograph. Why might you expect hydrogen at a higher level?

Note the flares of excited hydrogen. Is there evidence that these may rise far above the sun's atmosphere?

SCIENTIFIC HONESTY. These photographs illustrate differences obtained by special photographic methods. Such photographs may be misinterpreted by a person who thinks they represent what the eyes can see. Why is it scientifically dishonest to present special photographs without information about the ways they were taken? What are some other requirements for scientific honesty in the presentation of photographs, diagrams, models, and similar visual aids?

Niels Bohr

REVIEW QUESTIONS

1. Which spectrum above is a bright line spectrum? A continous spectrum? A dark line (absorption) spectrum?
2. Matter is usually in what state when bright line spectra form? When continuous spectra form?
3. What is different about the matter which forms bright line and dark line spectra?
4. How does the wave theory of light explain the difference between red light and blue light?
5. Why is the name, Niels Bohr, an important one in science?
6. Why does a fluorescent light give off two types of spectra?
7. Use each term correctly in a sentence: angstrom, spectrum, photon, fluorescence, energy level.

THOUGHT QUESTIONS

1. How could you design two different experiments to prove that nonvisible light, such as infrared and ultraviolet, really exist?
2. What questions would you ask a "scientist" who claimed to have discovered a new color of visible light that had never been seen before?
3. Why is the Bohr model of an atom sometimes called the solar system model of an atom?
4. If planets behaved like electrons, what would happen to the earth's motion around the sun if more energy were added to the earth?
5. On page 70 of this chapter, the artist shows the ultraviolet and infrared spectra as black. Why?
6. The spectra on page 75 show the burglary suspect's "jimmy" and a reference spectrum. Why is the reference spectrum used?

Continuous plus bright line spectra from a fluorescent lamp

The Periodic Table of the Elements

People can recognize scientists by the tools they use. For example, when people see a microscope they think of biologists. The photograph above shows one of the tools often seen where chemists work or where chemistry is taught. This tool is the chart called the Periodic Table of the Elements.

On this chart the elements are organized in a special way. By studying the positions of the elements, chemists can quickly predict many things about each element. The chart greatly simplifies the job of remembering each element and its properties. In this chapter you will learn some of the ways chemists use the Periodic Table.

THE DEVELOPMENT OF THE PERIODIC TABLE

Some people think that progress in science occurs at a steady rate. This picture is not accurate in many cases. One such example was the problem of organizing all that was known about the chemical elements. Until the middle 1800's, chemists had discovered new elements and studied their physical and chemical properties. But relatively little progress was made in organizing all this information into patterns. However, during the 1860's (about the time of the American Civil War), many of the ideas about the chemical elements began to fit into place. By 1870, all the known elements were organized into the Periodic Table which is used today. Since that time, many new elements have been discovered and new properties of known elements have been studied. All these new observations fit into the Table proposed in 1869.

Dimitri Mendeleev

Atomic Number and Physical Properties. In 1863, John Newlands, an English chemist, first used the term atomic number. Newlands arranged the chemical elements from lightest to heaviest. He gave the lightest known element, hydrogen, the atomic number 1. The second lightest known element, lithium, was given the atomic number 2, and so on.

In the early 1860's, several different chemists published their ideas about ways of organizing elements. Their ideas for organizing the elements usually arranged the elements by increasing atomic number, from lightest to heaviest. When another property was compared with atomic number, a repeating pattern was observed. Note the three graphs on the next page. The top graph compares the atomic number with the size of the atoms. Describe the pattern.

The next graph compares the atomic number to the density of each element. In what ways is it similar to the pattern of the upper graph?

The lower graph shows how the atomic number relates to the melting point of each element. Does this pattern have any similarities to the other graphs? Explain your answer.

Chemists have graphed many other properties of elements such as hardness, how much they expand when heated, solubility, how well they conduct electricity, color, and boiling point. All these properties, when graphed, produced the same repeating pattern or *periodicity*.

In 1869, Dimitri Mendeleev developed an idea for a Periodic Table of the Elements which is the type widely used today. Mendeleev, a Russian chemist, was so sure of his ideas that his first periodic table had several places for elements which were not yet discovered! Even though no one had ever seen these elements, Mendeleev predicted over 20 different properties for each of three undiscovered elements. Three new elements were discovered in 1874, 1879, and 1885. These elements had properties identical with those Mendeleev predicted! How does this affect the value of Mendeleev's scientific ideas?

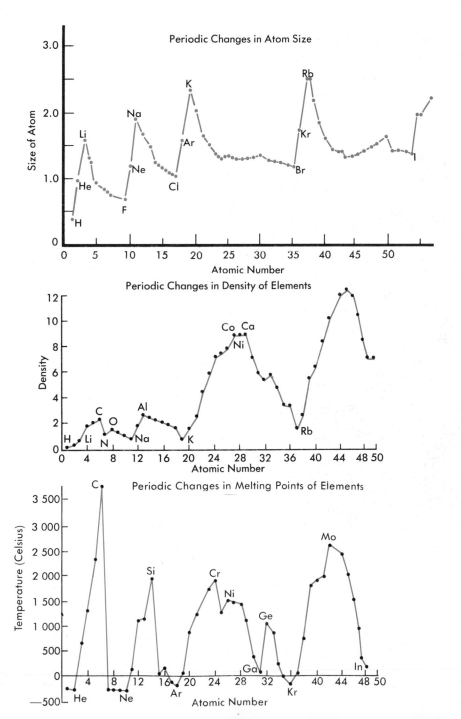

	1	
1	**1** **H** 1.008	

Light Metals

1 Hydrogen
2 Helium
3 Lithium
4 Beryllium
5 Boron
6 Carbon
7 Nitrogen
8 Oxygen
9 Fluorine
10 Neon
11 Sodium
12 Magnesium
13 Aluminum
14 Silicon
15 Phosphorus
16 Sulfur
17 Chlorine

18 Argon
19 Potassium
20 Calcium
21 Scandium
22 Titanium
23 Vanadium
24 Chromium
25 Manganese
26 Iron
27 Cobalt
28 Nickel
29 Copper
30 Zinc
31 Gallium
32 Germanium
33 Arsenic
34 Selenium

35 Bromine
36 Krypton
37 Rubidium
38 Strontium
39 Yttrium
40 Zirconium
41 Niobium
42 Molybdenum
43 Technetium
44 Ruthenium
45 Rhodium
46 Palladium
47 Silver
48 Cadmium
49 Indium
50 Tin
51 Antimony

52 Tellurium
53 Iodine
54 Xenon
55 Cesium
56 Barium
57 Lanthanum
58 Cerium
59 Praseodymium
60 Neodymium
61 Promethium
62 Samarium
63 Europium
64 Gadolinium
65 Terbium
66 Dysprosium
67 Holmium
68 Erbium

69 Thulium
70 Ytterbium
71 Lutetium
72 Hafnium
73 Tantalum
74 Tungsten
75 Rhenium
76 Osmium
77 Iridium
78 Platinum
79 Gold
80 Mercury
81 Thallium
82 Lead
83 Bismuth
84 Polonium
85 Astatine

2

| 2,1 **3** **Li** 6.940 | 2,2 **4** **Be** 9.013 |

3

| 2,8,1 **11** **Na** 22.991 | 2,8,2 **12** **Mg** 24.32 |

Transition Heavy Metals

4

| 2,8,8,1 **19** **K** 39.100 | 2,8,8,2 **20** **Ca** 40.08 | 2,8,9,2 **21** **Sc** 44.96 | 2,8,10,2 **22** **Ti** 47.90 | 2,8,11,2 **23** **V** 50.95 | 2,8,13,1 **24** **Cr** 52.01 | 2,8,13,2 **25** **Mn** 54.94 | 2,8,14,2 **26** **Fe** 55.85 | 2,8,15,2 **27** **Co** 58.94 |

5

| 2,8,18,1 **37** **Rb** 85.48 | 2,8,18,2 **38** **Sr** 87.63 | 2,8,18,9,2 **39** **Y** 88.92 | 2,8,18,10,2 **40** **Zr** 91.22 | 2,8,18,12,1 **41** **Nb** 92.91 | 2,8,18,13,1 **42** **Mo** 95.95 | 2,8,18,13,2 **43** **Tc** [99] | 2,8,18,15,1 **44** **Ru** 101.1 | 2,8,18,16,1 **45** **Rh** 102.91 |

6

| 2,8,18,18,1 **55** **Cs** 132.91 | 2,8,18,18,2 **56** **Ba** 137.36 | 57–71 See Lanthanide Series | 2,8,18,32,10,2 **72** **Hf** 178.50 | 2,8,18,32,11,2 **73** **Ta** 180.95 | 2,8,18,32,12,2 **74** **W** 183.86 | 2,8,18,32,13,2 **75** **Re** 186.22 | 2,8,18,32,14,2 **76** **Os** 190.2 | 2,8,18,32,15,2 **77** **Ir** 192.2 |

7

| 2,8,18,32,18,1 **87** **Fr** [223] | 2,8,18,32,18,2 **88** **Ra** 226.05 | 89–103 See Actinide Series |

6A
LANTHANIDE SERIES
(Rare Earth Elements)

| 2,8,18,18,9,2 **57** **La** 139.92 | 2,8,18,19,9,2 **58** **Ce** 140.13 | 2,8,18,21,8,2 **59** **Pr** 140.92 | 2,8,18,22,8,2 **60** **Nd** 144.27 | 2,8,18,23,8,2 **61** **Pm** [147] | 2,8,18,24,8,2 **62** **Sm** 150.35 |

7A
ACTINIDE SERIES

| 2,8,18,32,18,9,2 **89** **Ac** [227] | 2,8,18,32,18,10,2 **90** **Th** 232.05 | 2,8,18,32,20,9,2 **91** **Pa** [231] | 2,8,18,32,21,9,2 **92** **U** 238.07 | 2,8,18,32,22,9,2 **93** **Np** [237] | 2,8,18,32,23,9,2 **94** **Pu** [242] |

2 Electrons in 1st orbit
8 Electrons in 2nd orbit
1 Electron in 3rd orbit

| 2,8,1 **11** **Na** 22.991 |

← ATOMIC NUMBER is the number of protons in nucleus of atom.

← ATOMIC MASS is the average number of protons and neutrons in nucleus of atom.

11P 12N

Diagram of Sodium Atom

THE PERIODIC TABLE OF THE ELEMENTS

86 Radon
87 Francium
88 Radium
89 Actinium
90 Thorium
91 Protactinium
92 Uranium
93 Neptunium
94 Plutonium
95 Americium
96 Curium
97 Berkelium
98 Californium
99 Einsteinium
100 Fermium
101 Mendelevium
102 Nobelium
103 Lawrencium

☐ Gas at room temperature   ☐ Liquid at room temperature

Nonmetals

Noble Gases

							2 / 2 / He / 4.003

2 3 / 5 / B / 10.82	2 4 / 6 / C / 12.011	2 5 / 7 / N / 14.008	2 6 / 8 / O / 16.000	2 7 / 9 / F / 19.00	2 8 / 10 / Ne / 20.183
2 8 3 / 13 / Al / 26.98	2 8 4 / 14 / Si / 28.09	2 8 5 / 15 / P / 30.975	2 8 6 / 16 / S / 32.066	2 8 7 / 17 / Cl / 35.457	2 8 8 / 18 / Ar / 39.944

2 8 16 2 / 28 / Ni / 58.71	2 8 18 1 / 29 / Cu / 63.54	2 8 18 2 / 30 / Zn / 65.38	2 8 18 3 / 31 / Ga / 69.72	2 8 18 4 / 32 / Ge / 72.60	2 8 18 5 / 33 / As / 74.91	2 8 18 6 / 34 / Se / 78.96	2 8 18 7 / 35 / Br / 79.916	2 8 18 8 / 36 / Kr / 83.80
2 8 18 18 / 46 / Pd / 106.4	2 8 18 18 1 / 47 / Ag / 107.880	2 8 18 18 2 / 48 / Cd / 112.41	2 8 18 18 3 / 49 / In / 114.82	2 8 18 18 4 / 50 / Sn / 118.70	2 8 18 18 5 / 51 / Sb / 121.76	2 8 18 18 6 / 52 / Te / 127.61	2 8 18 18 7 / 53 / I / 126.91	2 8 18 18 8 / 54 / Xe / 131.30
2 0 18 32 17 1 / 78 / Pt / 195.09	2 8 18 32 18 1 / 79 / Au / 197.0	2 8 18 32 18 2 / 80 / Hg / 200.61	2 8 18 32 18 3 / 81 / Tl / 204.39	2 8 18 32 18 4 / 82 / Pb / 207.21	2 8 18 32 18 5 / 83 / Bi / 209.00	2 8 18 32 18 6 / 84 / Po / [210]	2 8 18 32 18 7 / 85 / At / [210]	2 8 18 32 18 8 / 86 / Rn / [222]

2 8 18 25 2 / 63 / Eu / 152.0	2 8 18 25 9 2 / 64 / Gd / 157.26	2 8 18 26 9 2 / 65 / Tb / 158.93	2 8 18 28 2 / 66 / Dy / 162.51	2 8 18 29 2 / 67 / Ho / 164.94	2 8 18 30 8 2 / 68 / Er / 167.27	2 8 18 31 8 2 / 69 / Tm / 168.94	2 8 18 32 8 2 / 70 / Yb / 173.04	2 8 18 32 9 2 / 71 / Lu / 174.99
2 8 18 32 24 9 2 / 95 / Am / [243]	2 8 18 32 25 9 2 / 96 / Cm / [247]	2 8 18 32 26 9 2 / 97 / Bk / [249]	2 8 18 32 27 9 2 / 98 / Cf / [251]	2 8 18 32 28 9 2 / 99 / Es / [254]	2 8 18 32 29 9 2 / 100 / Fm / [253]	2 8 18 32 30 9 2 / 101 / Md / [256]	102 / No / [254]	103 / La / [257]

SUMMARY QUESTIONS

1. How many elements at room temperature are solids (white)? Liquids (dark blue)? Gases (light-blue)?
2. Where are the nonmetals on this table? Name five nonmetals.
3. Where are the metals on this table? Name five metals.

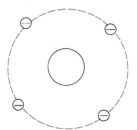
Diagram of Beryllium Atom

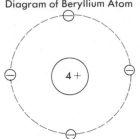
Diagram of Beryllium Atom

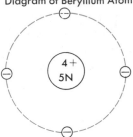
Diagram of Beryllium Atom

CONSTRUCTING ATOMIC MODELS

The Periodic Table is shown on the previous pages. Perhaps you are wondering why it is important. When you finish with this section you should be able to describe the atom of any element. The position of each element on this table aids you in this task.

Atomic Number. Part of the Periodic Table is shown at the right. The large number over the element symbol is the *atomic number*. The atomic numbers for the elements go from 1 for hydrogen, the lightest element, to over 100 for the heavy elements made in laboratories.

Note the atomic number for beryllium (Be) at the right. This number tells more about the atom today than it did when Newlands invented the term. A beryllium atom has the same number of electrons as its atomic number. Note the top left diagram showing a beryllium atom. How many electrons are shown? How did the artist know how many electrons to draw?

Electrons are negatively charged particles. These similarly charged particles tend to repel each other. This repelling force, as well as the effect of motion, would tend to cause the electrons to move away from the atom. But electrons do not readily leave their atom. Thus, there must be a strong attracting force in the nucleus. This attracting force is caused by the positively charged *protons* in the nucleus. Neutral atoms have the same number of positively charged protons in the nucleus as electrons. Note the middle drawing above. How many protons are shown? How did the artist know how many protons to draw?

Atomic Mass. The decimal number under each chemical symbol on the Periodic Table is called the *atomic mass*. The mass of an atom is almost the same as the mass of the nucleus because electrons

1	**1** **H** 1.008		
2 1	**3** **Li** 6.940	2 2	**4** **Be** 9.013
2 8 1	**11** **Na** 22.991	2 8 2	**12** **Mg** 24.32
2 8 8 1	**19** **K** 39.100	2 8 8 2	**20** **Ca** 40.08
2 8 18 8 1	**37** **Rb** 85.48	2 8 18 8 2	**38** **Sr** 87.63
2 8 18 18 8 1	**55** **Cs** 132.91	2 8 18 18 8 2	**56** **Ba** 137.36
2 8 18 32 18 8 1	**87** **Fr** [223]	2 8 18 32 18 8 2	**88** **Ra** 226.05

have so little mass. The mass of the nucleus is made up of protons (postively charged) and *neutrons*. Neutrons have no charge and have the same mass as protons.

Note the upper right drawing on the previous page. How many neutrons are shown? How did the artist know how many neutrons to draw?

Electron Positions. Note the small column of numbers to the left of each chemical symbol on the opposite page. These numbers add up to the atomic number. These numbers also indicate how many electrons each atom has, but they tell more. They indicate how far from the nucleus each electron orbits. The upper number tells how many electrons are in the orbit closest to the nucleus. Each number below describes the number of electrons in each orbit further out.

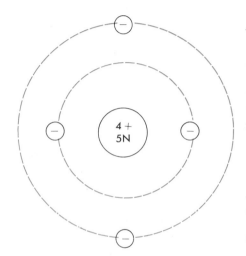

Note the drawing here. How many electrons are shown in each orbit? How did the artist know how to arrange the electrons this way?

Study the complete Periodic Table on pages 84 and 85. How does the number and position of electrons differ within each row of the table? How does the number and position of electrons differ within each column?

Study the upper graph on page 83. Note the right side of each "hill" on the graph. What does each right side represent on the Periodic Table?

What happens to the number of protons across each row of the Periodic Table? How might this change cause heavier atoms within a row of the table to be smaller in size?

Looking for Patterns in Electron Arrangements. Use the Periodic Table on pages 84 and 85 to study the arrangement of electrons in the first 20 elements. How does the arrangement change? Is there a pattern?

Now study the arrangement of electrons in elements 21–36. Does the same pattern occur? Discuss in class whether there is a pattern at all in this row. Test your decision by looking for a similar pattern in the next row of elements (37–54). Is this pattern similar? Compare these arrangements with the next row (elements 55–86).

Transition Elements. Metals, particularly lighter ones, are found at the left of the Periodic Table. Nonmetals, such as oxygen and sulfur, are found at the right. The wide group of elements in the center of the table, which includes metals with greater masses such as iron (26) and mercury (80), are called *transition elements*.

Compare the arrangement of electrons of transition elements with the lighter metals such as sodium (11) and magnesium (12). Why might transition elements have properties more similar to light metals than to nonmetals?

The Bottom Rows. Nearly all students will note the similarity in electron arrangements in rows 2 and 3. Most students will detect a different pattern in rows 4, 5, and 6. Now, how many students noticed the strange way the table-maker arranged the sixth row so it looked like rows 4 and 5? Note that about half the elements in this row are missing. Element 56 in the table is followed by element 72! The missing elements (57–71) are found in row 6a at the bottom of the table. These two bottom rows of atoms are called the *inner transition elements*. The elements in each row have very similar chemical properties. These elements, however, are not common. Another name for the elements of row 6a is the *rare earth series*.

The three-dimensional photograph below shows another way of representing the Elements of the Periodic Table. Where are the transition elements and inner transition elements on this table?

Isotopes. At the time the Periodic Table was invented, chemists believed all elements were composed of the same particles. Therefore, if hydrogen, the lightest element, had an atomic mass of almost exactly 1.0, then all other elements should have atomic masses which also were very nearly whole numbers. Study the table on pages 84 and 85 to decide if this idea is correct.

Chlorine (atomic number 17) does not have an atomic mass of nearly 35 or 36. Instead, its atomic mass is close to 35.5. This problem remained unsolved for many years.

In 1914, chemists reported that lead (Pb) did not always have the same atomic mass. Some minerals consisted of lead of atomic mass 206, other minerals contained lead of atomic mass 207 and still others of mass 208. But chemically, all the lead was the same!

Such an observation is explained if lead (atomic number 82) exists in nature with either 124, 125, or 126 neutrons. These three atomic forms of the same element are called *isotopes*.

The idea of isotopes also explains why atomic masses are not whole numbers. Chlorine, for example, has two isotopes, one of mass 35 and the other of mass 37. The atomic mass given on the Periodic Table is the average mass for chlorine atoms of both isotopes. Which isotope is more common?

The table below lists the isotopes of some of the elements. Look for patterns to these data.

Key

1	less than 0.0%
2	0.01%-1.0%
3	1.0%-10%
4	10%-90%
5	greater than 90%

Isotopes of Some Elements

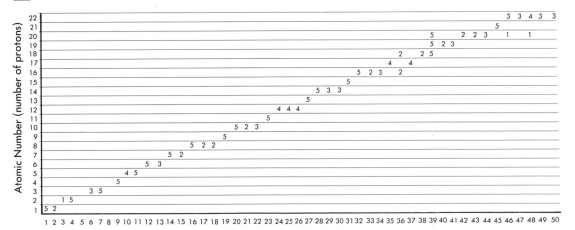

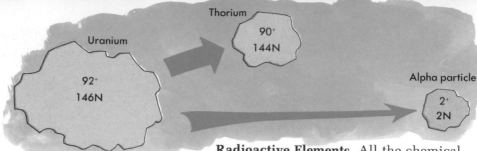
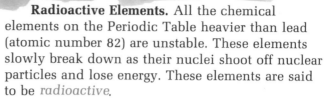

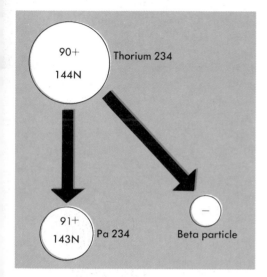

Radioactive Elements.

Radioactive Elements. All the chemical elements on the Periodic Table heavier than lead (atomic number 82) are unstable. These elements slowly break down as their nuclei shoot off nuclear particles and lose energy. These elements are said to be *radioactive*.

Study the diagram above. In this case, a uranium atom (isotope 238) breaks down as an *alpha* particle leaves the nucleus. An alpha particle consists of 2 protons and 2 neutrons (the same as a helium nucleus). The remaining nuclear particles now consist of 90 protons and 144 neutrons (isotope 234 of thorium).

The breakdown of uranium 238 to thorium 234 occurs at a slow rate. If you had 10 grams of U^{238} today, you would still have 5 grams left after 4.5 billion years. You would have 2.5 grams left after 9 billion years. How much would be left after 13.5 billion years?

Thorium 234 nuclei decay differently from Uranium 238, as shown at the left. In this case, a *beta* particle resembling an electron leaves the nucleus. You will recall that electrons contribute little mass to an atom. However, the negative charge of the beta particle results in a neutral neutron changing into a proton. The nucleus which had 90 protons and 144 neutrons now has 91 protons and 143 neutrons. The new element is protactinium 234.

Thorium 234 decays much more quickly than Uranium 238. If you had 10 grams of Thorium 234 today, in 3.5 weeks only 5 grams would be left. How much would be left in 7 weeks? In 10.5 weeks?

Gamma Rays. A third type of radiation from the nuclei of radioactive elements is called *gamma rays*. Gamma radiation, unlike alpha and beta, has no charge or mass. Gamma rays are like rays of visible light but with much greater energy. As a result, gamma rays (like X rays) can pass through matter such as people or thin metal.

Study the diagram at the left. Discuss the

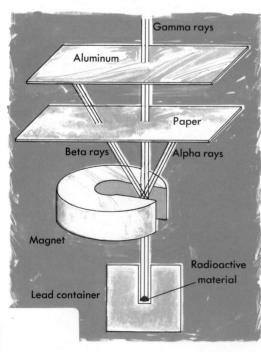

behavior of alpha, beta, and gamma radiation in terms of their masses, charges, and energies.

Laboratory-Made Elements. All the elements with atomic numbers greater than 82 (lead) are radioactive. Elements of atomic number greater than 92, however, do not occur in nature. These radioactive elements are produced in laboratories. Special equipment called particle accelerators are used. A beam of electrons, protons, neutrons or alpha particles is focused on a target, such as uranium, in a particle accelerator. Some of the accelerated particles smash into the nuclei of the target. As a result, the amount of matter in the nuclei change. For example, if a neutron combines with the nucleus of an uranium 238 atom, the uranium 239 isotope forms.

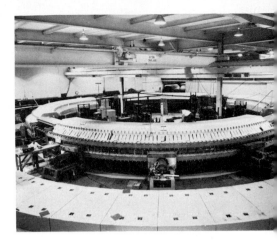

Cyclotron at Brookhaven National Laboratory, Long Island, New York

$$_{92}U^{238} + {}_0n^1 \longrightarrow {}_{82}U^{239}$$

The *half-life* of U^{238} (time to decay to half the original amount) is 4.5 billion years. The half-life of U^{239} is 23 days. Uranium 239 emits a beta particle forming element 93.

$$_{92}U^{239} \longrightarrow {}_{-1}e^0 + {}_{93}Np^{239}$$

This element was named Neptunium. Neptune is the planet beyond Uranus (Uranium). Why is element 94 named Plutonium?

Producing the artificial elements became possible with the invention of particle accelerators such as the cyclotron shown above. The cylcotron was invented by Dr. E. O. Lawrence in 1931. Most of these elements have been produced by Lawrence and his co-workers in their laboratory at the Berkeley campus of the University of California. Now study the names of the elements above 92. Suggest why these names were chosen.

Dr. E. O. Lawrence

SUMMARY QUESTIONS

1. What are isotopes?
2. Describe an atom of oxygen which has an atomic number of 8 and an atomic mass of 16.
3. What are transition elements?
4. Where are the naturally radioactive elements on the Periodic Table?
5. How do scientists make new elements?

FAMILIES OF ELEMENTS

Chemists since Mendeleev have always recognized that elements are chemically very similar to other elements just above or just below them on the Periodic Table. For this reason, elements in the same column have been given a special name. They are said to be elements of the same family.

Thousands of compounds, including many different compounds of oil and rubber, consist of molecules containing the element carbon. Note carbon's position on the Periodic Table. What other element could be most easily substituted for carbon to make synthetic rubber and oil?

The Noble Gases. One interesting family of elements is called the noble gases. Note their position on the upper graph on page 83. How does the position of noble gases on this graph differ from the position of all other elements?

The six noble gases were all identified between 1895 and 1900 as elements of the earth. Spectroscopic study of gas-filled tubes was used to identify each gas. In the case of helium, its discovery on earth only confirmed this element's existence. The helium spectrum was observed 30 years earlier as a part of the sun's atmosphere.

Early chemists observed that each element in this family did not combine with any other element to form compounds. (This is one reason they were not discovered sooner.) Chemists say that these elements are very stable. Some information about the noble gases is given in the table below.

Halogen Family	Noble Gas Family	Alkali Metals Family
	(2) · 2 He · 4.003	(2,1) · 3 Li · 6.940
(2,7) · 9 F · 19.00	(2,8) · 10 Ne · 20.183	(2,8,1) · 11 Na · 22.991
(2,8,7) · 17 Cl · 35.457	(2,8,8) · 18 Ar · 39.944	(2,8,8,1) · 19 K · 39.100
(2,8,18,7) · 35 Br · 79.916	(2,8,18,8) · 36 Kr · 83.80	(2,8,18,8,1) · 37 Rb · 85.40
(2,8,18,18,7) · 53 I · 126.91	(2,8,18,18,8) · 54 Xe · 131.30	(2,8,18,18,8,1) · 55 Cs · 132.91
(2,8,18,32,18,7) · 85 At · [210]	(2,8,18,32,18,8) · 86 Rn · [222]	(2,8,18,32,18,8,1) · 87 Fr · [223]

Symbol	Name	Melting Point (°C)	Boiling Point (°C)	Uses
He	Helium	—	−269	Balloons, dirigibles, blimps
Ne	Neon	−249	−246	Neon signs
Ar	Argon	−189	−186	Light bulbs; 1% of air
Kr	Krypton	−157	−153	Greenish-yellow luminous signs
Xe	Xenon	−112	−107	
Rn	Radon	−110	− 62	Treatment of cancer

Neighboring Elements of the Noble Gases. The noble gases are chemically stable. You might wonder if the same is true for the elements on either side of these gases. These elements are shown on page 92. Compare each noble gas with its neighboring elements. For example, compare the atomic structure of neon with sodium and fluorine. How do their atoms differ?

The chemical family to the left of the noble gases is called the *halogens*. The halogens are all nonmetals. The top one, fluorine, is the most active nonmetal of all nonmetals. The acid of fluorine, hydrofluoric acid (HF), is so strong that it quickly dissolves glass. Light bulbs are "frosted" by pouring this acid into clear glass bulbs. As the atomic number of the halogens increases, their chemical activity decreases. Thus, hydrochloric acid (HCl) is not as strong as hydrofluoric acid (HF).

The elements to the right of the noble gases are called the *alkali metals*. The term *alkali* means the opposite of an acid. Another name for an alkali is a base.

These elements are chemically very active. As a result, they are never found uncombined in nature. Active nonmetals make strong acids; active metals make strong bases. All the alkali metals make strong bases such as sodium hydroxide (NaOH - lye) and potassium hydroxide (KOH). Strong bases such as these are just as dangerous as strong acids. Acids and bases should be handled only by experienced adults trained in chemistry.

Study the table below. What are some physical and chemical differences between halogens and alkali metals?

HALOGEN FAMILY OF ELEMENTS

Name	Melts (°C)	Boils (°C)	Density (Sol.)	Year Discovered	Found in Nature
Fluorine	−223	−187	1.7(liq)	1771	Mineral fluorite (CaF_2)
Chlorine	−102	−35	1.9	1774	Mineral halite (NaCl)
Bromine	−7	59	3.1	1826	Part of sea water
Iodine	114	184	4.9	1811	Part of sea water and sea deposits
Astatine	(Radioactive-very short half life)				Synthesized in 1940 like other laboratory-made elements

ALKALI METALS FAMILY OF ELEMENTS

Lithium	186	>1200	0.5	1817	Numerous minerals and spring water
Sodium	98	880	1.0	1807	Halite, NaCl; feldspar
Potassium	62	760	0.9	1807	In many minerals: mica, feldspar
Rubidium	38	700	1.5	1861	In mineral spring water (rare)
Cesium	28	670	1.9	1860	Numerous minerals and spring water
Francium	Natural radioisotope			1939	Alpha disintegration of actinium

The Noble Gas Neighbors as Ions.

Consider the following observations: (1) Chlorine, the gas, has very different properties from the chlorine in table salt (NaCl). (2) The most active elements are very similar in atomic structure to the inactive noble gases. (3) Atoms of two different elements (such as sodium and chlorine) form compounds. These are important observations in chemistry. A good theory of why chemicals react should be able to explain these observations. The theory of chemical ions does.

Study the table at the left. Note that there are changes from the previous table. For example, how many electrons are shown in each orbit of fluorine (F), neon (Ne) and sodium (Na)? For chlorine (Cl), argon (Ar) and potassium (K)? Note that fluorine, atomic number 9, has 9 positively charged protons and 10 negatively charged electrons. As a result, the atom has a net charge of -1. Such charged atoms are called *ions*. The fluorine ion has a charge of -1 or a *valence* of -1. This is usually written F^-. Why is sodium written as Na^+ on this table?

The Ionic Theory.

The ionic theory states that when atoms react, they gain or lose electrons to either fill or empty their outer orbit of electrons. Metals consist of atoms which lose electrons and become positively charged. Nonmetals consist of atoms which gain electrons and become negatively charged. Oppositely charged metal and nonmetal ions are attracted to each other, forming a molecule of two oppositely charged ions. Note the example below.

Halogen Ions	Noble Gases	Alkali Metal Ions
	2 / 2 He	2 / 3 Li⁺
2,8 / 9 F⁻	2,8 / 10 Ne	2,8 / 11 Na⁺
2,8,8 / 17 Cl⁻	2,8,8 / 18 Ar	2,8,8 / 19 K⁺
2,8,18,8 / 35 Br⁻	2,8,18,8 / 36 Kr	2,8,18,8 / 37 Rb⁺
2,8,18,18,8 / 53 I⁻	2,8,18,18,8 / 54 Xe	2,8,18,18,8 / 55 Cs⁺
2,8,18,32,18,7 / 85 At⁻	2,8,18,32,18,8 / 86 Rn	2,8,18,32,18,8 / 87 Fr⁺

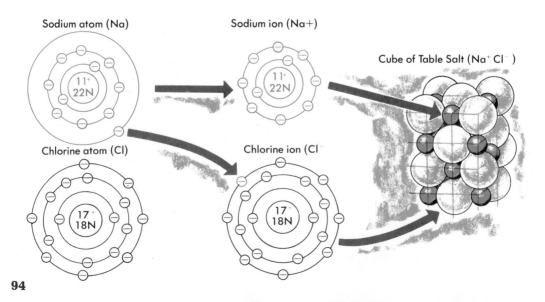

Sodium atom (Na)

Sodium ion (Na+)

Cube of Table Salt (Na⁺ Cl⁻)

Chlorine atom (Cl)

Chlorine ion (Cl⁻)

Name	Symbol	Valence
Aluminum	Al	+3
Ammonium	NH_4	+1
Barium	Ba	+2
Bicarbonate	HCO_3	−1
Bromine	Br	−1
Calcium	Ca	+2
Carbonate	CO_3	−2
Chlorate	ClO_3	−1
Chlorine	Cl	−1
Copper	Cu	+2
Fluorine	F	−1
Hydrogen	H	+1
Hydroxide	OH	−1
Iodine	I	−1
Iron	Fe	+2
Lead	Pb	+2
Magnesium	Mg	+2
Nitrate	NO_3	−1
Oxygen	O	−2
Potassium	K	+1
Silver	Ag	+1
Sodium	Na	+1
Sulfate	SO_4	−2
Sulfur	S	−2
Zinc	Zn	+2

Valences of Some Elements. The table at the left lists the valences for many common elements. What kinds of elements have positive valences, metals or nonmetals? What kinds of elements have negative valences, metals or nonmetals?

The table also includes some electrically charged groups of atoms. These groups frequently act in chemical reactions as though they were a single charged atom or ion. They are called *radicals*.

Using Valence to Write Chemical Formulas. The idea of valence helps chemistry students to write the chemical formulas for many compounds. The rule for writing compounds is simple; each compound must have enough metal and nonmetal ions to be electrically neutral, or the number of pluses and minuses must be the same. For example, copper ion has a valence of +2; nitrate ion has a valence of −1. Thus, for copper nitrate to be electrically neutral, its formula must be $Cu^{++} (NO_3^-)_2$ or $Cu(NO_3)_2$.

Write the correct chemical formulas for magnesium sulfate, barium hydroxide, potassium iodide, and aluminum oxide.

INCOMPLETE KNOWLEDGE: Chemistry is a very interesting but complicated subject. Whenever a complicated subject is presented in a brief or simplified form, the knowledge gained necessarily must be incomplete. Two examples of incomplete information on this page are: (1) Many ions sometimes have other valences than those listed in the table. These other valences are not easily explained with your model of an atom. (2) Although compounds such as salt form in the way described in this chapter, other compounds such as sugar do not appear to consist of ions. Such compounds (there are thousands) require other theories to explain their formation.

Many of the topics in this chapter can be studied for a lifetime. People with an understanding of scientific knowledge should recognize that a knowledge of atoms will never be complete.

$\begin{smallmatrix}2\\8\\8\end{smallmatrix}$	18 Ar 39.94

REVIEW QUESTIONS

1. What does periodic mean?
2. Explain what the numbers 2, 8, 8, 18, and 39.94 mean in the box at the left?
3. What do the numbers 2, 8, 8 tell you about the chemical properties of this element?
4. The element just before element 18 is chlorine. Describe the chlorine atom.
5. The element just after element 18 is potassium. Describe the potassium atom.
6. Which of these three atoms readily forms a positive ion? How?
7. Which of these three atoms forms a negative ion? How?
8. Define the following: transition elements, isotopes, radioactivity, alpha particles, beta particles, gamma rays, laboratory-made elements, half-life, element families.

THOUGHT QUESTIONS

1. How can there be millions of different substances when there are only 92 naturally occurring elements?
2. If a person claimed he or she discovered a new nonradioactive element, how would you react (no pun intended)?
3. How can you explain that magnesium has an atomic mass of about 24.3 instead of 24 or 25?
4. The particle accelerator in the photograph cost millions of dollars of taxpayers' money. Do you think it is wise to spend tax money this way? Why?

The Stanford University particle accelerator

BALANCED
AND
UNBALANCED FORCES

Atmospheric Pressure

We live at the bottom of an air mass many kilometers thick. This air is pulled downward by gravity and presses against all objects within it.

We rarely notice the great weight of the atmosphere. Pressures outside our bodies are nicely balanced against those from within. Only when we climb or drop suddenly, such as riding in elevators,

1

do we feel pressure changes against our eardrums.

However, the weight of the atmosphere influences our lives in countless ways. The atmosphere produces winds and storms, helps us breathe and drink, and operates hundreds of devices ranging from medicine droppers to automobile engines. We rarely escape from the effects of the huge mass of air above us.

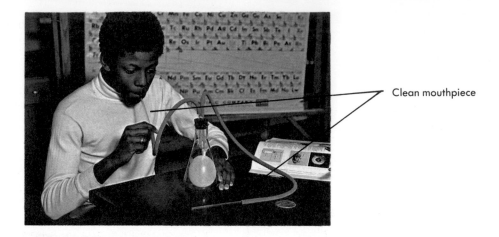

Clean mouthpiece

UNDERSTANDING ATMOSPHERIC PRESSURE

It is difficult to study atmospheric pressure directly because air cannot be seen and is not always felt. Indirect approaches must be used when studying the effects of atmospheric pressure upon other things.

Any indirect approach to science is awkard and may lead to misunderstandings. Therefore, special care must be used in this chapter when drawing conclusions and making general statements.

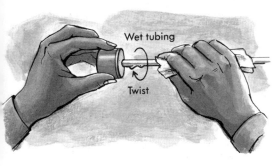

Wet tubing

Twist

PUTTING TUBING IN RUBBER STOPPERS. Dry glass tubing does not slip easily into a rubber stopper. It may break and cut the palm of the hand.

Always wrap glass tubing in a towel and wet it before pushing it into the hole of a rubber stopper. Then push it gently with a twisting motion.

Do not leave glass tubing in rubber stoppers. After a few days, the glass cements to the rubber; then it is difficult and dangerous to separate them.

Pressures on a Balloon. Everyone knows what happens when the pressure inside a balloon is increased. Fewer people, however, have any idea of what happens when the pressure outside a balloon is changed. They have never tried the following experiment.

Push glass tubes through the holes of a two-hole stopper (*see margin*). Attach a balloon to one of the tubes. Then put the stopper in the mouth of a bottle.

Blow on the tube leading to the balloon. (**CAUTION:** Blow through a glass mouthpiece that has been washed in hot, soapy water to remove bacteria.) What happens?

Decrease the pressure outside the balloon by sucking air from the second glass tube as shown in the picture above. Note what happens to the balloon.

What happens when air is sucked from the first tube? When air is blown into the second tube? Keep a record of the results.

With the help of another person, increase the pressure inside and outside the balloon at the same time. Decrease the pressure inside and outside the balloon at the same time.

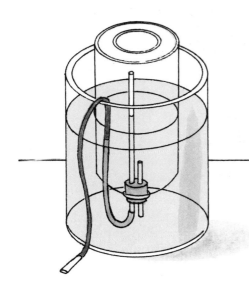

Discuss the behavior of a balloon in terms of both inside and outside pressures. When does a balloon grow larger? When does it grow smaller? What happens if the inside and outside pressures are equal?

Pressures on Water. Set up the apparatus shown at the left. Note the level of the water when the pressure in the bottle is the same as that outside.

Increase the pressure inside the bottle by blowing through the rubber tube. (**CAUTION:** Use a clean mouthpiece.) Decrease the pressure inside the bottle by sucking on the tube. Describe the behavior of the water level and discuss the reasons for any changes.

Balanced and Unbalanced Pressures. Diagrams can be helpful in understanding things that cannot be seen. The diagrams below use arrows to represent pressures. The directions of the arrows show the directions in which the pressures act. The sizes of the arrows show how great the pressures are.

Diagram A describes conditions when pressures inside and outside the bottle are equal. The two arrows are equal in size, showing that the pressures are balanced. The water level does not change.

Diagram B shows conditions when the air pressure inside the bottle is increased. One arrow is larger than the other. Pressures are no longer balanced. What will happen to the water level in the bottle?

Describe the conditions represented by diagram C. What will happen? Explain your answer in terms of balanced and unbalanced pressures.

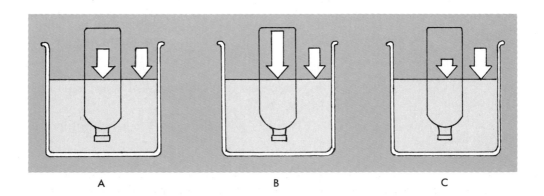

A   B   C

Fill a glass with water. Lay a card in place with one hand and turn the glass upside down with the other. The card should stay on and the water should stay in the glass.

An Old Trick and What It Means. Probably everyone has seen the trick described at the top of the page. This trick is often called an experiment and some people claim that it proves that the atmosphere pushes on things.

However, consider the trick carefully. Note that there has been no attempt to discover possible variables and check them. Note that there is no real evidence that atmospheric pressure has anything at all to do with the card and the glass of water.

The trick is not an experiment and it proves nothing. No conclusions about atmospheric pressure can be drawn from it.

The Scientific Approach. Scientists do not jump to conclusions when they work on problems. They look for all possible explanations and test them to find out which ones apply.

Suppose that you want to find out for certain why the card stays on the glass. Your first step is to list as many possible explanations as you can. Perhaps atmospheric pressure pushes up on the card. Perhaps the card sticks to the wet glass. Perhaps the water exerts a pull between the card and the glass. What other possible explanations can you suggest?

WHAT IS SCIENTIFIC THINKING? *Scientific thinking consists of many things. Among them are the habit of criticizing explanations and the habit of not jumping to conclusions.*

Testing Suggestions. The next step in the scientific approach is to test each suggestion. For example, bore a small hole in the bottom of a plastic drinking glass. Keep a finger over the hole while you fill the glass and turn it upside down. Then uncover the hole.

What is the independent variable in this experiment? What do the results seem to show? Do not draw any conclusions until you can prove that air enters the hole. Set up experiments to test other suggestions.

Drawing Conclusions. One of your experiments may prove that a certain suggestion is correct. Sometimes, however, it is not possible to prove or disprove all possible suggestions. Then you should accept temporarily the suggestion that seems best. A final decision must wait until you have more knowledge.

Pressure Diagrams. The diagrams below help explain the experiment described above. The pressure diagram below shows the glass before a hole has been bored in it. The weight of the water causes a small downward pressure on the card. The atmosphere causes a larger upward pressure on the card. Why does the card stay in place?

The other diagram shows the glass after the hole has been opened. Explain this diagram.

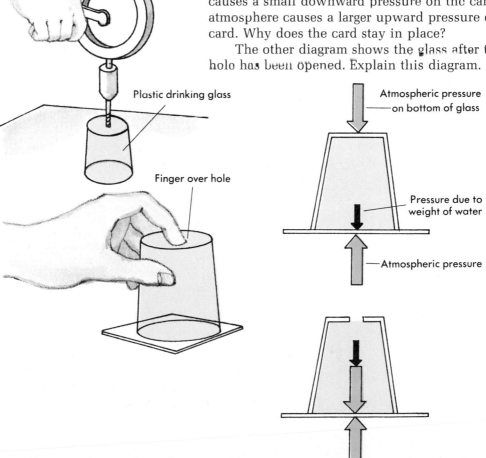

Plastic drinking glass

Finger over hole

Atmospheric pressure on bottom of glass

Pressure due to weight of water

Atmospheric pressure

Explaining a Soda Straw. It is often said that a soda straw operates because of *suction*. This explanation seems simple unless someone asks what suction is; then it may be seen that the explanation is far from simple. Suction is a word that is not always well understood.

For many years scientists explained that suction was caused by the pull of a vacuum, even though it was difficult to see how a vacuum could pull anything. Today, scientists say that suction is the result of two unbalanced pressures, and that suction is a push rather than a pull.

Suction in a soda straw results when the pressure is lowered within the mouth, thus producing a difference between the inside and outside pressures. Liquid rises up the straw because air pressure acting on the surface of the liquid is greater than the pressure in the person's mouth.

Testing the Explanation. An experiment for testing the above explanation should have only one independent variable. The atmospheric pressure in the bottle is probably the most easily controlled variable when simple equipment is used.

Use a glass tube in place of a soda straw. The ends of the glass tube should be smoothed by fire-polishing (*see* margin) and the tube should be washed in hot, soapy water to remove bacteria.

Push the glass tube through a two-hole stopper into a bottle as shown below. Fill the bottle completely with clean water.

Close the second hole of the stopper with one finger while sucking on the tube. What happens? Open the hole and suck. What happens? What conclusions can you draw from this experiment?

Sharp edges remain on a glass tube after it has been cut. These edges can give unpleasant cuts.

To smooth the edges, hold the tube in a hot flame, such as that of a gas stove. Turn the tube slowly so that all sides become hot. In a short time the thin edges will melt and flow. Then let the glass cool. (**CAUTION:** Glass stays hot for a long time. Do not touch it for five minutes.)

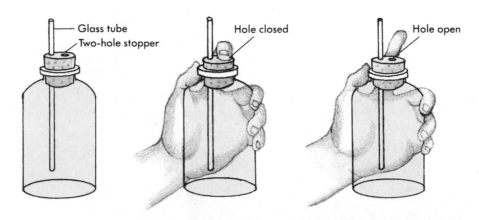

Glass tube
Two-hole stopper

Hole closed

Hole open

Explain the results of the experiment in terms of pressures. Make pressure diagrams using arrows to show why the water rises in the tube.

Experiments with Unbalanced Pressures. The pictures on this page describe the apparatus used in four simple experiments. For experiment *A*, compare what happens while the cap is on the bottle of water with what happens when the cap is removed. For experiment *B*, find out what happens when the upper hole in the jar lid is uncovered.

In experiment *C*, discover what happens when the full glass of water is lifted upwards from the bottom of the jar. For experiment *D*, suck air from the bottle, pinch the tube tightly and put the end under water as shown, and then open the tube.

Explain the results of each experiment in terms of unbalanced pressures. Make pressure diagrams with arrows which show where the pressures act.

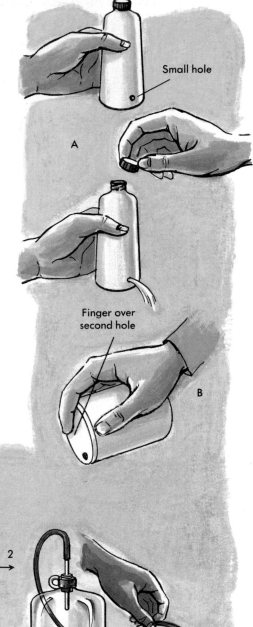

SUMMARY QUESTIONS

1. What happens to a balloon when the pressure outside the balloon is increased?
2. What are two ways of making a balloon larger?
3. Use arrows to show how water travels up a straw when you suck on it.

USING ATMOSPHERIC PRESSURE

The pressure of the atmosphere plays a bigger part in our lives than many people realize. It makes breathing possible and it helps us drink liquids. A large number of important devices ranging from drinking straws to automobiles require atmospheric pressure to function.

Medicine Droppers. Several devices use a rubber bulb to increase and decrease pressure. The medicine dropper is one of the most common of these devices.

Hold the tip of a medicine dropper in water and squeeze the bulb. Note what happens. Describe the way the pressure inside the dropper changes as the bulb is squeezed. Make a pressure diagram.

Let the bulb spring back again. Describe the pressure changes inside the dropper, and explain why water enters. Make a pressure diagram to help in your explanation.

Squeeze the bulb of a medicine dropper, place a finger against the tip, and then release the bulb. Describe what happens and explain the results using a pressure diagram.

Invent experiments to show that atmospheric pressure is needed to make a liquid rise into a medicine dropper.

Vacuum Cleaners. Dust is carried into a vacuum cleaner by a current of air which rushes through the cleaner. Study the picture of the vacuum cleaner and describe the way it operates. What causes air to enter the cleaner? What causes

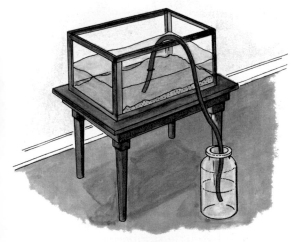

air to leave the cleaner? Why does dust stay in the cleaner?

Examine other types of vacuum cleaners. Discover how they operate and explain their action.

Making and Using a Siphon. At the left is shown a siphon being used to remove water and dirt from an aquarium. Such a siphon is a simple device—just a length of rubber tubing.

To start a siphon operating, fill the tube with water, pinch the ends shut, and put one end in a jar of water as shown. Note that the jar is lower than the aquarium.

Experimenting with a Siphon. Set two equal-sized jars side by side, one higher than the other and full of water and one empty. Siphon water from one jar into the other. When does the siphon stop? How much water flows from the full jar?

Try the same experiment with two unequal-sized jars as shown at the left. When does the siphon stop operating? Does the flow of the water depend upon the height of the liquid or upon the size of the jar?

Fill a large jar and set it on the edge of a table. Start a siphon operating and raise the lower end of the tube until the water stops flowing. What conclusion do you draw from this experiment?

Measuring Rate of Flow. Collect and measure the water that flows through a siphon during one minute. Repeat the measurements for larger and smaller siphon tubes. Does the rate of flow depend upon the size of the siphon tube?

Measure the flow of water per minute when one container is a little higher than the other. Repeat the measurements when the first container is much higher than the second. What conclusion can you draw from this experiment?

Measure the rate of flow when the level of the water in one container is always 30 centimeters above the level of the water in the other container. Raise or lower one of the containers as necessary to keep the difference between water levels constant.

Repeat the measurements keeping the distance between the water levels exactly 60 centimeters apart. Does the water flow twice as fast? If not, how do the rates of flow compare?

Repeat the measurements again keeping the water level of one container 120 centimeters above the other. What happens to the rate of flow?

120 cm

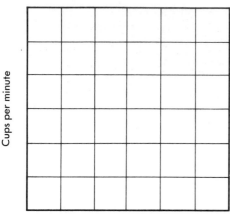

Cups per minute

Height

Graph the results of your experiments by comparing the height of the water level with the flow rate.

Explaining the Siphon. A siphon is easy to operate. It is not so easy to understand.

Many books say only that a siphon is operated by atmospheric pressure. A student of science should not be content with this statement unless it is supported by good evidence.

An experiment provides the best evidence. Make the pressure in the upper container the independent variable and the flow of water the dependent variable.

Use a large bottle for the upper container. Push a long glass tube through a two-hole stopper and attach a rubber tube to the glass tube. The two tubes serve as a siphon.

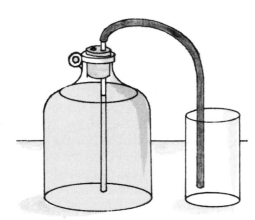

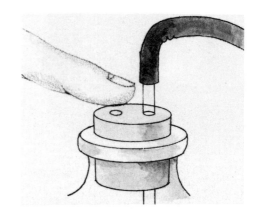

Start the siphon by sucking on the tube. After the water begins to flow, put a finger over the second hole in the stopper. Note what happens. Open the hole. What happens? What conclusions can you draw from this experiment? How can you increase the pressure inside the bottle? Try this and see if your conclusion remains the same.

SUMMARY QUESTIONS

1. How does a vacuum cleaner work?
2. How do you start water flowing through a siphon?
3. How does a medicine dropper lift water?
4. What effect does the level of water above the level into which the siphon flows have on the rate of flow?

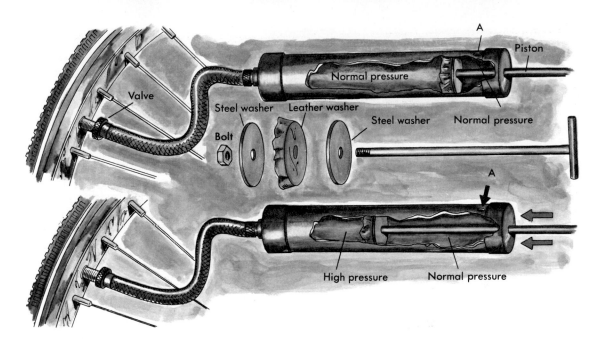

Labels on diagram: A, Piston, Normal pressure, Valve, Steel washer, Leather washer, Steel washer, Bolt, Normal pressure, A, High pressure, Normal pressure

CHANGING AIR PRESSURE

When additional air is squeezed into a space, the pressure increases. When air is removed from a space, the pressure decreases. Air pressure can also be changed by changing the size of the container. If the container is made smaller without air escaping, the pressure rises. If the container is made larger without air entering, the pressure decreases.

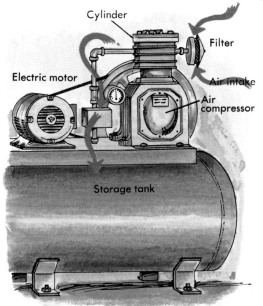

Labels on diagram: Cylinder, Filter, Electric motor, Air intake, Air compressor, Storage tank

Describe the air's path as it is pumped into the compressor's storage tank.

Air Pumps. The diagrams above show the parts and the operation of a bicycle pump. The part that slides back and forth is called the *piston*. The tube in which it slides is called the *cylinder*. Locate these parts in a bicycle or automobile tire pump.

Note the way the piston is made. The leather washer spreads out on the push stroke and fits tightly against the walls of the cylinder.

What happens to the air in front of the piston during the push stroke? Where does the air go? What happens during the back stroke? Why is there a hole in the cylinder at A?

Compressors. Air pumps used for raising the pressure of air are often called *compressors*. Why?

Air compressors are used in garages and auto service stations. For what purpose is compressed air needed?

Compressors are often seen (and heard) where streets are being dug up and where buildings are being constructed. How is compressed air used in these places?

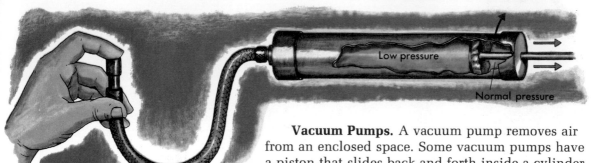
Low pressure

Normal pressure

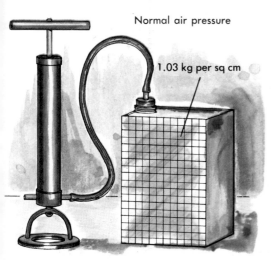
Normal air pressure

1.03 kg per sq cm

Vacuum Pumps. A vacuum pump removes air from an enclosed space. Some vacuum pumps have a piston that slides back and forth inside a cylinder, thus changing the pressure of the air inside.

Take the piston from a bicycle pump, put the leather washer on backwards, and replace the piston in the cylinder. Hold a finger over the hose of the pump while someone pulls quickly on the piston. Describe what you feel. What happens?

A bicycle pump cannot be used as a vacuum pump because it contains no valve to prevent air from returning into a container. If your school possesses a vacuum pump of the piston type, locate this valve.

Crushing a Can with a Vacuum Pump. Plug the opening of a metal can with a one-hole stopper. Connect a vacuum pump to a glass tube in the stopper. Pump out some of the air. Make a pressure diagram to explain what happens.

Measure off the number of square centimeters on one side of the can. If there were a perfect vacuum in the can, what would be the force pushing the side inward?

Testing a Vacuum Pump. Pump all the air possible from a bottle, pinch the connecting tube shut, and put the end of the tube in water.

Measure the amount of water that went into the

Air

Air

Pinch closed

Open

A   B   C

Heating

Cooling

bottle, and measure the amount of water that the bottle can hold. If the bottle was half full of water after the test, half of the air had been taken out and the pressure inside was about 0.5 kg/cm². How good is your pump?

Heating and Cooling Air. Attach one end of a rubber tube to the opening of a metal can. Put the other end of the tube in water. The can should be dry at the beginning of this experiment.

Slightly heat the can until bubbles come out of the tube. Let the can cool and note what happens. Explain the effect of heating and cooling on the pressure of air.

Measure the amount of water that went into the can. How much air was driven from the can? What part of the air was left in the can?

Heat the can again, this time without the tube. While the can is still hot, plug the opening with a solid stopper. Let the can cool for a few minutes. How do you know that the pressure inside the can has been reduced?

Egg-in-a-Bottle Trick. Peel a hard-boiled egg and place it on the top of a milk bottle. Set fire to a piece of tissue paper, lift the egg, and drop the burning paper into the bottle. Replace the egg immediately. Explain what happens.

The same method can be used to peel a banana. Invent other tricks. For example, put an inflated balloon on top of the milk bottle. Explain what happens in each case.

How Reasoning Can Be Wrong. For many years the following demonstration has been used to show that air contains 20% oxygen.

Set a candle on a small block of wood. Float the block in a tray of water. Light the candle and cover it with a straight-sided glass jar as shown here. After the candle has gone out, measure how much the water level has risen.

The conclusion that many people draw from this activity is that the burning candle used an amount of oxygen in the air equal to the amount of water that rises into the jar. This seems reasonable until we think about it carefully.

The conclusion is based upon three assumptions: (1) no air escapes from the jar; (2) all the oxygen is taken from the air by the candle; (3) no new gases are formed. If any of these assumptions is wrong, the conclusion may be wrong.

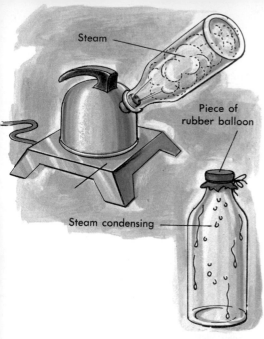

Steam

Piece of
rubber balloon

Steam condensing

Scientists do not like to draw conclusions until they have tested as many of their assumptions as possible. We can easily test two of the assumptions previously mentioned.

Repeat the experiment, putting the jar over the candle very quickly. Look for bubbles coming out from under the edge of the jar. Is any air escaping? If so, how does it affect the conclusions.

Repeat the experiment again. Notice the time when the water starts to rise and when it stops rising. If the water rises because oxygen is being used up, the water should rise all the time the candle is burning, and the water should stop rising when the flame goes out. Does it?

After you have made these tests, you may wonder how anyone could draw conclusions from the activity. But pupils have done so for years, which shows how easy it is to be wrong if a person has not learned to think critically.

Low Pressure by Condensation. Steam takes up much more space than the liquid to which it condenses. Therefore, a partial vacuum is produced in a closed space if steam condenses in that space.

Hold a large fruit juice bottle over the spout of a teakettle until the glass is hot and steam is coming from the mouth of the bottle. (**CAUTION:** Hold the bottle with gloves or a cloth.) Quickly stretch a piece of rubber balloon across the opening and tie it with a string or rubber band.

What happens? Explain why there is so little air in the bottle.

There are many experiments like this one, such as pressing the palm of your hand across the mouth of the bottle while the steam condenses. Invent some of these experiments.

Put a few tablespoonfuls of water into a metal can and boil the water. When steam begins to come from the opening, take the can from the heat and plug the opening with a tightly fitting cap. Explain what happens as the steam condenses.

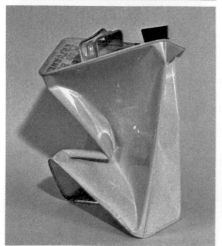

ASSUMPTIONS. All reasoning in science is based upon assumptions. However, we should always recognize the assumptions we are making, we should test them whenever possible, and we should be willing to change our minds if our assumptions are ever proven wrong.

Lifting Water with Steam. Steam the inside of a glass bottle for one minute as described on page 112. Close the mouth of the bottle with a one-hole stopper that is connected to a long glass tube. (Several short pieces may be joined with rubber tubing if necessary.) Dip the lower end of the glass tube into a jar of water. How high does the water rise in the tube? Explain why it rises.

The more perfect the vacuum, the higher the water rises. Atmospheric pressure can push water up to a height of about 10 meters.

Try different ways of making a vacuum. Which method raises water the highest?

How Food Jars Are Sealed. Food that is canned in glass jars is protected from bacteria by very tightly fitting lids. The following experiment shows what holds these lids in place.

Use three glass jars of the type made for home canning. Fill each jar with cool water. Put on the lids but do not clamp them.

Set one jar aside for comparison. Put the other jars in a kettle of cool water and heat them slowly. After the water has boiled for 15 minutes, remove the jars and clamp down the lids (**CAUTION:** Use gloves). Also clamp down the lid of the cool jar to be used for comparison. When the jars have cooled to room temperature, unclamp the lids. Which lid is tightly sealed and which comes off easily?

Note the level of the water in the jars. What has happened?

Open one jar by slowly pulling out the rubber ring. Listen as you do so and explain what is happening.

Test your explanation by opening another jar under water. Does air come from the jar or does water go into it when the ring is pulled out?

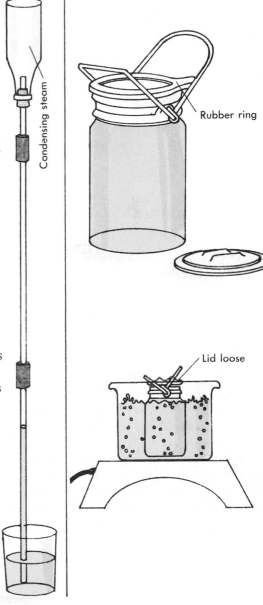

Condensing steam

Rubber ring

Lid loose

SUMMARY QUESTIONS

1. Explain two methods of decreasing the air pressure in a space.
2. How does an air pump work?
3. Why is it not possible to use a bicycle pump as a vacuum pump?
4. What conclusions can you draw from the activity using the can, the tube, and the beaker of water?

PRESSURES IN MOVING AIR

Air in motion is under pressure just as quiet air is, but the pressures may be different, often surprisingly so. Understanding the pressures in moving air has made possible the development of many important devices, including electric fans, high-speed trains, and airplanes.

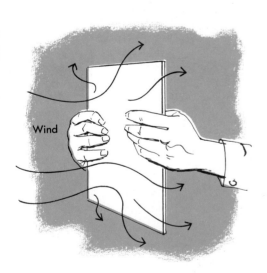

Wind

Effect of Slowing Down Air. Hold a sheet of cardboard in quiet air and then in a stream of moving air. Note the difference in the pressures acting on the cardboard. In the first case, atmospheric pressure acts equally on all sides of the cardboard; the forces are balanced. In the second case, there is an unbalanced force pushing the cardboard away from the source of air.

Imagine that air can be seen. Picture it striking the cardboard and piling up in front of it, especially at the center. The pressure of the air in the piled-up region is greater than normal atmospheric pressure.

Study the picture of a sailboat traveling before a high wind. Imagine air striking the sail, piling up behind it, and producing an unbalanced force which drives the boat ahead.

Discuss other devices that are operated by moving air. Decide where the air piles up and which way the pressures act. In the case of a windmill, study each blade separately.

The Electric Fan. An electric fan produces a wind instead of being driven by the wind. However, the pressures around the blades are much the same as those around the blades of a windmill. Each blade shoves its way through the air, piling air up in front of it. Thus, the air in front of the blades is under high pressure and streams forward into regions of normal atmospheric pressure.

Effect of Speeding Up Air. It was shown on the opposite page that slowing down air increases its pressure. A person might now reason that speeding up air should decrease its pressure.

This type of reasoning is very useful and has led to many discoveries. However, the conclusions drawn from such reasoning are actually no more than guesses. Experiments are needed to test the guesses.

The pictures on this page describe four experiments in which air is speeded up. Try each of these experiments and decide whether the pressure in the moving air is higher or lower than in the quiet air nearby. What conclusions can you draw from these experiments?

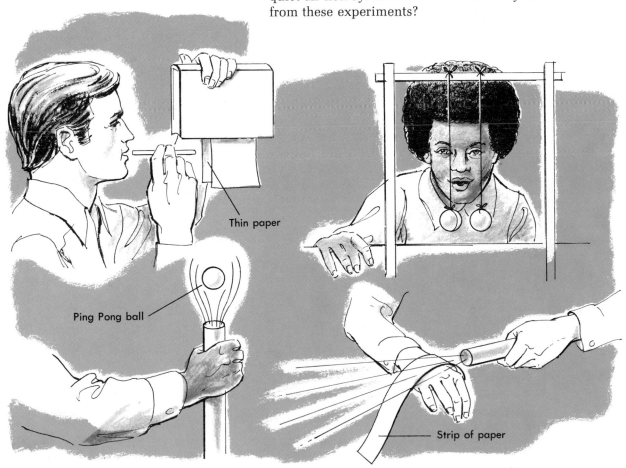

Thin paper

Ping Pong ball

Strip of paper

Pressures on Airplane Wings. When an airplane moves through the air, the pressures on the wings are the same as when the airplane stands still in a strong wind. Airplanes are commonly tested in wind tunnels where they are held in a rapidly moving stream of air.

To study air pressures around airplane wings, make a section of a wing from a strip of stiff paper about 5 centimeters wide and 20 centimeters long. Double the paper and press the fold lightly to make a slight crease. Tape the two ends together with an overlap as shown in the picture above.

Hold the wing section on a round pencil and blow against the bottom. Explain what happens in terms of air pressure. Now blow across the top. Again explain what happens in terms of pressures.

This study shows that both streams of air, one across the bottom of the wing and one across the top of the wing, produce an upward force on an airplane. Precise measurements indicate that the air flowing across the top of the wing provides about 70% of the lifting force.

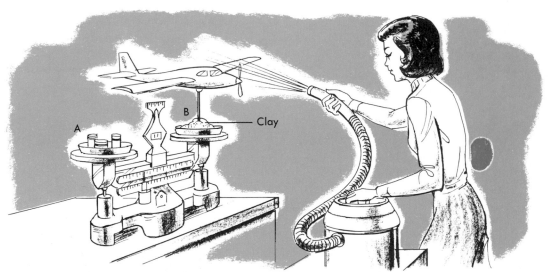

Clay

Measuring the Lifting Force. Set up the apparatus shown above. Add enough weights to pan A so that the weights balance the airplane and its support. Start the blower. Measure the lifting force acting on the airplane by adding weights to pan B until the pans balance.

Set up an experiment to find out what effect the distance between the blower and plane has on the lifting force acting on the plane.

Studying Air Currents. Air tends to move from a region of high pressure to a region of lower pressure. Therefore, a study of air currents may show where high pressures and low pressures can be found.

Stand a board on edge. Put a lighted candle on one side of the board, and direct a stream of air against the other side. Which way does the candle flame point? Which way is the air moving? Locate an area of low pressure. Try the same experiment with a round object such as a pail or a large glass jug.

Plotting Air Currents. A map of the air currents around an object gives a picture of the high and low pressure areas that are formed. Instead of a candle flame, use a small wind vane for locating the currents. The smaller the vane, the more detailed the map that can be produced.

Set the object to be tested on a large sheet of paper. Direct the air from a vacuum cleaner hose against the object. Move the tiny wind vane into different positions, and draw arrows showing the wind direction at each place. The sheet of paper becomes a map of the air currents around the object.

Make a map of the air currents around objects of several shapes and sizes, such as round objects, long and short boards, and large round can covers.

From vacuum cleaner

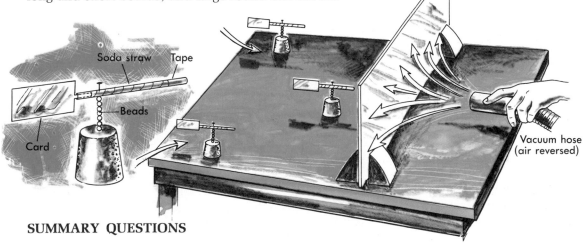

Soda straw Tape

Beads

Card

Vacuum hose (air reversed)

SUMMARY QUESTIONS

1. Use arrows to explain how the wind makes a sailboat move.
2. How can you demonstrate that speeding up air decreases its pressure?
3. What are the pressures acting on an airplane wing in flight?

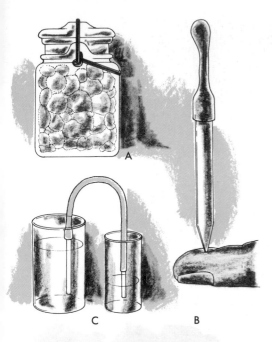

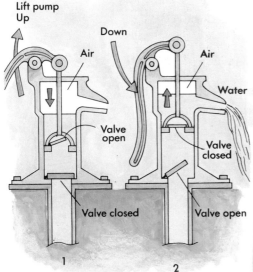

REVIEW QUESTIONS

1. What holds the lid on in *A* at the left?
2. Why does the bulb stay flat in *B* at the left?
3. When will the water stop flowing in *C* at the left?
4. Why does water rise in a soda straw?
5. Why are two holes desirable when pouring juice from a can?
6. What causes the lifting force on an airplane wing?
7. Why does a metal can collapse when air is removed from it?

THOUGHT QUESTIONS

1. Explain why a table is not crushed flat to the floor by atmospheric pressure acting on the table top.
2. Why do leaves and papers tend to fly into the space behind a rapidly moving automobile?
3. Why does increasing the speed of an airplane increase the lifting force of the wings?
4. Why is a flat signboard more apt to blow away than a round tank of the same width, height, and weight?
5. When can a siphon be used to take water from a rowboat?
6. How does the pump pictured at the left operate?
7. When does water flow upward in the coffee-maker shown below? When does the water flow downward?

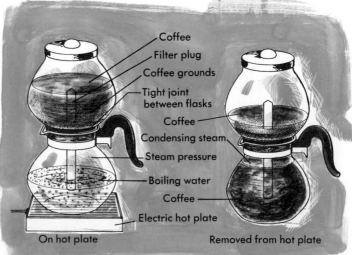

118

Starting and Stopping

  Our world is a restless place. Everywhere forces act to set objects in motion or to bring them to a stop. The quiver of a leaf, the jolt of an earthquake, the tinkle of a mechanical toy, and the roar of a rocket suggest forces together with the motions the forces produce.

  Such forces never seem more important than when they act on our own bodies. We learn at an early age to take advantage of common forces, such as gravity and friction, to control our bodies. We may even develop special skills in using these forces during athletic contests. Field hockey, track, and football players all must learn special skills for producing forces on objects. These forces are often used to stop moving objects or to start objects moving.

FORCES FOR STARTING AND STOPPING

Our modern theory of motion states that a force must always act upon an object to set it in motion. Likewise, no moving object slows down by itself; a force must always act to slow it down.

Gravity is the most common force that sets objects in motion. Friction is the most common force that brings objects to a stop. List as many other forces as you can, and give an example of an object that is set in motion or stopped by each force.

Acceleration and Deceleration. When a person pushes on the pedals of a bicycle, the bicycle will begin to speed up. By squeezing the brake handles, a force is exerted on the wheels of the bicycle, slowing them down. An object that is increasing its speed is *accelerating*. An object that is slowing down is *decelerating*.

Effect of Friction. Cover a flat table top with a large towel or other soft cloth. Give a roller skate a gentle push to start it rolling across the table top. Notice any change in speed as the skate rolls. What force is acting on the skate? Is the force an accelerating force or a decelerating force?

Effect of Gravity. Prop up two legs of a table so that the top is slightly tilted. Place a roller skate near the high end of the table. What happens? What force is acting on the skate? Is the force an accelerating or a decelerating force?

Raise the end of the table. Note the effect on the skate and the change in the force acting on the skate.

Acceleration of Falling Objects. Fill two cans with sand, making sure that they weigh the same.

Bath towel

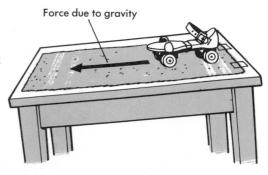

Force due to gravity

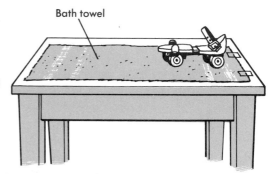

Hang the cans from the opposite ends of a cord placed through a clothesline pulley.

Give one can a downward push and then watch its motion. Does the can accelerate or decelerate? Explain what happens in terms of (1) the downward pull of gravity, (2) the upward pull of the cord, and (3) the friction of the pulley. Which of these forces acts to accelerate the can, and which force acts to decelerate it?

Add some weights to one of the cans. Then explain the changes in the acceleration of the can in terms of accelerating and decelerating forces.

Add sand to one can until it moves downward at a steady speed (without acceleration) after being given a slight push. Explain what has happened to all the forces acting on the can.

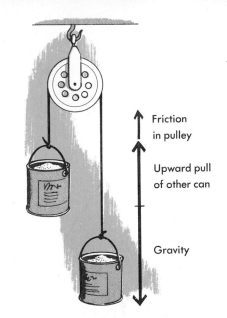

Friction in pulley

Upward pull of other can

Gravity

Parachutes. Two forces act on a parachute as it descends. What are these forces? In which direction does each of the forces act? Which of the forces is an accelerating force? Which of the forces is a decelerating force?

What happens to the parachute when the downward force is much greater than the upward force? What happens to the parachute when the downward force is slightly greater than the upward force? What happens to the parachute when the upward force equals the downward force?

Free Fall. The term *free fall* is used when an object has only the force of gravity acting on it. Such an object accelerates constantly and its speed soon becomes very great.

Free fall is impossible in air because of air resistance. However, objects that are streamlined or nearly streamlined are not greatly affected by air resistance until they are going very rapidly. For example, a ball accelerates almost as much during the first 16 meters of its fall as it would if it were in free fall.

Acceleration During Free Fall. Acceleration during free fall takes place so rapidly that the eye cannot observe it very well. However, by using a buzzer and strip of paper you can observe and measure the changing speed of a falling object.

To observe and measure changing speeds, use a device such as shown on page 122. The clapper on an electrical buzzer moves up and down with a regular beat so it can be used as a timer. The clapper should be made to leave a mark on a strip of paper being pulled by a rolling object. Then you

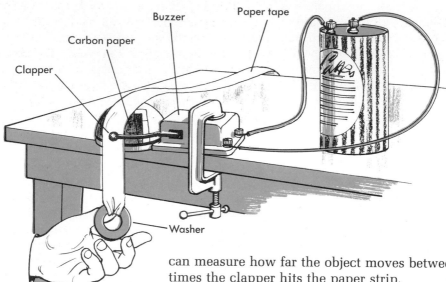

Clapper · Carbon paper · Buzzer · Paper tape

Washer

can measure how far the object moves between the times the clapper hits the paper strip.

Fasten a strip of paper to a large washer. Run the paper tape through the buzzer. Start the buzzer and drop the washer. Where on the strip of paper are the dots close together? When the dots are far apart, what do they show you about the speed of the washer? How did the speed of the washer change as it fell?

Upward Acceleration. Tie a string to a can of sand and hold the can by the string. Note the force required to support the can.

Pull upward suddenly on the string. What happens to the can? Compare the force needed to accelerate the can with the force needed to support the can.

Accelerate the can upward at different rates. Compare the forces required in each case. Explain why the forces are different.

Can you make the force great enough to break the string? How?

Raising Your Body. Stand on a bathroom scale with your knees bent. Note the reading of the scale. Straighten up suddenly. What happens to the pointer of the scale? Do not try to make an accurate reading because the pointer will probably swing beyond the correct value. What does the scale tell you about the act of straightening up?

Repeat the experiment, sometimes straightening up slowly and sometimes rapidly. How do the accelerating forces compare in each case? What method of straightening up requires the least muscular effort? Why?

Climbing. Place a bathroom scale on the first step of a stairway. Note the readings as you climb the stairs, first as you raise your body onto the scale, and then as you raise yourself onto the next step. Is the required force equal to, greater than, or less than your weight? Explain why.

Try to go up the stairs in a way that requires the least effort. Describe what you do in terms of acceleration. Mountain climbers try to climb in this way. Discuss the advantages of this method.

Behavior of a Tossed Ball. An old saying states that "what goes up must come down." Study the motion of a tossed ball to see how it behaves.

The photograph below was taken in a dark room with the shutter of the camera fixed open. When the ball was tossed up, a flashing light was switched on. Each time the light flashed, the ball appeared in a different place. How many times did the light flash in this photograph?

The photograph shows that the ball starts from rest at A. It is pushed upward with an accelerating force from A to B. Two other forces also act on the ball: gravity and air resistance. Which way do these forces act? Are they accelerating or decelerating?

The force causing upward acceleration stops acting when the ball leaves the hand. Only gravity and air resistance continue to act. What do these two forces do to the ball? As the ball goes from B to C, is it accelerating, decelerating, or rising at a steady speed?

What is the upward speed of the ball when it reaches C? Only one force acts on the ball at C. What is this force and what does it do to the ball?

What forces act on the ball as it moves from C to D? Is the ball accelerating, decelerating, or moving at a steady rate?

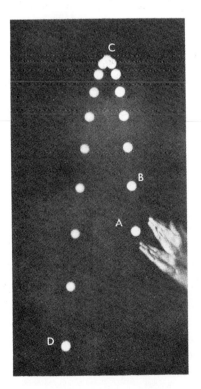

Study the motion of a tennis ball as someone tosses it upward. Watch from a second floor window to study the motion of the ball while it is at the top of its climb.

SUMMARY QUESTIONS

1. What do accelerate and decelerate mean?
2. What common force brings objects to a stop?
3. What two forces act on a parachute?
4. Why is free fall impossible in air?
5. Why is the force required to climb one step on the staircase greater than your own weight?

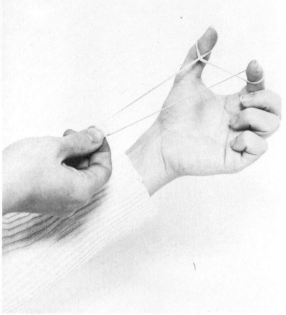

STARTING DISTANCES

A golf ball hit with a powerful blow seems to reach its full speed instantly. Acceleration is so rapid that our eyes cannot follow the golf ball changing speed. Nevertheless, cameras show that objects never reach full speed instantly. Objects always accelerate through a distance which may be long or short, depending upon the force applied.

Study the photograph above. How do you know that the ball was hit with great force? Was the ball in motion or at rest when the photograph was taken? How do you know? The dent in the ball shows that the club has been moving more rapidly than the ball. It also shows that the ball will pick up more speed as the dent straightens out. The ball is still accelerating.

Measuring Starting Distances. The girl is about to shoot a paper wad with a rubber band. She will apply a force through a certain distance to accelerate the paper wad. This *starting distance* begins at the point where the paper wad begins to move and ends at the point where the rubber band stops pushing on the wad. The starting distance is the distance through which the wad accelerates.

Bows and Arrows. The picture on page 125 shows an archer ready to shoot an arrow. Where does the starting distance of the arrow begin and where does it end? Compare the starting distance of the arrow with its length.

At what instant does the arrow have its greatest speed? Why does it not pick up speed after it leaves the bow?

Ask someone to demonstrate the action of a bow and arrow. Experiment with the effect of pulling the arrow back different lengths, measuring the distance the arrow travels each time.

Varying Starting Distances. Make wads of damp paper towel to fit inside a soda straw. Blow a wad from the straw and note the path it takes.

What force sets the paper wad in motion? At what point does the wad begin to accelerate? At what point does the acceleration end? How long is the starting distance?

Test the effect of using straws of different lengths. Measure the distance a paper wad can be blown from a full-length straw, a half-length straw, and a quarter-length straw.

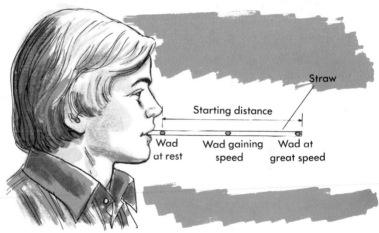

Several variables must be controlled in this experiment. How can you be sure that the paper wads are all the same size? How can you be sure that the air pressure is nearly the same each time? What is the advantage of averaging the results of ten or more trials?

Blowguns. In certain parts of Central and South America, Indians have developed great skill with blowguns. They blow darts through long, hollow tubes with high speed and accuracy. Small birds are easy prey. Larger animals are often killed by using poisoned darts.

Blowguns are sometimes 1.5 meters long. What is the advantage of a long blowgun? Why is a long, steady puff more effective than a short, hard puff?

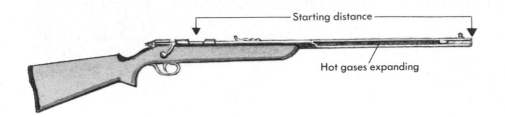

Starting distance

Hot gases expanding

Firearms. Rifles, shotguns, and other firearms are blowguns that use the pressure of hot gases instead of lung pressure. Burning gunpowder produces a large quantity of gases. These gases are compressed in the small space behind the bullet in the barrel of the gun. These gases push the bullet along the barrel, accelerating it to a very high speed.

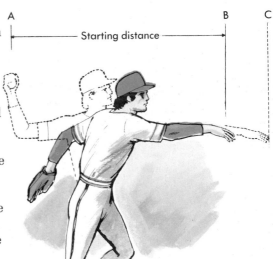

Starting distance

Discuss the action of firearms in terms of the length of the barrel, gas pressure, bullet speed, and accuracy.

Pitching a Baseball. Study the diagram of a pitcher throwing a baseball. Through what distance does the pitcher exert an accelerating force on the ball? When does the ball reach its greatest speed? Why does the ball not gain speed after it leaves the pitcher's hand?

Suppose that the pitcher leaned forward at the end of the pitch and released the ball at C instead of B. Why would the ball be going faster? What advantage do long arms give a pitcher?

Below is a series of pictures taken 1/100 of a second apart. Measure the distances between

images of the baseball. Make a graph plotting distance against time. When was the ball accelerating? Did it accelerate after it left the pitcher's hand?

Driving a Golf Ball. The picture above shows how a golfer drives a golf ball. The golfer continued the swing until the club made almost a complete circle.

Measure the distances between images. When was the club going most rapidly? Note that the club appears to be bent backward at one point. Explain this bend.

The purpose for the long swing can best be understood in terms of acceleration. A ball can pick up speed only while it is being pushed by the club. It cannot gain any more speed after it has left the club. Therefore, a player tries to make the starting distance as long as possible. Golfer's do not let the club slow down after first striking the ball.

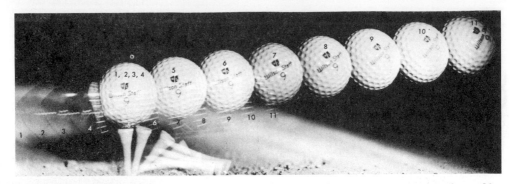

Study the series of close-up exposures of a golf club striking a ball. In which position does the club first touch the ball? In which position has the ball just left the club?

Describe the differences in exposures if a player let the club slow down after first striking the ball. What would be the effect on the speed of the ball?

Describe the differences in exposures if a player could make the club move faster after first striking the ball. What would be the effect on the speed of the ball? Explain why.

Follow-Through. Athletic coaches often speak of *follow-through*. They explain that a golf player, for example, should continue the swing after striking a ball, as shown on the previous page. The purpose of follow-through is to lengthen the starting distance as much as possible.

Examples of follow-through are pictured on this page. Discuss each example in terms of the starting point of the ball, the distance through which the accelerating force acts, and the possible effect of not using follow-through.

Ask an athletic coach to demonstrate proper follow-through in several sports. Practice some of the methods. Compare your results from using and not using follow-through.

SUMMARY QUESTIONS

1. When studying starting forces, where does the starting distance begin and end?
2. What is the effect of the length of a straw on the distance it will shoot a wad of paper?
3. How is a rifle or shotgun like a blowgun?
4. Why are most baseball pitchers taller than most of the players on the team?
5. When hitting or throwing a ball, what is meant by follow-through? Why is it important?

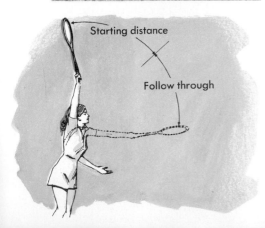

Starting distance

Follow through

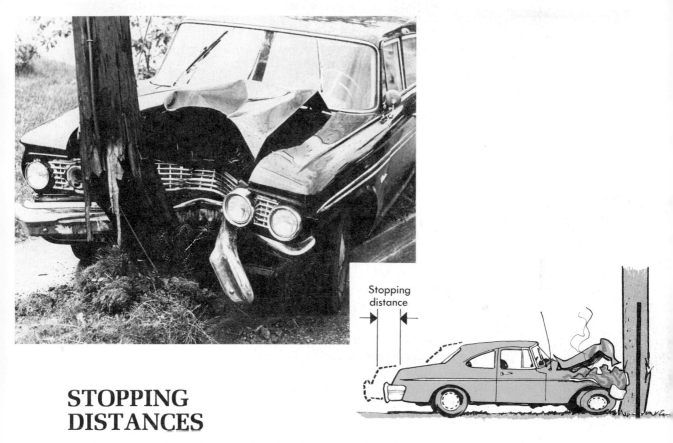

Stopping
distance

STOPPING DISTANCES

Just as an object is set in motion by an accelerating force, an object is brought to rest by a decelerating force. Just as there must be a starting distance to set an object in motion, there must be a stopping distance to bring an object to a stop.

Nothing stops instantly. There is always a period during which an object is slowing down. Sometimes the deceleration is too rapid for the eye to follow, but high-speed cameras can reveal a stopping distance.

Hitting Immovable Objects. Some people argue that if a moving body hits an immovable object, motion is stopped instantly. These people do not consider what happens to the moving body.

The photograph shows an automobile that hit an immovable object. The diagram shows what happened in terms of the stopping distance.

Dashed lines show the position of the car when it struck the tree. The car did not stop instantly, but continued to move forward as the front end crumpled. The car came to rest in the position here after decelerating through the stopping distance shown. What evidence is there that the decelerating force was enormous?

Stopping
distance

Changes in Stopping Distance. Stand on a low step or box not over 10 centimeters high. (**CAUTION:** Do not use anything higher.) Jump from the step and compare the effect of landing (1) with knees stiff, and (2) with your knees free to bend. Test also the effect of letting your knees bend different amounts. Compare the decelerating forces that were produced.

The diagrams above show how the stopping distance can change. At the left, the girl's body stops in less than 3 centimeters. At the right, her stopping distance is more than 15 centimeters as measured by the motion of her head. How does the change in stopping distance affect the force needed to stop her body?

Landing from a Jump. Below is a series of illustrations showing an athlete landing from a 5-meter pole vault. Measure 5 meters on a wall and imagine dropping this distance.

Note how the vaulter lands. What is the vaulter landing on? Why does he land on his back and not on his feet?

Catching a Ball. A hard ball traveling at high speed can hurt a person's hand seriously if the ball is not caught properly. The ball should be decelerated slowly through a long distance rather than quickly through a short distance.

Study the way a baseball catcher slows down a fast pitch. Notice where the hands are held before the ball arrives. Notice what the catcher does when the ball strikes the mitt. Does the catcher ever try to stop a fast pitch without moving both hands? How does the padding in the mitt help slow down the ball?

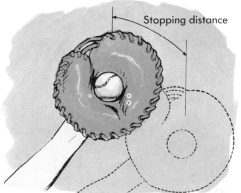

Stopping distance

The other players of a baseball team cannot wear such heavy mitts as the catcher does. They have to catch fast balls by decelerating them through a longer distance. Watch an outfielder catch a ball. Notice the stopping distance of the ball.

Experiment with the catching of a softball. Catch it without moving your hands. Catch the ball again, letting your hands "give" to increase the stopping distance. Compare the forces that you feel.

Boxing. Sometimes a boxer seems to have been hit with a very hard blow, but he shows no signs of being hurt. Probably the boxer was able to stop the blow with very little force.

The picture below shows how he can do this. When the boxer sees a blow coming, he tries to move his body away from it. Thus the stopping distance of the blow is increased and the decelerating force is decreased. Boxers call this trick "rolling with the punch."

Sometimes a boxer makes a mistake and moves toward the blow as it lands. What does this do to the stopping distance? To the decelerating force? To the boxer?

Stopping distance

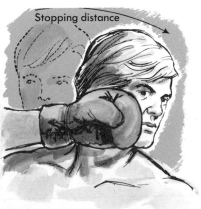

Stopping distance

Diving. It is helpful to know something about stopping distances when diving. A diver is usually going rapidly when he or she strikes the water. If the diver stops suddenly, the decelerating force can be painful and even dangerous.

The underwater photograph shows a girl landing from a high dive. Note that her stopping distance is several meters long. Why is this helpful? Discuss what is likely to happen to the boy in the photograph at the left.

The Action of a Hammer. A common hammer, though simple in appearance, is one of our most amazing tools. With a hammer, a person can use a small muscular force to develop an enormous force on the head of a nail.

All members of the pounding tool family, from the midget tack hammer to the giant drop forge, make use of long starting distances and short stopping distances. A small accelerating force is applied over a long distance, thus providing a tool with enough power to produce a huge force when stopped suddenly.

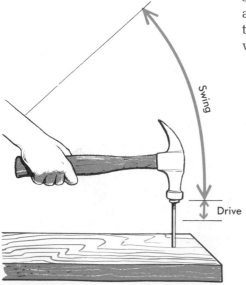

Swing

Drive

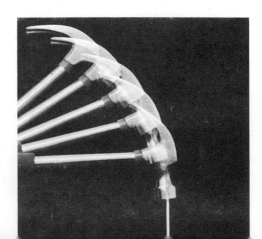

Importance of Stopping Distance. Lay a thin board across two blocks as shown above. Try to drive a nail into the board halfway between the blocks. Place a third block under the nail and try again. Which is the better way?

What happens to the board when there is no support under the nail? How far does the board bend? Compare the stopping distance of the hammer when the board bends with the stopping distance when the board does not bend. What is the relation between the stopping distance and the force produced on the head of the nail?

Purpose of Hammer Handles. The handle of a hammer increases the starting distance of the head as shown in the diagram below. Therefore, the head is going faster when it strikes.

Try to drive a nail while holding a hammer by its head. Try again, holding the hammer by its handle.

Hold a hammer by its head and hit a board with all your strength. Note the dent in the wood. Hold the hammer by its handle and swing it just hard enough to produce another dent as deep as the first. What effect does the handle have?

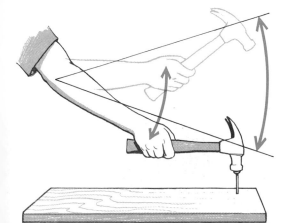

SUMMARY QUESTIONS

1. What kind of force causes an object to stop?
2. Why is it a good idea to bend your knees when jumping down from a high place?
3. How does a catcher prevent a hard-thrown ball from hurting his or her hands?
4. What action does a boxer take to prevent a blow from hurting him?

SELF-PROPULSION

All the objects studied thus far in this chapter have been set in motion by outside forces pushing or pulling on them. However, automobiles, airplanes, and people are able to set themselves in motion. They are self-propelled.

This does not mean that outside forces do not act on self-propelled objects. Usually, outside forces are necessary. For example, the truck in the picture is unable to move. Its wheels spin, but there is no forward motion.

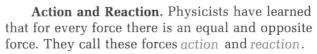

Action and Reaction. Physicists have learned that for every force there is an equal and opposite force. They call these forces *action* and *reaction*.

Suppose that a block rests on a table top. The block is being pulled down on the table by the force of gravity. The table pushes back with an equal force.

Suppose that a pitcher pushes forward on a baseball with a certain force as the ball is thrown. The baseball pushes back on the pitcher's hand with an equal and opposite force.

Wind up a toy car and set it on a thin board that rests on rollers. What happens when you hold the car still? What happens when you hold the board still? What happens when you do not hold either the car or the board? Discuss the forces applied in each case.

A force is needed to start a car moving forward. Which way do the wheels push? What provides the forward push? Why might the car fail to move when the road is covered with ice?

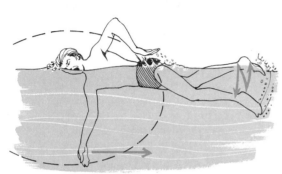

Importance of Friction. A person on a skate board needs a forward push to drive ahead. The person can get this forward push by pushing *backward* on the sidewalk. The sidewalk then pushes forward on the person and drives the skate board ahead.

The ground is able to provide a forward push because of friction. The forward push is really a frictional force. What would happen if a person tried to ride a skate board on very smooth ice? How could the person increase the forward push on smooth ice?

Discuss what happens when a person walks. Compare walking with riding a skate board. What happens when a person tries to walk on very smooth ice, on marbles, or on anything that does not provide much friction?

Self-Propulsion in Water. How do swimmers move themselves through the water? Upon what do they push? What pushes them forward? Why do they keep their fingers together instead of spread apart?

Discuss the forces that move canoes, kayaks, and rowboats through the water. Why are paddles and oars shaped the way they are?

135

Dowels

Propellers. Propellers exert a backward push on the air. This can be noticed at an airport when an airplane's brakes keep it from moving even though the propellers are spinning rapidly. A strong wind is produced behind the airplane.

As the propellers push backward on the air, the air pushes forward on the propellers. The forward force is the force of reaction. It is this force that drives the airplane ahead.

The force of reaction on a propeller can be demonstrated with an electric fan. Place the fan on a light board that rests on round sticks. Hold the board from moving and start the fan. Note that the fan is pushing the air away from it. Release the board. What happens? Explain the behavior of the fan and the board in terms of action and reaction.

Motor boats and ships also use propellers. Discuss the forces that act as the propellers turn. Explain why boats move.

Jets and Rockets. Jets and rockets are self-propelled. Unlike automobiles and propeller-driven airplanes, however, they do not depend upon outside forces for movement. They are driven by forces that act within themselves.

A balloon full of air can drive itself without the aid of outside forces. Set up the rocket car shown below. Control the rate of air flow with a tube made

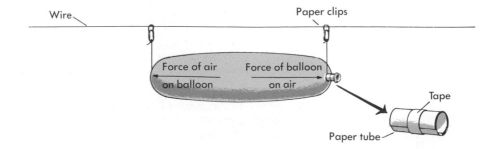

by wrapping a strip of paper around a pencil. The size of the opening can be changed by pinching the end of the tube.

The rubber of the balloon is stretched and it pushes on the air in the balloon, driving this air out through the opening at the back. Because the balloon pushes on the air, the air pushes back on the balloon. This reaction force drives the balloon forward.

Such a rocket car can operate anywhere. It could operate in a vacuum because it does not depend upon outside forces.

Jet planes and rockets operate in much the same way. Gases are heated to a very high temperature in the engines. Heating increases the pressure of the gases and they are forced out the back of the engines. As the gases are pushed in one direction, they push back with an equal and opposite force. The reaction force propels the plane or rocket.

SUMMARY QUESTIONS

1. What is a self-propelled machine?
2. Give three examples to show that for every force there is an equal and opposite force.
3. How does friction make self-propulsion work?
4. What forces cause a propeller to move a boat or an airplane?
5. What forces cause a jet or rocket to move an airplane?

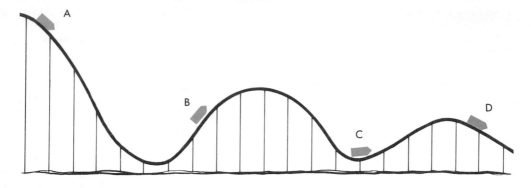

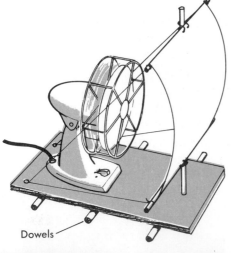

Dowels

REVIEW QUESTIONS

1. Which of the cars on the roller coaster are accelerating and which are decelerating?
2. Which car probably has the greatest speed and which probably has the slowest speed?
3. Why is a hammer such an effective tool?
4. Why is there a tread on a bicycle tire?
5. What provides the forward push on a rocket?
6. What happens when a moving object, such as a ball, strikes an immovable wall?
7. Why does a good driver stop the car slowly instead of slamming on the brakes?
8. Why does a ball player let his or her hand "give" when catching a ball?
9. Why is a long-handled ax more effective than a short-handled ax or hatchet?
10. What force or forces act on a rocket after it has left the earth's atmosphere and has burned up its fuel?

THOUGHT QUESTIONS

1. Why do track runners need spiked shoes?
2. Why do automobile wheels spin when drivers try to make too fast a start?
3. Why are bleacher seats under a special strain when everyone stands up at once?
4. Why are seat belts a necessity for automobiles?
5. How are small jets used to steer a rocket outside the earth's atmosphere?
6. What is the increase in speed per second of a baseball that is going 19.5 meters per second two seconds after it was dropped?
7. Which way will the device shown at the left move when the fan is turned on?

Moving Objects

The study of how things move and what causes them to move was advanced by Sir Isaac Newton. He realized that motion could only be brought about or changed by a force. He defined the relationships among mass, changing motion (acceleration or deceleration), and force. He also recognized that gravity is a force which exists throughout the universe.

Since earliest times we have planned and built structures. A great improvement in methods of building was the discovery that we could use simple machines. These devices can be used to lift greater weights than we alone can lift or to exert greater forces than we alone can exert. For example, the wheelbarrow above allows workers to lift and move a load that they alone could not lift. Even the shovel makes work easier by allowing the load to be moved faster or to be lifted higher than could be done without this simple machine. In what other ways do the simple machines above make work easier?

MOTION OF OBJECTS

About 250 years ago, Isaac Newton stated that any object in motion will continue in constant motion along a straight line unless some force acts upon it. Perhaps you might not agree with that statement. Your experience has been that anything moving on the surface of the earth slows down and soon stops unless a force acts upon it. But maybe Newton is right because friction is a force all over the earth causing things in motion to slow down.

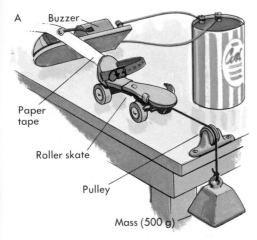

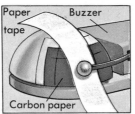

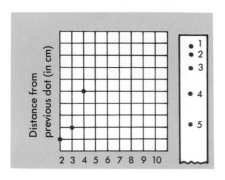

Effect of Force and Mass on Motion. In order to move or change the motion of an object, Newton stated that a force must be applied to the object. Furthermore, Newton said that more force is needed to move an object with greater mass than an object with less mass.

Set up an experiment similar to the one shown here. Use an electric buzzer to measure the speed of the roller skate. An electric buzzer operates in such a way that the clapper goes up and down at a steady rate.

Tape a piece of carbon paper with the carbon side up on the buzzer where the clapper hits. Cut a long piece of paper tape and thread it between the carbon paper and the clapper. If you pull the paper tape while the buzzer is on, a series of dots will appear on the tape. If you pull the tape through quickly, will the dots be close together or far apart? If you pull the tape at a constant rate, how will the distance between the dots appear? How would the distance between the dots change if you pulled the tape faster and faster?

Attach paper tape to the back of the roller skate. Thread on the paper tape through a buzzer with the carbon paper in place. Tie a string with a mass on it to the front of the roller skate. Let the mass fall off the edge of the table while the buzzer is on. Examine the paper tape and select ten dots in a row. Number the dots and measure the distance between each one. Plot the distance between each dot on a graph similar to the one in the margin. This graph will show how the motion of the skate was changed by the force of a falling mass.

In this experiment you are keeping all variables the same except for varying the mass of the skate. By dropping the same weight each time, you are using the same force and investigating the change in motion resulting from the change in mass of the skate.

Determining the Mass of the Skate. Add weights to the skate until you have doubled its mass. Attach a new piece of paper tape to the skate, thread the tape through the buzzer, turn on the buzzer, and drop the same weight as before, letting the force of gravity move the skate. Again examine the paper tape and select ten dots in a row to study. Number the dots and measure the distance between each one. Plot these distances on the graph you used earlier. How did the motion of the skate this time compare with the motion before? What effect did doubling the mass of the skate have on its motion? How does the acceleration of the skate change when mass is added?

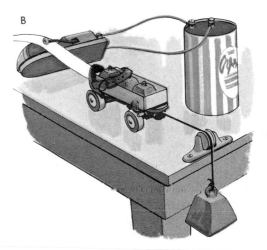

B

If you assume that the clapper on the buzzer goes up and down regularly, then the time between each dot is the same. Therefore, each dot marks a period of time which for convenience you can call a "dot." To calculate the actual acceleration of the skate, determine how much farther it traveled with each succeeding dot. For example, if the skate went 1 cm farther between succeeding dots, then you can say that the skate accelerated by 1 centimeter per dot faster between each dot. This acceleration can be written as 1 cm/dot/dot or 1 cm/dot^2. In your experiment, what was the actual acceleration of the skate without the added mass? What was the acceleration with the mass doubled?

Force Equals Mass Times Acceleration. Newton stated the relationship among the three variables: force, mass, and acceleration. Force is directly proportional to the mass times the acceleration ($F = m \times a$). If the force on a mass were doubled, its acceleration would double.

Another way of expressing this relationship is that acceleration is directly proportional to force but inversely proportional to mass $\left(a = \dfrac{f}{m}\right)$. In this equation, what happens to the value of a when m is increased? According to this equation, what should have happened to the acceleration of the skate when its mass was doubled?

$$F_{gravity} = G \frac{m_1 \times m_1}{d^2}$$

where G = gravitational constant
m₁ = one mass
m₂ = another mass
d² = distance squared

where G = gravitational constant
m_1 = one mass
m_2 = another mass
d^2 = distance squared

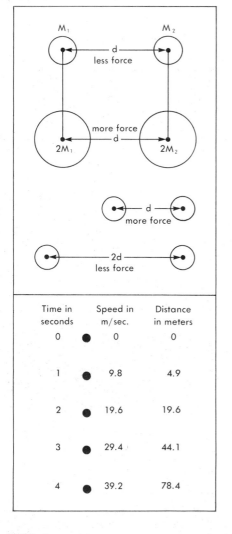

Time in seconds	Speed in m/sec.	Distance in meters
0	0	0
1	9.8	4.9
2	19.6	19.6
3	29.4	44.1
4	39.2	78.4

The Newton: A Unit of Force. Based on the experiments you have done, how can a force be described? Newton described a force as the acceleration the force causes on a certain mass. If one force accelerates a mass twice as much as another force, then the first force must be twice as strong. If one force accelerates a 2-kg mass just as much as another force accelerates a 1-kg mass, then the first force must again be twice as strong.

In honor of having discovered these relationships, the measurement of a force is expressed in units called newtons. One *newton* is the force needed to accelerate a 1-kg mass one meter per second faster every second. If a 2-kg mass were accelerated 1 m/s², what would the force be? How much force would it take to accelerate a 1-kg mass to 2 m/s²?

The Force of Gravity. As you well know, everything is attracted toward the center of the earth by a force called gravity. Again, we can thank Newton for clearly describing this force. He realized that the earth is not the only piece of matter having gravity. Newton stated that any bit of matter has a force of gravity which attracts all other matter.

The strength of the force of gravity between any two pieces of matter is found to depend upon two variables. One variable is the mass of each piece. The greater the masses, the stronger the force of gravity between them. The other variable is the distance between the centers of the two pieces. The farther apart the two pieces are, the weaker the force. However, this relationship is not a simple one. When the distance between the two objects is doubled, the force of gravity between them becomes one fourth as great as it previously had been. If the distance between the two objects is tripled, the force becomes one ninth as great. In other words, the force is inversely proportional to the square of the distance.

Study the equation in the margin where G stands for a constant value which is the same throughout the universe. What happens to the value of F when a mass is made greater? What happens to the value of F when the distance is increased?

Gravity Around the World. The force of gravity on the earth is slightly different from place to place for several reasons. At the top of a mountain the force of gravity is less than at sea level. Why do you suppose that this is so? Anybody weighing

142

themself on a bathroom scale at the top of a mountain would find that they weigh less there than at sea level. This is so partly because on the mountain top one is farther away from the center of the earth. How does distance from the earth's center affect gravity?

At the equator the force of gravity is slightly less than at the poles. Study the diagram of the earth as seen from above the North Pole. Since the earth is spinning, a person at any point on the earth spins around once every 24 hours. How much faster does a person on the equator travel than a person at 60° north latitude? Why?

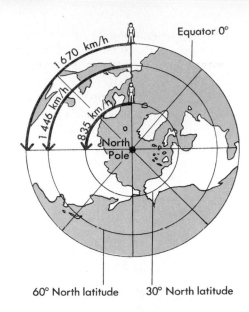

Tie a paper clip to a piece of thread and swing it around over your head. What can you feel when you make the paper clip go faster? Can you feel it tugging at your finger? What happens to the amount of tugging as you make it go faster? According to Newton, the clip would continue in a straight line but the thread is pulling it around.

Gravity is keeping people from flying off the earth. The people on the equator are spinning faster than others closer to the poles, so they are tugging with a greater force against gravity. Any person or object, therefore, will weigh slightly less on the equator than at the poles.

Scientists have also discovered that the earth is not a perfect sphere. The earth bulges a little at the equator. Explain how this might happen.

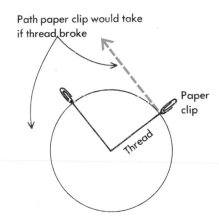

Path paper clip would take if thread broke

Paper clip

Thread

In some places on earth there are large underground formations which have densitites that are different from those of the surrounding formations. If a formation is more dense, the force of gravity over that formation is greater.

The force of gravity is measured in newtons. Your weight is actually a measurement of the force of gravity between the earth and you. The force of gravity on a 1-kg mass is about 9.8 newtons. What is your mass in kilograms? What is the force of gravity between you and the earth in newtons?

Effects of Gravity. Previously you did an experiment in which you used a mass with a long piece of string attached and dropped it off a table. The tape ran between a clapper and a bell. The clapper left dots on the paper tape which allowed you to study the motion of the falling washer. No doubt you found that the washer had accelerated during its fall. This acceleration was caused by the force of gravity acting continuously during the fall.

Careful measurements have shown that if we disregard air resistance, objects at sea level will fall with an acceleration of 9.8 m/s². Study the chart in the margin of a falling object.

The earth's gravity attracts all matter toward its center. Therefore, any object which is suspended by a string will point to the center of the earth. To be certain that a structure is not leaning, carpenters use a *plumb bob*.

To determine whether a structure is level, carpenters use a *level*. Study the photograph of a level and explain how it can be useful.

Time in s	Speed in m/s	Distance in m
0	0	0
1	9.8	4.9
2	19.6	19.6
3	29.4	44.1
4	39.2	78.4

SUMMARY QUESTIONS

1. How can the motion of a mass be changed?
2. If the force on an object is kept constant but the mass of the object is increased, what will happen to the rate of acceleration of the object?
3. How does a buzzer help you to study motion?
4. Why did the skate accelerate while it was being pulled by the falling weight?
5. What is the meaning of F = m × a?
6. How can a force be described?
7. What is a newton?
8. How does mass effect the force of gravity?
9. How does the distance between two objects affect the force of gravity between them?
10. Give three reasons why the force of gravity might be slightly different at different places on the earth.
11. Why will a person weigh slightly less at the equator than close to the North Pole?

THE FORCE: FRICTION

The photographs above show people on two different slopes. In one case, the person slides freely down the slope. In the other, however, the person is held in place by a force between the shoes and the slope. This force is called friction.

Friction opposes movement of one surface along another. Compare the frictional force in each photograph with the force pulling the person down the slope. In which case is friction undesirable?

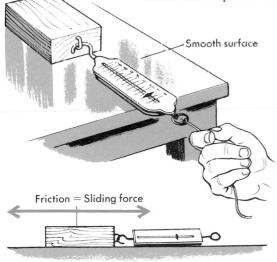

Smooth surface

Friction = Sliding force

Wooden block moving at constant speed

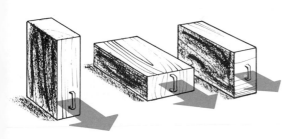

Measuring Friction. Pound a staple into a block of wood about $5 \times 10 \times 15$ cm. Set the block on a level desk top or other smooth surface and measure the force which keeps the block moving at constant speed. Compare this force with the force required to keep the board moving on other surfaces, such as cloth, paper toweling, sandpaper, aluminum foil, and glass.

A moving object travels at constant speed if no forces act on it. However, a sliding block slows down quickly, indicating that some force acts to slow it down. This slowing down force (friction) can be determined by measuring the force required to keep the block sliding, as shown here.

Which surfaces in contact with the table top produced high friction? Which produced low friction?

Effect of Contact Area. Study the effect of contact area on friction. Compare the force necessary to slide the block when on its face, edge, and end, as shown here. What conclusions can you make?

Decreasing Friction. Push three thumbtacks into the bottom of the block used in the last

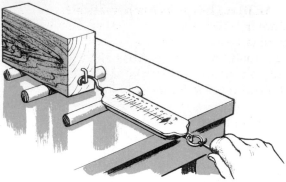

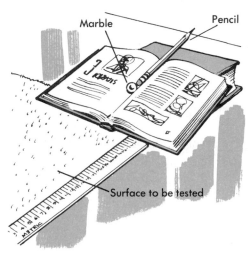

Marble   Pencil

Surface to be tested

activity. Now measure the force required to slide the block across the level desk top. How did the thumbtacks affect friction?

Remove the thumbtacks and use three rods as shown above. How does rolling friction compare with sliding friction?

Rolling Friction. Set up two books on a level table as shown here. Release a marble in the fold of the open book. Explain what happens in terms of the forces acting on the marble. Why does the marble slow down as it travels across the table?

Determine where to release the marble along the fold so that it will travel about one meter before stopping. Mark the point of release with a pencil.

Cover the table with a cloth and record the average distance the marble travels. Test other materials which were used in the first experiment in this chapter.

List the materials tested so that the material which produced the highest friction with the marble is first, and the material which produced the lowest friction with the marble is last. Compare this list with the list obtained from the first experiment in this chapter.

Mass and Friction. Study the effect of the block's mass on the force needed to slide the block. Use a spring scale to measure the force needed to move the block at constant speed across the level table. Use a balance to calculate the mass of the block.

Add a book or other object to the block to change its mass. Then measure the force needed to slide the block. Test at least four different loads on the blocks. Record the data in a table as shown here.

Study the data. How does the mass of the block affect the force required to slide it? Graph the data.

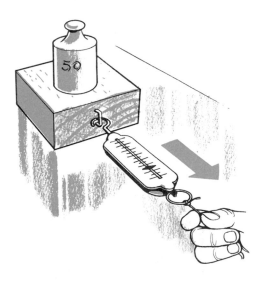

Force Needed to Slide Block and Load (S)	Mass of Block and Load (W)

Analyzing Graphs Mathematically. A graph of a similar experiment with friction is shown below. The data forms a straight line starting from the origin (point 0, 0) of the graph. Such a straight line graph can also be expressed as a mathematical statement or equation.

The mathematical statement which describes this straight line can be determined by analyzing some of the points which make up the line. Four points (A, B, C, and D) have been marked at random along the line. The values for the force required to slide the block (abbreviated as S) and the values for the masses of the block (abbreviated as W) can be determined for each of these four points. For example, at point A, S is 5.0 grams and W is 16 grams as shown in the table.

Prepare a table like the one shown here. Determine S and W for points C and D on the graph. Then calculate the ratio of S/W for points B, C, and D.

Is the ratio S/W about the same at these four points? What would be the ratio if it were calculated for all the other points making up the line?

A straight line graph such as the one below can be expressed mathematically as:

$$\frac{S}{W} = M$$

where M is a constant (about 0.32). Is the statement S = MW correct, too?

Point	Force to Slide Block and Load (grams)	Mass of Block and Load (grams)	$\frac{S}{W}$
A	5.0	16.0	0.31
B	12.0	38.0	
C			
D			

Mathematical Predictions. The equation S = MW can be used to predict the amount of force necessary to slide this block with any load. For example, to calculate the force needed to slide this block with a mass of 70 grams, you would use the formula:

$$S = MW$$

and substituting for M and W,

$$S = 0.32 \times 70$$

and multiplying,

$$S = 22.4 \text{ or } 22 \text{ grams.}$$

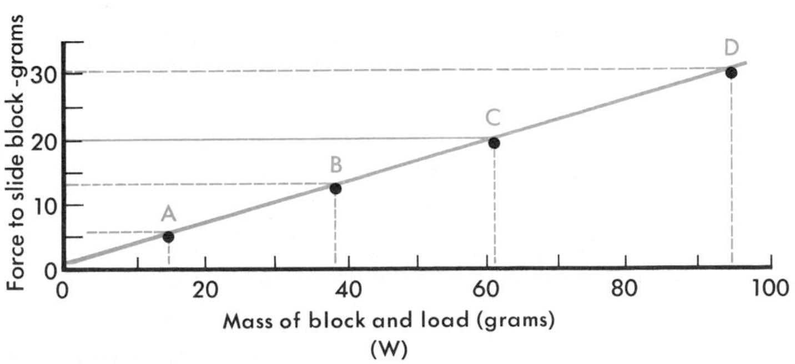

Calculating the Coefficient of Friction. Do your data from the friction experiment form a straight line as shown on the opposite page? If so, calculate M, called the *coefficient of friction.* Use the equation S = MW to predict the force needed to slide the block for some mass not yet tried. Check this prediction. How accurate is this mathematical method of predicting?

Discuss the usefulness of a formula such as S = MW compared to the same information on a graph.

Calculate the coefficient of friction for the block on other surfaces. Also determine some coefficients of rolling friction.

Shoes and Friction. Plan an experiment to study the frictional forces produced by several different shoes. What variables should be controlled during this experiment? What evidence do you have to support your belief about the effect of shoe size?

Conduct the experiment by putting a shoe on a board as shown here. Increase the slope until the shoe begins to slide. Record each shoe material tested and the angle at which the shoe began to slide.

Repeat the experiment using different surface materials, such as linoleum, carpeting, and a waxed smooth board. How do different shoe materials behave on these surfaces?

What type of shoe soles would be best for mountain climbers, steeplejacks, and roofers? How do shoes in sports such as track, golf, football, basketball, and baseball provide large frictional forces?

Effect of Slope. Set up an experiment to study what effect slope has on the force needed to move a block. Study both uphill slopes and downhill slopes. Plot the results on a graph like the one here.

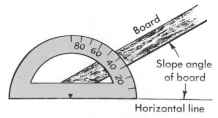

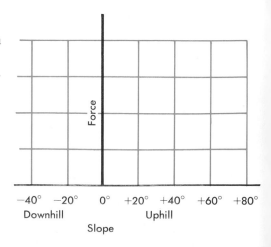

At what downhill slope does the sliding force become zero? Why is the force zero at this angle? When is the force greatest? Compare this greatest force with the mass of the block.

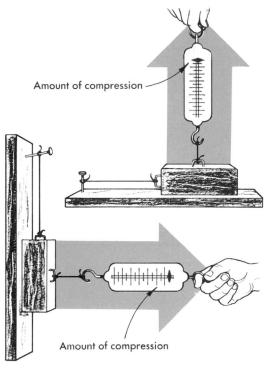

Amount of compression

Amount of compression

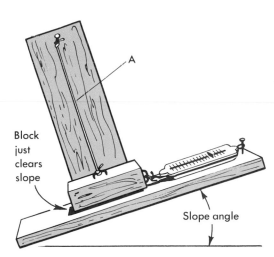

Block just clears slope

A

Slope angle

Angle of Slope and Compression. An earlier experiment in this chapter demonstrated that friction depended on the mass of the block. The last experiment, however, showed that friction changed with the angle of slope even though the mass of the block did not change. These experimental results can be better understood by studying the effect of the angle of slope on the force pressing the block against the surface. This force is called *compression*.

Measure the compression caused by the block pressing against the surface, as shown in the two examples here. Also measure the compression when the surface is at other angles. Graph your findings.

How does the angle of slope affect compression? How will the force needed to slide the block change as the compression changes.

Slope and the Effect of Gravity. Set up the device shown below. The tension on string A should be great enough to lift the block slightly off the surface of the slope. Measure the effect of gravity on the block as the slope changes. Graph the results.

How is the force tending to slide the block affected by changes in the slope? What is the cause of this force? How does compression between the block and surface change as the slope changes? What happens to the block when the slope becomes so large that the force tending to slide the block is greater than the frictional force?

At what angle of slope does the block first begin to slide (the critical angle)? Predict how the critical angle would be affected if the pressing force were increased by adding a larger mass to the board. Test your prediction and explain the results. How would the critical angle be affected if the coefficient of friction were increased? Test your prediction by gluing sandpaper to the bottom of the block and repeating the experiment.

SUMMARY QUESTIONS

1. What is friction?
2. How does the area of contact between two objects affect the friction between them?
3. How does the mass of an object affect the friction it has with the surface upon which it is resting?

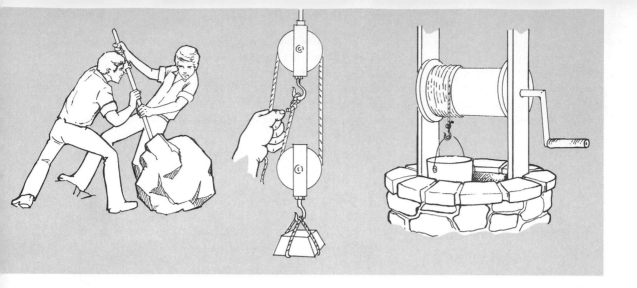

SIMPLE MACHINES

An inclined plane is used to raise objects which are too heavy to lift directly. Other devices such as pulleys, levers, and wheel and axles also help make work easier. These devices which can be used to exert large forces are called simple machines.

In order to describe simple machines and how they function, it is important to understand the scientific definition of work. Work is done whenever a force causes something to move. The advantage found in using simple machines is that they can make work easier.

Measuring Work. The girl shown below is accomplishing some work because she is exerting a force on the wooden block which is pulling the block up the board. The force she is exerting is measured by the spring scale, and she can easily measure the distance she has moved the block. The amount of work she has done can be calculated from the amount of force she exerted and the distance through which the force moved. These relationships can be stated in the form of an equation: Work = force × distance.

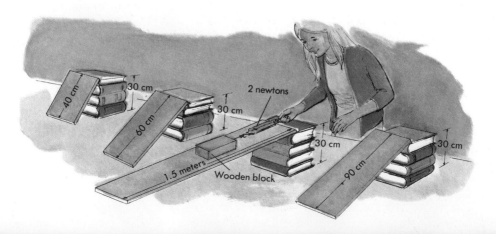

Measurements. *In this section on simple machines, the measurements you make will be used for making comparisons. You will compare the force exerted by the machine to the force you exerted. You will also compare the work done by the machine to the work you did. Since the answers you are looking for are comparisons, the units of work and force are eliminated.*

However, engineers calculate work by measuring the forces in newtons and the distance in meters. The unit of work is called a joule, in honor of James Prescott Joule whose studies did much to advance our knowledge of machines. One joule is the work done when a force of one newton acts through a distance of one meter.

Inclined Planes. Set up a pile of books about 30 cm high. Use a spring scale to measure the force needed to lift the block directly upward to this height. Place one end of a long board on the pile of books as shown here. Measure the force needed to slide the block up this inclined plane. In which case did the block require less force to reach the height of 30 cm? What is the advantage of using the inclined plane?

Actual Mechanical Advantage. Obviously there is an advantage to using an inclined plane because, as shown in the drawing, the force to pick up the block is 6 newtons and the force to slide it up is only 2 newtons. These two forces are given names for convenience. The force required to pick up the block is called the *resistance*. The force required to slide the block up the inclined plane is called the *effort*. The advantage of this simple machine can be expressed by comparing the resistance with the effort. This comparison is called the *actual mechanical advantage*. Notice that the value for the actual mechanical advantage has no units. The units cancel themselves out in the formula.

Use the shorter boards as shown at the right. Make a chart like the top one in the margin. Enter the measurements you made in each case and calculate the actual mechanical advantage. Make a chart like the middle one. Calculate the work accomplished by lifting the block directly to the height of the pile of books. Complete a chart like the bottom one and calculate the work done in each case. Which case required the most work? The least? Which case had the greatest actual mechanical advantage? If you had to raise a heavy object, which case would you choose? Why?

$$\text{Actual Mechanical Advantage (AMA)} = \frac{\text{Resistance}}{\text{Effort}}$$

$$\text{AMA} = \frac{6 \text{ newtons}}{2 \text{ newtons}}$$

$$\text{AMA} = 3$$

Height × Resistance = Work

Length of Board | Resistance ÷ Effort = AMA

Length of Board × Effort = Work

151

Useful Work and Efficiency. The total work done sliding objects up inclined planes is equal to the sum of the work done lifting the object to the same height plus the work done in overcoming friction. Total work = Useful work + Work overcoming friction.

The girl on page 150 needs less pull to slide the block up the inclined plane than to lift it. But she exerts the pull through a greater distance, and the block has the force of friction slowing it down. So she does more work by using the inclined plane. Useful work is calculated by measuring the height the block is raised and the force needed to raise it. The ratio of useful work to actual work is called *efficiency*.

Pulleys and String Tension. Connect a 500-g block, paper clip, string, and spring scale like example 1. What is the force on the spring scale as the block is lifted?

The force which tends to stretch the string is called tension. What is the tension on the string in example 1?

Connect the string as shown in example 2. What is the tension on each half of the string? Lift the spring scale and compare the force exerted on the spring scale (resistance) and the mass of the block which is lifted (effort).

Calculate the mechanical advantage of this device by dividing the mass of the block by the force exerted on the spring scale. Or,

$$\text{A.M.A.} = \frac{\text{Resistance}}{\text{Effort}}$$

where M.A. is the mechanical advantage. Propose a reason why the M.A. is not exactly 2 if the tension on each spring is ½ the mass of the block.

Predict the tension in each part of the string when it is arranged as shown in example 3. Test your prediction.

Lift the spring scale and record the force exerted on the scale as the block lifts. Calculate the mechanical advantage of this device.

Decreasing Friction with Pulleys. Make two pulleys from coat hangers and spools as shown here. Replace the two paper clips in the last experiment with the spool pulleys.

Calculate the mechanical advantage. Why does the mechanical advantage increase when the paper clips are replaced with spool pulleys?

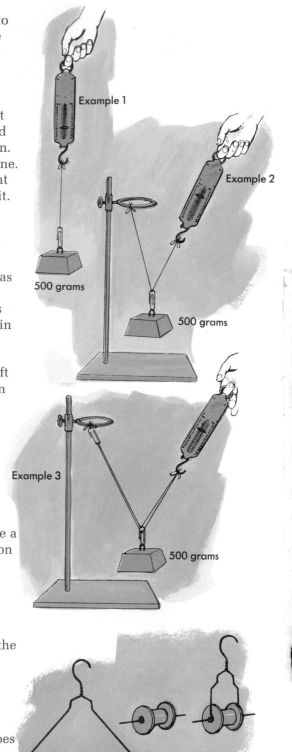

Example 1

Example 2

500 grams

500 grams

Example 3

500 grams

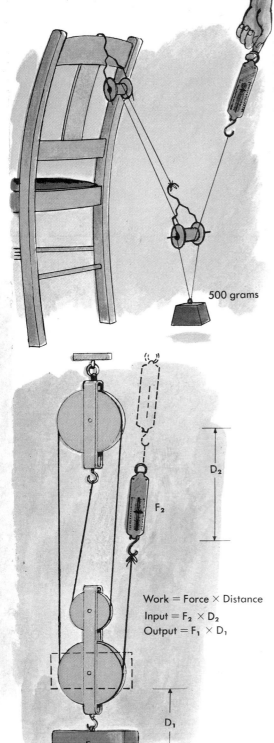

500 grams

Work = Force × Distance
Input = $F_2 \times D_2$
Output = $F_1 \times D_1$

Replace the spool pulleys in the apparatus with two single pulleys of the type shown below. Measure the force on the spring scale as it lifts the block. Calculate the mechanical advantage using these pulleys. Explain why the mechanical advantage is slightly larger.

Increasing the Mechanical Advantage. Make a drawing of a pulley arrangement which produces a mechanical advantage of two (if there were no friction). Make a second drawing of pulleys which produce a mechanical advantage of three. Use these two drawings as aids in planning a pulley set-up which has a mechanical advantage of four. Construct the apparatus according to your plan and test your prediction. Use pulleys which have little friction.

Plan and test an experimental apparatus or device which you predict will produce a mechanical advantage of five.

How can you estimate the mechanical advantage of pulleys without measuring the force acting on the pulley? Explain your answer.

Doing Work with Pulleys. Set up a four-strand pulley as shown at the left. Use the pulley to lift a large object 30 cm. Weigh the object and calculate the work done lifting it (output).

Measure the force on the spring scale and the distance through which this force is exerted while lifting the object 30 cm. Calculate the work you did lifting the spring scale (input).

How does the work you did on the spring scale (input) compare with the work the pulley did lifting the object (output)? Calculate the efficiency of the pulley:

$$\text{Efficiency} = \frac{\text{output}}{\text{input}}$$

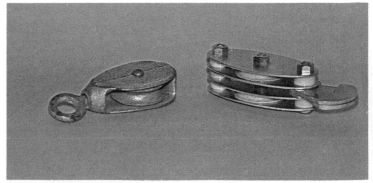

Single pulley   Double pulley

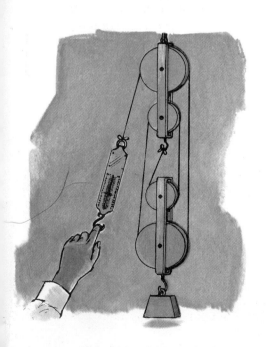

Changing Direction of a Force. Set up a fixed pulley as shown at the left. Use a spring scale to measure the force required to lift a book, using the fixed pulley. Compare this force with the force required to lift the book without the pulley. How do the forces differ? How are they similar? Where are pulleys of this type used?

Why would the first spring scale reading include a larger instrument error than the second reading?

The pulley shown below has the same mechanical advantage as the pulley on the previous page. However, this pulley is more common because it is more useful. Why is it more useful?

Large pulleys are called a block and tackle. These are often used to hoist large objects, such as scaffolding, pianos, ship's sails, and materials used in the construction of large buildings.

SUMMARY QUESTIONS

1. What is a simple machine?
2. Give two examples of simple machines.
3. What is actual mechanical advantage?
4. How can you calculate the actual mechanical advantage of an inclined plane?
5. How does a pulley system such as the one here make it possible to lift heavy objects by exerting only a small force?
6. How can you calculate the efficiency of a pulley system?
7. What is the scientific definition of work?
8. What is a joule?
9. Calculate the efficiency of the pulley system shown at the left.
10. Engineers claim that simple machines are not work savers because the work input is more than the work output. Why then do workers continue to use simple machines?
11. How would reducing friction in pulley systems and on inclined planes help increase their efficiency?

TORQUE

The simple machine above rotates when a force is applied. That is, the pedals rotate around the axis of rotation (labeled A) as the feet push against the pedals. The steering wheels of automobiles and the handles on fishing reels and pencil sharpeners are other examples of devices which produce rotation when forces are exerted on them. This turning action is called torque. *The amount of torque produced by such a device depends on the force applied, and the distance from the axis of rotation to the point where the force is exerted.*

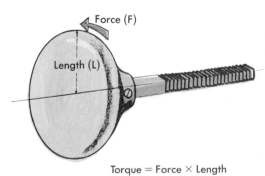

Torque = Force × Length

Changes in Torque. Remove the knob from a door and try to turn the shaft with your hand. Turn the shaft again with the knob in place. Explain why the knob now turns more easily.

Torque is defined mathematically as $T = F \times L$, where T is torque, F is the applied force, and L is the distance from the center of rotation to the point where the force is applied.

Measure the distance from the center of the door knob shaft to the edge of the shaft. Assume that five newtons are applied to the shaft. Calculate the torque (in newtons—centimeters) produced.

Measure the distance from the center of the door knob to its edge. Assume that 5 N are applied to the edge of the knob and calculate the torque produced. Why is the shaft easier to turn when the knob is in place?

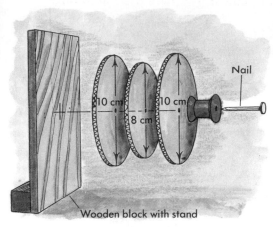

Wooden block with stand

Wheel and Axle. Construct a wheel and axle as shown here, using a wooden spool, two cardboard discs 10 cm in diameter, and a disc 8 cm in diameter. Apply glue between all surfaces when assembling the apparatus.

Tie a string around the wooden spool (axle) and tie a 100-gram block to the other end of the string. Tie another string around the cardboard wheel and attach a paper cup to the other end of the string, as shown below at the left. Add masses to the cup until the 100-gram block lifts. Then determine the mass of the cup.

Measure the distance the paper cup travels as the block rises 30 centimeters. How does the energy added to the block (output) compare to the work done by the paper cup as it falls? Calculate the efficiency of this simple machine.

Torque of Wheels and Axles. Adjust the two forces acting on the wheel and axle (spool) until they balance. Then measure each force, and measure the lengths from the axis of rotation to the edge of the cardboard wheel (L_W) and to the edge of the spool (L_a). Use these measurements to calculate the torque ($F \times L$) tending to turn the wheel in one direction and the axle in the other direction. Do the wheel and axle turn? Why?

Discuss the use of the equation $F_W \times L_W = F_a \times L_a$ for wheels and axles.

Use the equation to calculate the force needed on the edge of a wheel with a radius of 20 cm to balance 100 kilograms on an axle with a radius of 2 cm. What will happen if the wheel torque is less than the axle torque?

What is the advantage of the handles in the photograph? How do the handles aid the tennis player to exert force on the wheel?

Levers. Drill a series of holes through a stick, including one through a point where the stick will

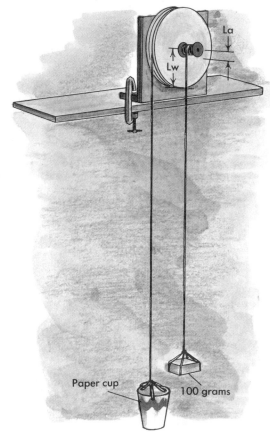

Paper cup   100 grams

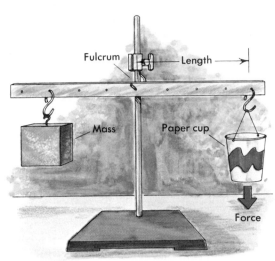

Fulcrum — Length →

Mass

Paper cup

Force

Force and Length Needed
to Lift Mass with Lever

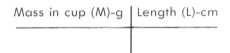

Mass in cup (M)-g	Length (L)-cm

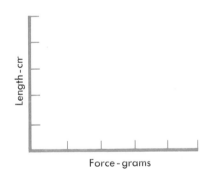

Length-cm

Force - grams

balance when suspended by a wire passing through the hole. Suspend the lever from this point and move one end of the lever down. What happens to the other end? Note the pivot point of the lever, called the *fulcrum*.

Use a bent paper clip to fasten a mass through one of the holes in the lever. Push down with your hand on the other end of the lever to lift the mass.

Sense the force against your hand as the lever moves. Move your hand closer to the fulcrum and again push down on the lever. How does the required force change? Where on the lever is the required force greatest? Least?

Lever Torque. Hang a mass from one side of the lever. Add masses to a paper cup hanging from the other side of the lever until it balances. Measure the mass in the paper cup (in grams) and the length (in centimeters) from the paper cup to the fulcrum. Record the data in a table as shown here.

Move the paper cup to a different position on the lever and again determine the mass and length when the lever is balanced. Repeat the procedure two more times. Make a graph of the results.

Analyzing the Graph. A graph of a similar experiment is shown below. Four points (A, B, C, and D) have been marked on the graph to aid in analyzing it. At point A, the length (L) is 18 centimeters and the mass (M) is 6 grams. The ratio L/M is 18/6, or 3.

Calculate the ratio, L/M, for points B, C, and D. Is this ratio about the same at all four points? Is L/M a constant for this line?

Calculate the product of M × L at the four points on the graph. Is this product about the same at all four points? What does the graph show about the torque (M × L) needed to balance the lever?

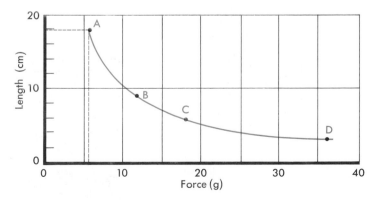

Calculate the product of M × L at several points along your graph. Describe the results. Calculate the torque on the other side of your lever and compare the torque on each side of the lever as it balances. Discuss similarities in the behavior of levers, and wheels and axles.

Mechanical Advantage of Levers. Use a lever to lift a 5-kilogram mass (M_1) with a 1-kilogram mass (M_2) as shown here. What is the M.A. of this lever (M_1/M_2)?

Measure the distance (L_1) from the fulcrum to the center of the 5-kilogram mass and the distance (L_2) from the fulcrum to the center of the 1-kilogram mass. Calculate the ratio L_2/L_1 for the lever and compare this ratio with the mechanical advantage of the lever.

Change the distance from the 5-kg mass to the fulcrum. Move the 1-kg mass until it again lifts the 5-kg mass. Calculate the ratio L_2/L_1 again. How does the ratio L_2/L_1 compare with the mechanical advantage of a lever?

Propose values for L_1 and L_2 which would produce a mechanical advantage of 3. Check your predictions. Predict and test a lever with a mechanical advantage of 10.

Types of Levers. Levers are divided into three classes. The levers studied previously were all of the first class. Levers of the second and the third classes are shown below. How do the classes differ?

Set up second and third class levers. Use a spring scale to determine if the equation, $M_1 \times L_1 = M_2 \times L_2$, holds for these levers. Discuss the advantages of each type of lever.

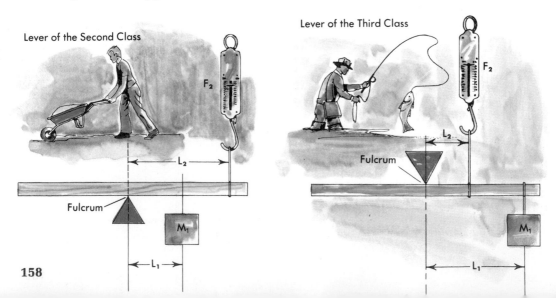

Lever of the Second Class

Lever of the Third Class

Efficiency of Levers. Set up a first class lever so that a 500-gram block can be lifted 30 cm. Add masses to a paper cup on the other side of the lever until the 500-gram block is lifted 2 cm. Record the mass necessary to lift the 500-gram block and the distance this mass travels to raise it 2 cm. Calculate the input (input = $Mg \times D_{cm}$) and the efficiency:

$$\text{Efficiency} = \frac{\text{output}}{\text{input}}$$

Repeat the measurements at three other places on the lever. Calculate the efficiency for each trial. Discuss the reasons for the high efficiency of levers.

What would happen to the efficiency if the lever bent?

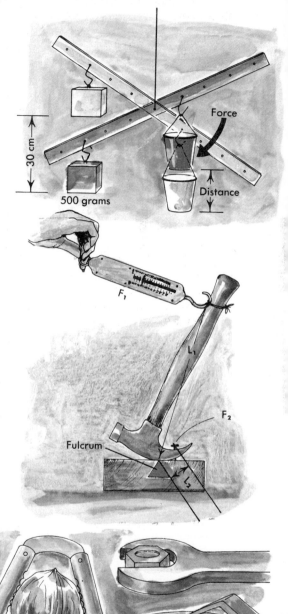

Efficiency of a Lever

Mass	Distance	Input (MxD)	Efficiency

Common Levers. Pound a nail into a board and use a spring scale as shown here to measure the force (F_1) needed to pull out the nail. Measure the distance from the fulcrum to this applied force (L_1) and from the fulcrum to the nail (L_2). Use the equation $F_1L_1 = F_2L_2$ to calculate the force which lifts the nail.

Use the same technique to calculate the forces produced while using other levers. Examples are shown at the right.

SUMMARY QUESTIONS

1. What is torque and how can it be calculated?
2. How much torque can a 50-kg person develop on a bicycle with a pedal 25 cm long?
3. How are wheels and axles used to do work?
4. What is the fulcrum of a lever?
5. How is a lever similar to a wheel and axle?
6. How can you calculate the mechanical advantage of a lever?

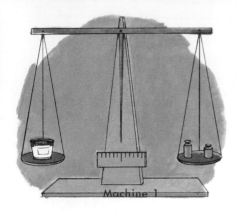
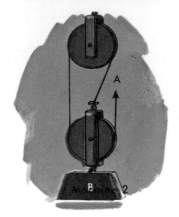
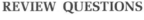
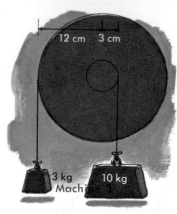

Machine 1

A

B
Machine 2

12 cm 3 cm

3 kg   10 kg
Machine 3

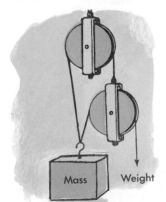

Mass   Weight

REVIEW QUESTIONS

1. What is friction and how is it measured?
2. How does the mass of a block affect friction?
3. Of what use are coefficients of friction?
4. What is work and how is it determined?
5. The farmers in the photo are using simple machines to make feeding the horses easier. What kind of a simple machine is the shovel? The wheelbarrow?
6. What simple machines are shown above?
7. Which simple machine above has the greatest mechanical advantage?
8. What is the mechanical advantage of machine 2?
9. How far must A be lifted to raise B 10 meters in machine 2?
10. What would happen to a wheel and axle if it were set up as it is above?
11. To which class does each lever belong?

THOUGHT QUESTIONS

1. What is the mechanical advantage of the pulley arrangement at the left?
2. How do the simple machines being used by the farmers make their work easier?
3. The lever at the left has only one mass hanging from it. Why does the lever balance?
4. If a 1-kg ball and a 10-kg ball fall from the same height at the same instant, will they hit the ground at the same time? Why?
5. Why is the force of gravity on the moon 1/6 for force of gravity on the earth?
6. How much would a person weighing 75 kg on the earth weigh on the moon?

Investigating On Your Own

The study of balanced and unbalanced forces offers you many opportunities for investigating on your own. Most of the projects suggested in this chapter require that you plan, carry out, and think about your results in the same way that research scientists, engineers, and technicians do. In this way you can consider whether a career in some area of science has interest for you.

Most of the activities suggested will require more time than one class period. Some projects require special materials. With your teacher's help, you can plan to obtain or make what you need. Plan for a time when you can share your experiences with your classmates.

EXPERIMENTAL RESEARCH WITH PRESSURE

1. Find out how much air you can withdraw from a bottle. Suck out as much air as possible and then hold the bottle under water as shown below. Measure the amount of water that enters. Is it equal to the amount of air that was removed?

2. Use the apparatus at the right to compare the weight of two liquids. Withdraw columns of the liquids into the tubes and compare the heights of the columns. A liquid that weighs half as much as water will rise twice as high.

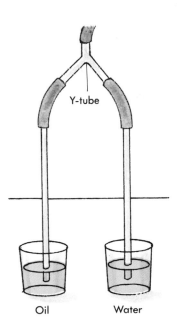

Y-tube

Oil   Water

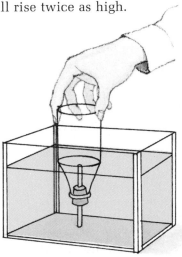

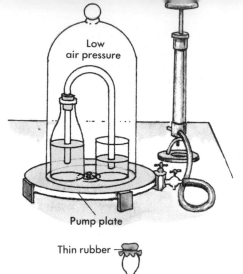

Low
air pressure

Pump plate

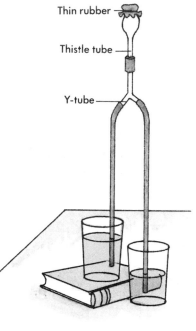

Thin rubber

Thistle tube

Y-tube

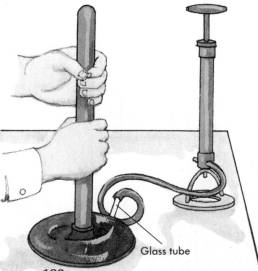

Glass tube

3. Use a small wind vane to make a study of the air currents around a building on a windy day. Plot the currents on a map.

4. See if round cans or flat-sided cans are more easily crushed by atmospheric pressure.

5. Bend a glass tube to make a siphon from a bottle like that shown at the left. Push the tube through a one-hole stopper into the bottle. Place the bottle and siphon under a bell jar and remove the air with a vacuum pump. Loosen the stopper in the bottle and make another trial.

6. Make a device for studying pressure in streams of air. Use a curved or U-shaped tube with a little colored water in it. Direct streams of air from vacuum cleaners, tire pumps, and electric fans across the open upper end of the tube.

7. Find out how good a vacuum can be produced with a vacuum pump. Connect the pump to a gallon jug, remove as much air as possible, and then open the jug under water. How much water enters? How much air remained in the jug?

8. Find out how high water can be siphoned with a garden hose. Fill the hose with water and put the ends in two pails of water. Slowly raise the middle of the hose until water stops flowing from one pail to the other. Measure this height.

9. Investigate the effect low pressure has on the boiling point of water. Put hot water and a thermometer in a bell jar. Start reducing the pressure. At what temperature does water boil?

10. Study the pressure in the top of a siphon tube. Join two rubber tubes, a Y-tube, and a thistle tube as shown above at the left. Cover the thistle tube with thin rubber and watch this rubber while the siphon is operating. Note what happens when the siphon tube is raised and lowered.

11. Connect a vacuum pump to the plumber's plunger as shown at the left. Pump out some air and note the load that can be lifted. Make a graph showing the load that can be lifted for varying numbers of strokes of the pump.

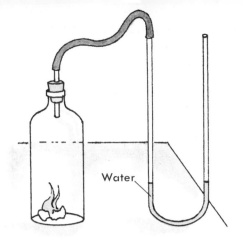

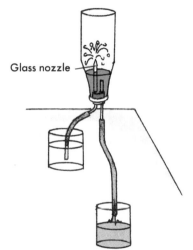

Glass nozzle

Water

12. Put a little water in a glass tube that is shaped like the letter U. Connect the tube to a bottle in which burning paper has been dropped. Note the pressure changes as shown by the water.

OTHER INVESTIGATIONS AND PROJECTS WITH PRESSURE

1. In the picture below, an airplane model is being tested in a wind tunnel. Read about wind tunnels and how they are used. Prepare a report on your findings.

2. Prepare an exhibit of common devices that make use of atmospheric pressure.

3. Make a flying model of an airplane and demonstrate it to the class.

4. Set up the "fountain in a bottle" apparatus shown at the left. Fill the bottle about one-third full of water and raise it to the position shown. Ink or food coloring in the water makes the action clearer.

5. Make a picture collection showing types of early airplanes. Prepare a bulletin board exhibit which illustrates the changes that have taken place in the shapes of wings.

6. Examine an aneroid barometer and find out how it operates. Prepare a chart which describes your findings.

7. Make models of automobiles or trains to show the development of streamlining.

8. Investigate the operation of a milking machine. Find out the advantages and disadvantages of machine milking as compared with hand milking.

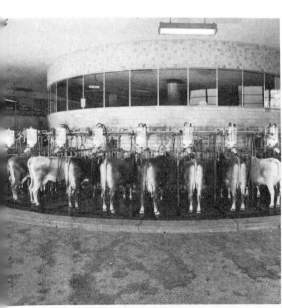

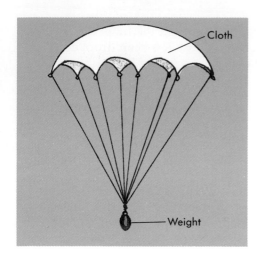

Cloth

Weight

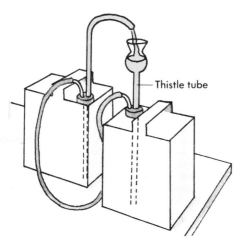

Thistle tube

9. Read about the famous experiments performed by Von Guericke with his Magdeburg hemispheres shown in the print below.

10. Collect pictures of racing cars to show the development of streamlined bodies.

11. Make a parachute by tying strings to the corners of a square of cloth. Demonstrate the action of this parachute.

12. Make a chart which shows how lift pumps and force pumps operate.

13. Read about the experiments of Torricelli, the Italian physicist who proved that there is atmospheric pressure and invented a way to measure it.

14. Study the valve of a tire and find out how it lets air into the tire without letting air escape.

15. Make a chart that shows the air currents around the wing of an airplane.

16. Set up the mystery apparatus shown at the left. Fill the upper can with colored water. Pour a cupful of water into the thistle tube to start it operating. Ask your classmates to explain what makes the water keep flowing.

17. Examine the construction of a toilet bowl to find out how it flushes. Make a chart to illustrate your findings.

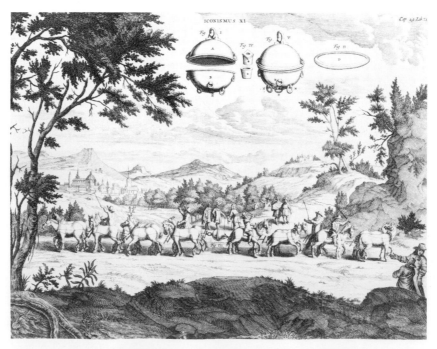

EXPERIMENTAL RESEARCH WITH FORCES

1. Study the relation between stopping distance and decelerating force with the apparatus shown at the right. Use your sense of touch to determine the lowest points reached by the weight and the pointer of the scale. Change the stopping distances by using rubber bands of different thicknesses, but keep the over-all length of string and slack rubber band the same so that the weight falls the same distance before being checked. Plot the results on a graph.

2. Arrange a series of weights on a long string so that when the string is dropped from a height, the weights land at equal intervals of time and produce a steady series of sounds.

3. Use a watch or clock with a sweep second hand to measure the approximate time needed for a heavy object to fall 1 meter, 5 meters, and 20 meters.

4. According to some books, Galileo dropped two balls of different weights from the Leaning Tower of Pisa, and discovered that the balls landed at the same time. Check this experiment. Drop two objects of equal size and shape but of greatly different weights from a height of 6 meters or more. (Suggestion — use large paint cans, one empty and one full of sand.)

5. Stand on a bathroom scale in an elevator. Note the readings of the scale as the elevator accelerates, decelerates, and travels at a steady rate of speed.

6. Place a bathroom scale on the first step of a flight of stairs. Note the readings as you climb the stairs, first as you raise your body onto the scale, and then as you raise it onto the next step.

7. Make the device shown here and find out what happens when you burn the string. Test stones of different weights. What happens when the stone and the board weigh the same? Change the accelerating force by using rubber bands of different thicknesses. Test the effect of adding different loads to the board.

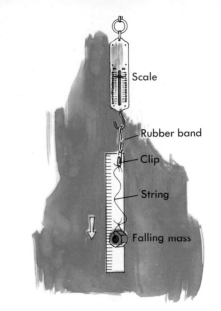

Scale

Rubber band

Clip

String

Falling mass

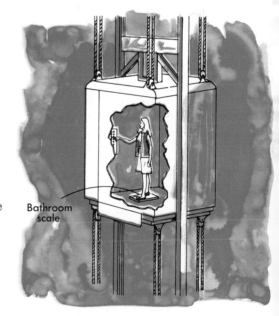

Bathroom scale

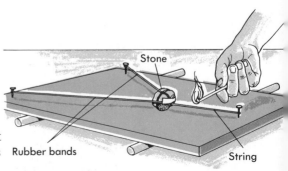

Stone

Rubber bands

String

8. Make an accelerometer like the one shown at the right. Find out when a vehicle is accelerating or decelerating. The accelerometer works best in a train or a plane in level flight, and it can be used in an automobile on a level stretch of road. The fishline sinker should be free to swing like a pendulum.

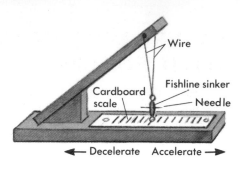

9. Use an accelerometer like the one shown at the right to find out when an elevator is accelerating or decelerating. A rubber band can be used in place of the spring. The zero point on the cardboard scale should line up with the pointer on the fishline sinker when at rest.

10. Put a short board on rollers under the front wheel of a child's tricycle and find out whether there is a backward force on the board as well as a forward force on the tricycle. Use two spring balances to find out if these forces are equal or nearly so.

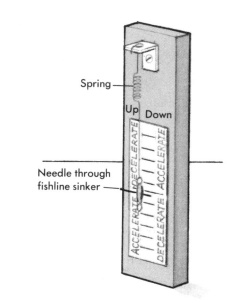

11. Compare the force that a hammer can produce on a nail with the force produced by jacking up one wheel of a car and letting it down slowly on the head of a nail as shown below. The force on the nail will be about one-quarter of the weight of the car (found on the registration certificate). Experiment to find the largest nail that can be pushed into the wood with this force. Ask an adult to help you with this. (Make sure that the hand brake is set in the car.)

OTHER INVESTIGATIONS AND PROJECTS WITH FORCES

1. Find out about the stopping distances of automobiles traveling at different speeds and under different road conditions. Make a chart that summarizes your findings.

2. Ask a scout leader or experienced forester to demonstrate the proper use of an ax. Have the person discuss the reasons why the ax is such a useful and at the same time, dangerous tool.

3. Visit a forging plant and report on the methods used to pound hot metal into various shapes.

4. Study the details of how the first "soft" lunar landings were made. Watch one of these operate and make a chart or a model that describes the way it works.

5. Prepare an exhibit of common pounding tools, including those used in the home, in small shops, and in athletics.

6. Ask a shop teacher or a carpenter to help you learn how to drive a nail properly. Demonstrate the method to the class.

7. Prepare a report on the way jet planes and propeller-driven planes are decelerated in the air and on the ground.

8. Most types of well-drilling rigs used for getting water pound their way into the earth. Watch one of these operate and make a chart or a model that describes the way it works.

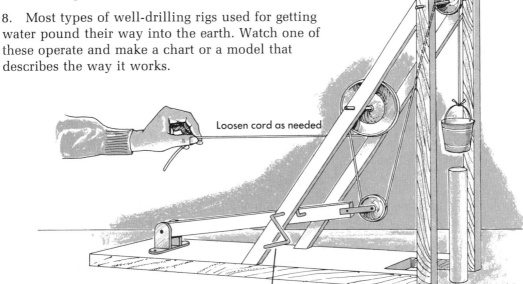

Loosen cord as needed

Model of well-drilling rig   Turn crank

EXPERIMENTAL RESEARCH WITH MACHINES

1. Ask an adult to put on the emergency brake of an automobile and to assist you in measuring the force exerted on the jack handle while the jack lifts the front of the car. Use a bathroom scale to measure the force exerted. Measure the distance the jack handle and car move when the front wheels are off the ground. Calculate the weight of the car, assuming the jack has an efficiency of 100% and that the car's weight on the jack is one half the total weight of the car.

2. Support a meterstick with two fingers. Slide the fingers slowly together. Explain the effect in terms of friction, force, and torque. Use the same procedure to find the balancing point of a baseball bat.

3. Determine the mechanical advantage of a bicycle by comparing the force exerted on the pedal with the force produced at the rim of the rear wheel. Also determine the M.A. of the pedal and sprocket, the sprocket and axle, and the axle and wheel.

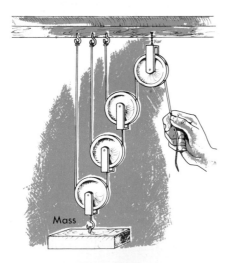

Mass

4. Set up a pulley arrangement as shown at the left. Estimate the force necessary to lift the weight. Check your estimate.

5. Saw a small, wooden spool in half crosswise and glue it to the center of a 15-cm disc of flat, stiff cardboard. Attach an inflated balloon to the spool. Bore a small hole through the bottom of the disc and place the disc on a smooth surface. Adjust the size of the hole so the device will move across a flat surface as though it were frictionless. Study the behavior of this device.

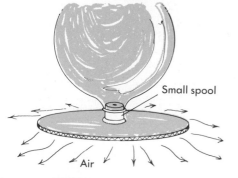

Small spool

Air

6. Measure the force required to turn a person around on a piano stool. Measure also the distance through which this force is exerted. Calculate the input and the output (the weight of the person times the height the person is lifted). Find the efficiency of the screw on the piano stool.

7. Determine the torque of a small electric motor, using the Prony brake device shown here. Suspend a rope from two spring scales and place the pulley of the electric motor in the rope loop. Pull up on one of the scales as the motor turns, and record the maximum difference in the readings of the spring scales.

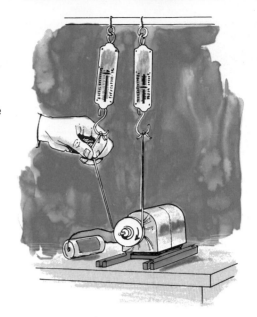

OTHER PROJECTS AND ACTIVITIES WITH MACHINES

1. Make and display examples of simple machines, such as levers, inclined planes, wheels and axles, and pulleys. Also display compound machines which are combinations of two or more simple machines.

2. Make a large chart listing the different forms of energy and devices which convert energy from one type into another.

3. Make a display of pictures of air vehicles like the one shown below. Make diagrams showing how these vehicles reduce friction, making their high speeds possible.

4. Visit a garage and examine the brakes of automobiles to learn how friction stops a car.

5. Make a graph comparing the standard of living and the amount of energy used per person in different countries. Obtain the data in a library.

6. Make a display of toys which utilize simple machines.

7. Set up a display of gears, pulleys, and belts, and explain their different uses.

8. Prepare a bulletin board display depicting our sources of energy through the ages.

9. Make a working model of a steam turbine. Explain to the class how the energy from heat can be changed into mechanical energy and then into electrical energy in an electric generating plant.

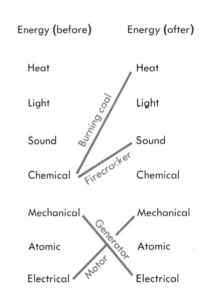

Energy (before)		Energy (after)
Heat		Heat
Light		Light
Sound		Sound
Chemical		Chemical
Mechanical		Mechanical
Atomic		Atomic
Electrical		Electrical

Burning coal — Firecracker — Generator — Motor

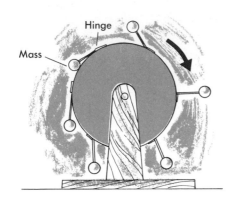

Hinge

Mass

10. One type of perpetual motion machine is shown at the left. Its inventor believed the falling weights would keep the wheel turning. Build such a wheel and explain why its efficiency is less than 100%.

11. Read about the use of nuclear reactors as energy sources. Prepare a model of a reactor and explain how it works.

12. Design and build a "Rube Goldberg" type of nutcracker (ridiculously complicated mechanical contrivance) composed of three different simple machines which produce a mechanical advantage of 5.

13. Use library references as an aid in constructing a model of a catapult which was used as a weapon of war centuries ago. Explain why the mechanical advantage of this device is less than 1.0.

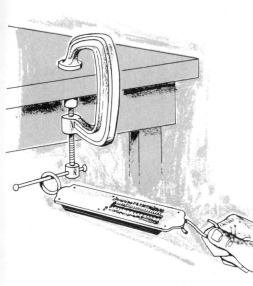

14. Using large weights, set up full-size pulleys and demonstrate their operation.

15. Read about the simple machines invented by Archimedes (287?–212 B.C.) to defend his home city.

16. Make a large poster showing energy changes which occur in an automobile.

17. Measure the force exerted and the force produced when tightening a C-clamp as shown at the left. Calculate its mechanical advantage.

18. Spin a hard-boiled egg on a smooth surface. Stop the egg momentarily and then release it. Repeat again with a raw egg. Explain the difference in behavior.

19. Swing a pendulum consisting of a spring scale connected between the string and weight. Explain the differences in the spring scale readings.

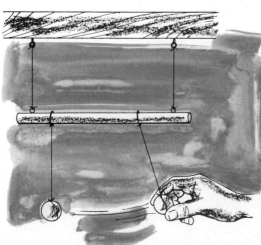

20. Set up two pendulums of exactly the same length as shown here. Start one pendulum swinging. Explain the behavior of both pendulums in terms of potential and kinetic energy.

21. Read about the life of Isaac Newton and report on his great contributions to the science of motion.

Machines

When compared with other creatures, humans are only average in size, yet we dominate the earth. We have been able to escape the gravitational pull of the earth and travel to neighboring bodies in space. How has this come about?

We have discovered that there is energy stored in the earth, and we have learned how to control this energy. The invention of machines which can convert stored energy into useful energy has made possible our modern civilization. We can communicate with almost any place on the earth. We can travel long distances in a short time, and heat or cool our houses with the flick of a switch. We can build enormous structures, move the ground around to suit our desires, and accomplish numerous other feats that were undreamed of just two generations ago.

The photograph of the New York City skyline is a living example of the extent of our capabilities. There are many machines which convert one form of energy into another form to do work in the city. List at least six of these machines.

HISTORY OF MACHINES

To better understand the tremendous increase in power available through the use of machines, examine the following statements.

1. The average worker can produce about one-twentieth of a horsepower. This is about 37 watts when converted to electrical energy. Working at this rate 8 hours each day for 240 working days in one year, the average worker can produce about 71 kilowatt hours of work per year.

2. Approximately 17 500 kilowatts of electrical energy are used for each employed worker for one year. In other words, each worker has the power of another 257 workers added to his or her own.

3. The average family uses more than 2600 kilowatt hours of electricity each year. In addition, the family uses an engine with about 150 horsepower for transportation. Adding up these figures, we find that the average family is being served by the energy equal to a labor force of 37 workers, and the family has at its command the energy of 3000 workers to transport them.

The history of how machines evolved to their present state shows how practical the minds of humans can be. Most inventions were in use before an acceptable theory had been proposed of the principles by which they operated. However, the refinements of machines in use today could not have been designed without an acceptable theory.

Pumping Machines. During the 1600's, many mines were forced to close because water seeped through the walls and flooded them. There were many imaginative ways for removing water from the mines. Horses were used to pull chains with buckets attached, or to turn gears which operated pumps. But in most cases, the horses could not keep up with the volume of water seeping into the mines. More power was needed than the horses could supply.

There was plenty of coal and wood to produce heat energy. However, the miners needed a way to turn the heat energy into mechanical energy to run the pumps.

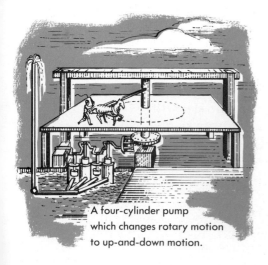

A "four-horsepower" pump for draining a mine.

A four-cylinder pump which changes rotary motion to up-and-down motion.

Early steam engine, 1698.

A Solution to the Problem. What happens to a can when you boil water in it with the cap off, replace the cap, and suddenly cool the can? When the can is filled with steam, there is enough pressure inside to balance atmospheric pressure. When the can is cooled and the steam condenses to water, the pressure inside the can is suddenly reduced. Atmospheric pressure then crushes the can because of the partial vacuum created when the steam condensed.

In 1659, experiments were made with the idea of lifting water by using the partial vacuum created when steam condenses. These experiments showed that atmospheric pressure will raise water to a height of 9.6 meters.

The first working steam engine was produced by Thomas Savery and patented in 1698. The device was called a "fire engine" because fire provided the energy for changing water to steam.

Water was heated in the boiler and the steam passed through a valve into another container. When the steam had forced most of the air from the container, the valve was closed by hand. A jet of cold water was then sprayed on the container to cool it, causing the steam to condense and create a partial vacuum. The vacuum allowed atmospheric pressure to force water up into the container where the steam had been. The pressure of the steam in the boiler was then used to force the water out.

Savery's steam engine was very inefficient because it wasted fuel. Savery had no way of testing the efficiency of his engine because there were no theories to test it against. The theory of heat engines actually was developed well after the engines were built and in widespread use.

Efficiency of Machines. We judge the efficiency of a machine by comparing the amount of energy put into the machine with the amount of work the machine produces. In other words, by comparing the power supplied with the power developed, it is possible to calculate the engine's efficiency.

Laws of Thermodynamics. The word *thermodynamics* describes all cases where heat energy is changed to another form of energy. The theory about heat engines, which change heat energy into mechanical energy, rests on two statements or laws.

The first statement includes the idea that energy is never lost even though it may be transformed. Whenever heat is changed to other forms of energy, any increase in heat energy is accompanied by an equal decrease in some other form of energy, and vice versa. When gasoline is burned in the cylinder of a gasoline engine, chemical energy is changed to heat energy. The hot gases do work on the piston by moving it, so heat energy is changed to mechanical energy. Some energy is not transformed into doing useful work. Friction between moving parts transforms mechanical energy into heat energy which is not used. In an automobile engine, so much useless heat energy builds up that the cylinders have cool water flowing around them to prevent overheating.

The second law states that heat naturally flows from a body at a higher temperature to a body at a lower temperature. Heat can be made to flow from a cooler body to a warmer body, but external work must be done. The refrigerator is a good example of driving heat from a cooler place (inside the refrigerator) to a warmer place (the room the refrigerator is in).

Steam Engines. The first heat engine to be widely used for transportation and industry was the steam engine. As pollution problems increase in today's world and the availability of gasoline decreases, the steam engine may once more regain its usefulness.

Study the diagram and read the following description of how this steam engine functions. A close-fitting piston moves in a cylinder that is connected to a steam chest by two passageways, "A" and "B." These passageways have valves that allow steam to enter the cylinder at one time as

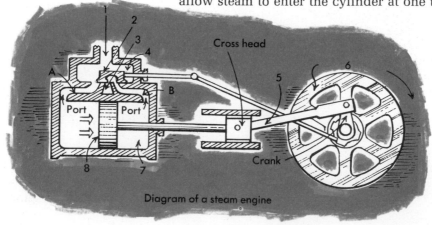

Diagram of a steam engine

well as to exhaust the steam at another time. As the piston moves to the right, steam enters through "A" and the used steam is forced out through "B." When the piston moves to the left, steam enters the cylinder at "B" and used steam is forced out at "A."

With this simple arrangement, steam would be exhausted from the cylinder at a high temperature, which means a loss of energy. In order to use this steam, a cut-off is provided. When the piston moves along its stroke, a slide valve automatically cuts off the incoming steam. After this cut-off, the steam in the cylinder expands and pushes the piston forward through the remainder of the stroke. While the steam is expanding, the piston does work and the pressure of the steam is reduced. In this way, the heat energy of the steam is transformed into useful work.

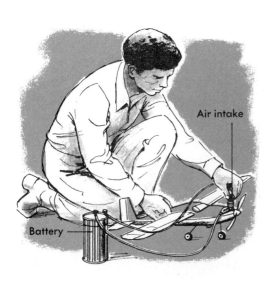

GASOLINE ENGINES

Gasoline engines make use of the pressures produced when gases are heated. Gasoline is ignited inside the engines, heating air and other gases to a very high temperature. The forces thus produced are used to drive lawn mowers, boats, automobiles, and airplanes. Why are gasoline engines called "internal-combustion engines?"

Model Airplane Engines. Study one of the small engines used for model airplanes. Locate the fuel tank. Trace the path of the fuel from the tank into the engine. Find the place where the fuel mixes with air before entering the engine. Can the flow of gasoline be regulated? If so, how?

The picture at the left shows a student starting a model airplane engine. Note that a battery is connected to a plug at the very top of the engine. Locate the plug which ignites on the engine you are studying.

Air intake

Battery

Inside an Engine. Use a small screwdriver or a wrench to take apart a model airplane engine. (**CAUTION:** Store the pieces in a small box so that they are not lost.) First, take off the part which holds the spark plug. Examine the end of the plug that is inside the engine. How does the plug use electricity to ignite the fuel?

Turn the propeller shaft (see the drawing) by hand. Note what happens to the *piston* inside the engine when the shaft turns.

Take apart the remainder of the engine and find out how the piston is connected to the propeller shaft. How is the *crankshaft* (see the drawing at the right) able to change the up-and-down motion of the piston into the turning motion that spins the propeller?

The Two-Stroke Cycle. The piston of a model airplane engine makes two strokes for each explosion. This type of engine is called a "two-stroke" or "two-cycle" engine.

Turn the propeller shaft until the piston is near the top of the engine. Imagine that the space above the piston is filled with a mixture of fuel and air. What does the spark plug do to the mixture? What does the explosion do to the piston? Why is the motion of the piston called the *power stroke*?

Move the piston to the position it has at the end of the power stroke. Look for holes in the side of the engine just above the piston. Which of these holes leads to the outside? The hot, burned gases escape through these holes.

Fuel and air enter the engine through the other

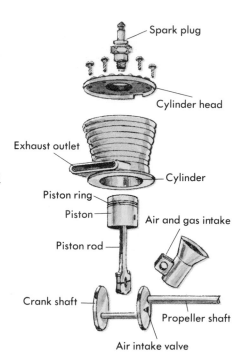

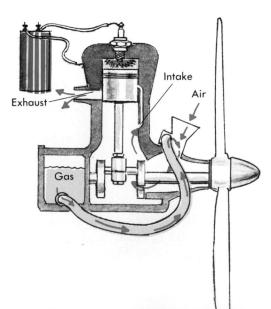

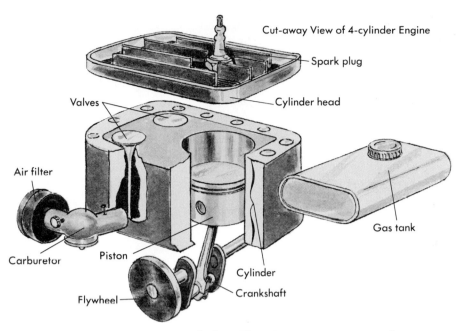

Cut-away View of 4-cylinder Engine

Spark plug

Cylinder head

Valves

Air filter

Carburetor

Piston

Flywheel

Cylinder

Crankshaft

Gas tank

holes. The piston moves up and compresses the fuel and air. Why is this stroke called the *compression* stroke? What happens at the end of each stroke?

The Four-Stroke Cycle. Most large gasoline engines have a four-stroke cycle, that is, the piston makes four strokes for each explosion. The diagram above shows the parts of a four-stroke (four-cycle) engine used for a lawn mower.

Remove the air filter and the head from a four-stroke engine. Do not tip the engine because oil will spill out. Remove the wire from the spark plug and attach it to any metal part of the engine.

Turn the flywheel slowly and watch the motion of the piston and the valves. How many strokes does the piston make while a valve opens once?

Blow through the hole from which the air filter was removed. Turn the flywheel slowly while you are doing this. Which valve allows air and fuel to pass from the carburetor into the engine? Why is the valve called the *intake valve?*

Blow through the muffler while someone turns the flywheel. Which valve allows burned gases to escape from the engine into the muffler? Why is this valve called the *exhaust valve?*

Rotate the flywheel slowly in the direction it normally operates. Which valve opens while the piston moves downward? Which valve opens while the piston moves upward?

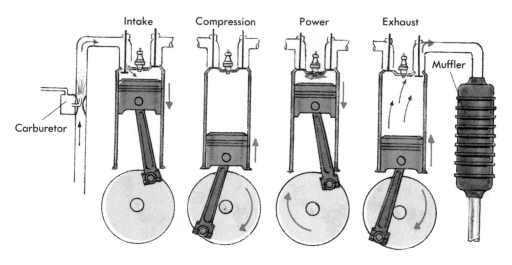

Intake Compression Power Exhaust

Muffler

Carburetor

The diagram above shows the four strokes of a four-stroke engine. Refer to this diagram while turning the flywheel of an engine from which the head has been removed. Note the action of the valves and the motions of the piston. Identify each of the strokes.

1. Intake Stroke. The piston moves downward, producing a partial vacuum inside the engine. The intake valve opens. Air flows into the carburetor, mixes with gasoline vapor, and goes to the engine.

2. Compression Stroke. The intake valve closes and the piston moves upward. The air and gasoline mixture is compressed above the piston.

3. Power Stroke. When the piston reaches the top of its compression stroke, the spark plug produces a spark and ignites the fuel. The resulting fire heats the gases to a temperature of about 1300°C. The gases expand and push downward on the piston, turning the crankshaft and flywheel.

4. Exhaust Stroke. The energy of the spinning flywheel starts the piston moving upward again. The exhaust valve opens and burned gases are forced out through the muffler. At the end of the exhaust stroke, the exhaust valve closes, the intake valve opens, and the four-stroke cycle begins again.

The Turbine. The turbine is a highly efficient engine because it delivers its mechanical energy as rotary motion. Steam engines and gasoline engines produce a to-and-fro motion which has to be converted into rotary motion by means of crankshafts and connecting rods. This implies that energy is being wasted in starting and stopping these parts, in addition to the energy lost in friction. Energy loss means less energy is available for useful work.

The illustration here shows a turbine designed by an Italian as early as 1628, although there is no record of it ever having been built. The boiler was built to look like a boy's chest, with the fire underneath it. The steam squirted out of a pipe in the boy's mouth and was directed against the rim of a wheel set with small paddles or vanes.

In 1889, a Swedish engineer made a successful turbine driven by steam. It is called an impulse turbine because the wheel or rotor is driven by the impulse of the steam against its vanes. The success of this turbine is due to the design of the nozzles.

When steam under pressure is released from a boiler it does two things. It speeds up and it expands, but it does not do these things at the same rate. When it is first released, its speed increases faster than its volume. Later its volume increases faster than its speed. For an efficient impulse turbine, the steam should leave the nozzle at the highest possible speed. When the steam hits the rotor, the steam loses some energy which is transferred to the rotor, making it spin.

In an ideal turbine, the jet of steam would stop when it hit the vanes and bounced backward. In practice, the best design is to make the steam strike against a series of vanes shaped as shown in the illustration. As the steam passes between the vanes, it follows a V-shaped path. This action produces a force which is nearly the same as it would be if the steam stopped and bounced backward.

The most widely used impulse turbine today is named after the American inventor, C. G. Curtis. In the Curtis turbine, the steam follows a path through alternate rings of moving and fixed vanes. The moving vanes are mounted on a single shaft, but the fixed vanes poke inward from the casing. The moving blades are shaped to reverse the direction of the steam. The fixed vanes are also curved, but opposite to the rotor's vanes.

Hero's steam engine

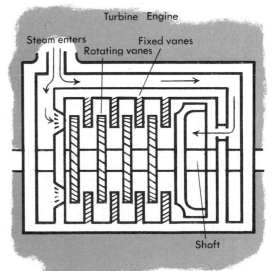

SUMMARY QUESTIONS

1. Explain how cooling steam in a closed container can cause the container to collapse?
2. What are the two laws of thermodynamics?
3. Why are turbines more efficient than gasoline or steam engines?

6

Objects in Orbit

The motions of objects in the sky have been a puzzle and a challenge to people since the beginning of recorded history. The questions raised have been of two types. What are the paths of objects in the sky? Why do objects follow these paths?

The paths of objects have been determined by making observations as accurately as possible over thousands of years. From these observations, astronomers developed theories to explain why objects followed these paths. As new instruments permitted improved observations, astronomers developed better theories to predict paths of objects in the sky.

MOTION

Some teachers in early Greece were among the first people to try to understand why objects moved. They noticed that objects needed to be pushed in order for the objects to move. If they stopped pushing, the object stopped moving. From these observations, they concluded that the natural state for all objects was for them not to be moving. The only things which moved continually were the sun, moon, and stars, but these heavenly objects did not have to obey the laws for normal things.

Observing Motion. The photograph above shows a tennis ball as it rolls across a table. As the ball moved, the camera lens was left open and a bright light flashed 6 times a second. The resulting multiflash photograph shows where the ball was every 1/6th of a second.

Measure the distance between the center of the ball at successive flashes. Are the distances the same? Use your answer to describe the speed of the ball? Is it increasing, decreasing, or traveling at a constant speed?

An Observation Which Doesn't Fit Early Theories. The multiflash photograph on the next page shows a bowling ball being rolled down a bowling alley. Does the speed of this ball increase, decrease, or remain the same? How does your answer fit the early Greek theory above?

Newton's First Law of Motion. The movement of the bowling ball differs from the tennis ball above in at least one important way. The smooth surface that the bowling ball rolls on does not offer much resistance to the ball as it moves forward. This resistance to motion is called *friction.* Today, scientists know that it is just as natural for objects to be moving as it is for them to be stopped. If an object is stopped, it will not move unless a force pushes hard enough against it. If the object is moving, it will move at constant speed in a straight line, unless a force acts on the moving object. This idea, published by Isaac Newton in 1687, is called Newton's First Law of Motion.

What force acts on a golf ball to get it to move? What force acts on a baseball to get it to stop moving in your hand?

Study the multiflash photograph below. How is the ball's speed changing? What causes the force which speeds up the ball? Where is the ball following a curved path? What force causes the ball's path to be curved? Where is the ball traveling at constant speed in a straight line? What tiny force is probably acting on the ball in this part of its path?

The Force Called Gravity. Newton formulated the First Law of Motion to explain the moon's orbit. Note the moon at M in the drawing. According to Newton's law, the moon would soon travel to A. But the moon does not travel to A. Instead, it goes to B. Thus, there must be a force acting in the direction shown by the black arrow. To explain the moon being at B, Newton believed a force called gravity acted between the earth and moon. Gravity is also the force which acts between the earth and the sun, thus keeping the earth orbiting around the sun.

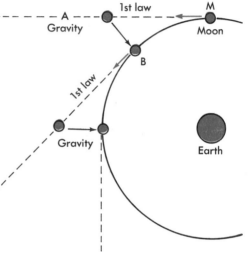

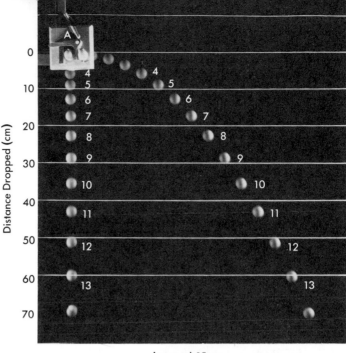

Distance Dropped (cm)

0 — A
— 4 4
10 — 5 5
— 6 6
— 7 7
20 — 8 8
30 — 9 9
— 10 10
40 — 11 11
50 — 12 12
60 — 13 13
70 —

├────┤ 10 cm

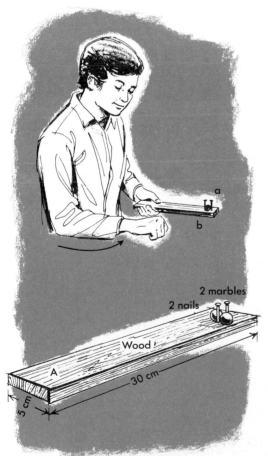

a
b

2 marbles
2 nails

Wood

A

30 cm

5 cm

An Experiment with Falling Bodies. Construct the apparatus shown at the left. Hold it level and predict which marble will hit the floor first when the device is hit horizontally. Check your prediction.

Repeat the experiment, changing the horizontal force. How does a change in the horizontal force affect the rate at which the marbles fall?

The photograph shows two balls which started at A and were propelled in the same manner as were the marbles here. After 4/30th's second, each ball was at 4; after 5/30th's second, each ball was at 5; and so on. Note the change in position of the left ball. Describe its motion. Is the ball accelerating, decelerating, or moving at constant speed? If this ball is accelerating downward, there must be (according to Newton) a force acting on it. What is this force?

Now study the right ball. Compare the amount this ball has dropped in 2/10's second (image 6) with how far the left ball fell in the same length of time. Compare the distance dropped after other time intervals. How do the dropping rates compare?

Now ignore the dropping of the right ball and consider only its motion to the right. Measure the distance each ball is to the right of the previous image. Is the ball's movement to the right increasing, decreasing, or remaining constant? Explain this motion in terms of Newton's First Law.

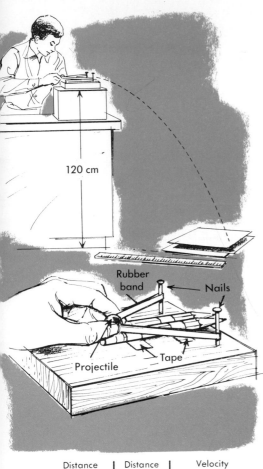

Projectile Experiments. Make the launcher shown here using three small nails, two pencils, a marble, a rubber band, tape, and a board about 20 cm long and 15 cm wide. Place the launcher in a level position 120 cm above the floor. Place a piece of carbon paper between two pieces of white paper. Tape the papers to the floor where you expect the marble to land when fired from the launcher. You will know where the marble first hit the floor by the mark it makes when it hits the carbon paper. Pull the marble back 3 cm against the tension of the rubber band and release it. Measure the horizontal distance the marble travels.

Record the accelerating distance of the marble (3 cm for this trial) and the horizontal distance the marble traveled. Carry out an experiment to learn how the accelerating distance of the marble affects the horizontal distance it travels.

Physicists have discovered that an object dropped from a height of 121 cm hits the ground in 1/2 second, regardless of its horizontal velocity. You can see this from the multiflash photograph on the previous page. Estimate from this photograph how far the ball would fall in one second. Thus an object that travels 2 meters horizontally before falling 121 cm, travels 2 meters in 1/2 second.

Use the graph to determine the horizontal velocity in kilometers per hour of each marble fired in the last experiment. For example, a marble that traveled 2 meters horizontally in 1/2 second was traveling 15 kilometers per hour.

Make a graph, comparing the accelerating distance of the marble with the speed of the marble in kph. What effect does the accelerating distance have on the speed of the marble?

Distance Through Which Accelerated	Distance Projectile Traveled	Velocity of Projectile (km/h)
2.5 cm		
5.0 cm		

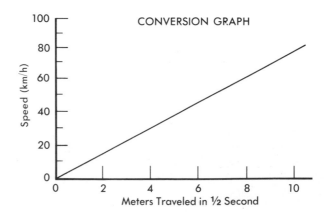

Effect of Force. Conduct a class experiment to determine how the amount of force exerted on a projectile affects its speed. Change the amount of force by adding more rubber bands to the launcher. Use the distances the projectiles travel and the graph on the previous page to determine the speed of the projectiles. Make certain that other variables, such as the accelerating distance of each projectile, are kept constant in this experiment.

Marble

121 cm

Glass marble

Wooden ball

Steel ball

Projectiles

Carbon paper between
2 sheets of white paper

Make a graph showing the effect of the projectile's accelerating force on its speed. What conclusions can you draw from the graph?

Effect of Mass. Obtain a steel ball and wooden ball of nearly the same volume as the glass marble. Launch each ball in level flight 121 cm above the floor. Use the same accelerating distance and force for each launching. Measure the distance each projectile travels. Compare the speeds of the projectiles. Compare the acceleration of the projectiles.

Physicists have noted that when the same force is applied to different objects, the objects may accelerate at different rates. The property of objects which cause differences in acceleration is called *mass.* Physicists say that steel has a greater mass per unit of volume than wood or glass. Most wood has less mass than either glass or steel.

Compare the masses of the three projectiles with the distances each traveled. How did mass affect the distance traveled? How did mass affect speed?

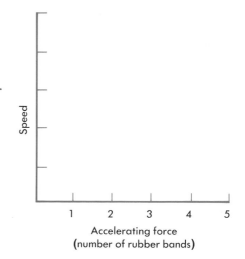

Speed

Accelerating force
(number of rubber bands)

1   2   3   4   5

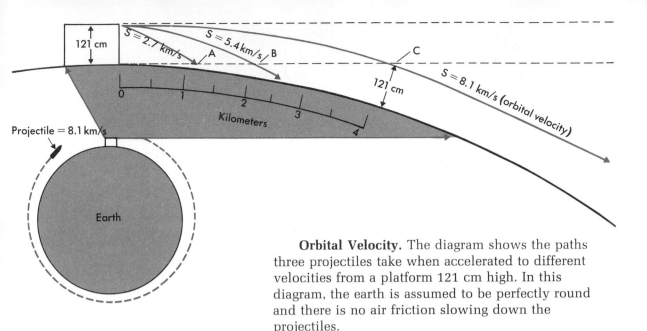

121 cm

S = 2.7 km/s

S = 5.4 km/s

A

B

C

121 cm

S = 8.1 km/s (orbital velocity)

0   1   2   3   4

Kilometers

Projectile = 8.1 km/s

Earth

Orbital Velocity. The diagram shows the paths three projectiles take when accelerated to different velocities from a platform 121 cm high. In this diagram, the earth is assumed to be perfectly round and there is no air friction slowing down the projectiles.

Note the path of the projectile traveling 2.7 kilometers per second. When this projectile has traveled 1.35 km, it has dropped 121 cm to point A, but it has not yet hit the ground. Why?

The projectile traveling 5.4 km/s falls 121 cm to point B after ½ second. Why does it not hit the ground after falling this far? How far will it travel before landing?

Note the path taken by the projectile traveling at the speed of 8.1 km/s (or 29 000 kilometers per hour). At this speed, the projectile travels 4 km before it falls 121 cm. However, the earth's surface curves down 121 cm in every 4 km, so, at point C, the projectile is still 121 cm off the ground. Describe the path of this projectile during the next ½ second. Why is 29 000 kilometers per hour (8.1 km/s) called the *orbital velocity*?

Reaching Orbital Velocity. Review your experiments with projectiles. What was the greatest velocity of any projectile you launched? Use the data from these launchings to describe the combination of mass, accelerating force, and accelerating distance which could be used to reach orbital velocity (29 000 km/h).

Compare the mass, accelerating force, and accelerating distance of the following projectiles: baseball, shotput, arrow, pistol bullet, and rifle bullet. Discuss what factors could be changed to accelerate these projectiles to orbital velocity.

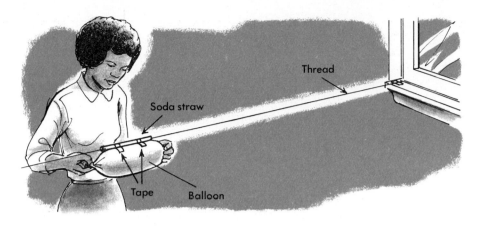

Thread

Soda straw

Tape   Balloon

THE MOTION OF ORBITING OBJECTS

Our knowledge of orbits and rocket engines has developed enough so that scientists and engineers now launch objects into orbit. However, long before scientists sent the first objects into orbit, they developed theories enabling them to predict the paths of these objects. The basis for these predictions was obtained by experimenting with falling bodies.

A Model Rocket. Stretch a thread across the classroom and set up the balloon rocket as shown in the illustration above. Inflate the balloon and release it. What is the force that acts on the balloon causing it to accelerate?

As gases rush out the back of a rocket, a forward push is exerted on the rocket. The amount of force with which the gases are pushed out the back is equal to the amount of force which accelerates the rocket forward. Thus, if the gases are pushed out the back with greater force, a greater force will accelerate the rocket forward.

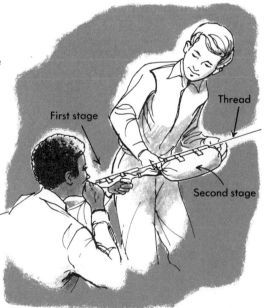

Thread

First stage

Second stage

Compare the accelerating distance of the balloon rocket with that of the projectile launcher used earlier. What advantage is a large accelerating distance? Why are artificial satellites accelerated with rockets instead of with giant guns?

Two-Stage Model Rocket. Put two balloon rockets on a string stretched across the classroom. Inflate the front balloon or *second stage* of the two-stage balloon rocket. Hold this balloon tightly about 3 cm from its mouth. Push about 2 cm of the *first stage* (rear balloon) into the mouth of the inflated second stage as shown here. Inflate the first stage. Then, while holding the two stages together, release the first stage. What happens?

Experiment with the connection between the

two stages of the rocket. What is the greatest distance the second stage travels?

What are some advantages to a rocket with more than one stage? What are some disadvantages?

Multi-Stage Rockets. The diagram below shows the path of a three-stage rocket accelerating an artificial satellite to a velocity of 29 000 kilometers per hour at an altitude of 500 kilometers. At what altitude does the first-stage rocket run out of fuel? At this point, the first stage breaks off and the second-stage engine starts. At what altitude does the third-stage engine start?

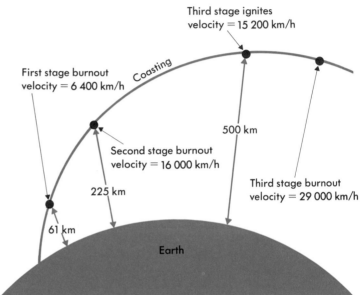

The photograph shows a space vehicle composed of three stages, or rockets. Which stage is accelerating the vehicle in this photograph? Which stage has to accelerate the greatest mass? Which stage has to accelerate the smallest mass?

Note in the diagram above that the first stage accelerates the vehicle to 6400 km/h. What *increase* in speed is caused by the second stage? What increase in speed is caused by the third stage? Why does the large first-stage engine increase the vehicle's velocity only 6400 km/h whereas the much smaller third-stage engine accelerates the vehicle much more?

Why are multi-stage rockets able to send objects into orbit more easily than large one-stage rockets?

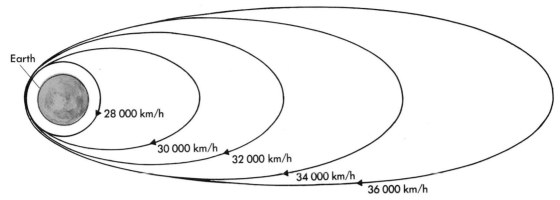

Earth

28 000 km/h

30 000 km/h

32 000 km/h

34 000 km/h

36 000 km/h

Shapes of Satellite Orbits. Any object in level flight at 28 800 km/h, and above nearly all the atmosphere, will travel in a curved path around the earth. Note the path above taken by a satellite traveling under these conditions. Such a satellite travels around the earth about every 90 minutes.

The diagram above also shows other possible orbits of satellites which, in level flight above most of the atmosphere, have velocities greater than 29 000 km/h. What shape are these orbits? How does changing the velocity of a satellite affect the shape of its orbit?

Many satellites have been designed to study different regions in space such as 5000, 10 000 and 20 000 kilometers from earth. An elliptical orbit is useful because regions at different distances from the earth can be sampled during each orbit.

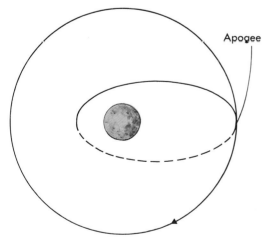

Apogee

Circular orbits are usually planned if satellites are to be kept in orbits for long periods. One method of attaining a circular orbit is shown at the left. First, the rocket is launched in an elliptical orbit. Then, as the last stage of the rocket reaches the point in its orbit farthest from earth, the rocket motor fires and accelerates the satellite into a circular orbit. The point in an orbit which is furthest from the earth is called *apogee*.

The closest point to earth of an orbit is called *perigee*. If perigee is less than 500 km, molecules of the earth's atmosphere will collide with the satellite. How does the force of these collisions affect the speed of the satellite? What happens when a satellite's speed falls below 28 800 km/h?

Space vehicles sometimes carry *retro-rockets* which are fired to slow down the vehicle as it orbits the earth. How does slowing down affect the vehicle's path?

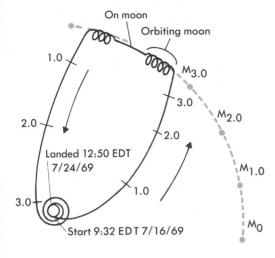

On moon

Orbiting moon

1.0

$M_{3.0}$

3.0

$M_{2.0}$

2.0

2.0

$M_{1.0}$

Landed 12:50 EDT
7/24/69

1.0

3.0

M_0

Start 9:32 EDT 7/16/69

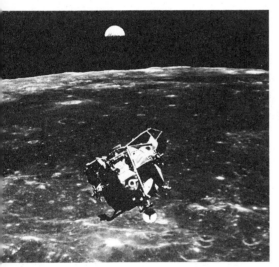

Moon Flights. A rocket at a velocity of 40 000 km/h in level flight above the earth's atmosphere travels in a highly elliptical orbit. This rocket travels about 400 000 kilometers (about the distance to the moon) from the earth, and each orbit lasts longer than a week. Astronauts Armstrong, Collins, and Aldrin traveled along such an orbit during their Apollo 11 flight to the moon.

The drawing at the left shows the path of Apollo 11. Note the distance traveled during the first three days. Note the position of the moon at M_0 on the scale drawing. The spacecraft left earth when the moon was in this position. Twenty-four hours after the spacecraft left earth, it was at 1.0, the moon was at $M_{1.0}$. Compare the distances the moon and spacecraft traveled in their orbits during the next 3 days.

The slightly curved path during the first 3 days was the result of earth's gravity acting on the spacecraft. What force caused the spacecraft to travel in a tight circle after three days?

After orbiting the moon, the spacecraft separated and the lunar module, shown at the left, used its rocket motor to decelerate so that it landed on the moon's surface in the "Sea of Tranquility."

Describe the forces which were needed in order to return the astronauts safely to earth.

Escaping the Earth's Gravity. If a rocket is accelerated to a velocity greater than 40 000 km/h, the rocket escapes from the earth's gravitational pull. Thus, the *escape velocity* from earth is 40 000 kilometers per hour. When a rocket escapes the earth's gravitational pull, the rocket may be captured by the sun's gravitational attraction and travel in an orbit around the sun.

Spaceflights to Other Planets. Spacecrafts that escape the earth's gravitation may travel in elliptical orbits around the sun. The shape and size of the craft's orbit around the sun depend on the speed and direction of the craft as it leaves the earth. Thus, spaceflights can be planned along elliptical paths which take the craft to other planets. For example, the path of the unmanned Mariner spacecraft as it traveled from the earth to Venus is shown on the next page. Note that the craft continued past Venus, completed its elliptical orbit around the sun, and returned back near where it started.

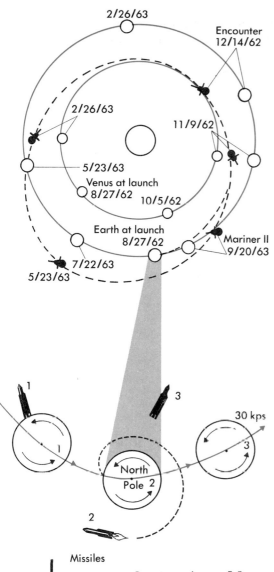

2/26/63

Encounter
12/14/62

2/26/63

11/9/62

5/23/63
Venus at launch
8/27/62

10/5/62

Earth at launch
8/27/62

Mariner II
9/20/63

7/22/63

5/23/63

The velocity of the moving earth must be considered when spaceflights to other planets are planned. The earth moves 30 km/s along its orbit around the sun. Thus, a spacecraft will be traveling at a speed greater than 30 km/s if it escapes from earth traveling in the same direction as the moving earth. A spacecraft which escapes from earth in the opposite direction travels less than 30 km/s.

Mariner II was launched from earth in the direction opposite the earth's motion. Note the diagram. After escaping from the earth, Mariner's speed around the sun was less than earth's. Compare the path of the earth with the slower-moving Mariner II between 8/27/62 and 11/9/62. How did the deceleration of Mariner II affect its path?

Spacecrafts that escape from earth with speeds greater than 30 km/s move further away from the sun. Propose how a launching rocket would be fired for a spaceflight to Mars.

The Paths of Projectiles. The graphs show the paths of projectiles accelerated to different speeds. The left graph shows the distance a missile travels if fired at speeds between 0 and 29 000 km/s. To what speed must a missile be accelerated to land 8000 km from its starting point?

The middle graph shows the height of a satellite at apogee if accelerated to speeds between 29 000 and 40 000 kph. To what speed must it be accelerated to reach a 390 000 km apogee?

The right graph shows the maximum distance a spacecraft travels from earth if accelerated to speeds between 40 000 and 60 000 km/h. Discuss the path of a spacecraft accelerated to a speed greater than 60 000 km/h.

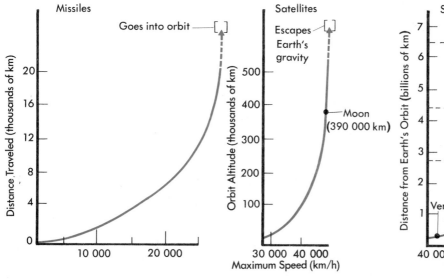

1

3

30 kps

North Pole 2

2

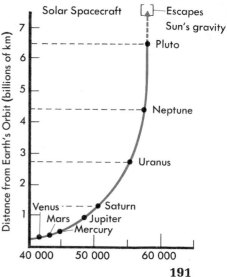

Missiles

Goes into orbit ⟶ [↑]

Distance Traveled (thousands of km)

20

16

12

8

4

0

10 000   20 000

Satellites

Escapes Earth's gravity

Orbit Altitude (thousands of km)

500

400 — Moon (390 000 km)

300

200

100

30 000 40 000
Maximum Speed (km/h)

Solar Spacecraft   [↑] Escapes Sun's gravity

Distance from Earth's Orbit (billions of km)

7

6 ● Pluto

5

4 ● Neptune

3

2 ● Uranus

1 Venus ● Saturn
  Mars ● Jupiter
  Mercury

40 000   50 000   60 000

Mariner Spacecraft

Uses of Space Flights. Orbiting space vehicles are of great value. They measure and keep track of conditions in the atmosphere. These observations have enabled meteorologists to gain more insight into the factors which affect the weather. Some satellites receive and transmit television and radio signals, allowing people all over the world to communicate with each other. Photographs taken from satellites help scientists predict the size of the harvest of certain crops. The photographs show quite clearly the effects and sources of pollution. Some photographs are useful in prospecting for various natural resources.

There are also indirect benefits derived from the space program. Certain fabrics designed for use in space have found their way into use by all of us. Techniques for preserving food, such as freeze drying, are now used extensively.

REVIEW QUESTIONS

1. How did Newton's First Law of motion disagree with the thoughts about motion in ancient Greece?
2. What is friction?
3. What force is responsible for keeping the moon in orbit around the earth?
4. Describe the path a missile takes when shot horizontally.
5. What is the apogee and perigee of an orbit?

THOUGHT QUESTIONS

1. What arguments can you think of in favor of continuing support for the space program?
2. What is an advantage of locating an observatory on an orbiting satellite instead of on earth?
3. What are the advantages of using a 2-stage rocket over a single stage rocket?
4. What can you infer about the "Sea of Tranquility" from the photograph at the left?

HEAT, SOUND, AND LIGHT

The Effects of Heat

Heat produces many changes, both physical and chemical. Changes in size and state are among the more common physical effects of heat.

Changes in state are especially noticeable. At almost any time we can see evidence of water evaporating, condensing, melting, or

1

freezing. Changes in size are generally less noticeable but are of equal importance. Winds, for example, are caused by the unequal expansion of air.

We have learned to use many of the physical effects of heat to our advantage. Among our numerous inventions are thermostats, refrigerators, and gasoline engines.

HEAT TRANSFER

If you have ever been warmed by a fire, you know that heat travels from one place to another. Heat always travels from a place that is hot to a place that is cool. Heat travels in several ways. Molecules may pass heat along to neighboring molecules. This way of traveling from molecule to molecule is called conduction. *Heat travels through air and water by being carried along in* currents. *When molecules of a fluid carry heat along with them, this process of heat transfer is called* convection. *Coals in a fire may be red hot and heat travels from the coals by the red light the coals give off. Heat traveling by means of light is referred to as* radiation.

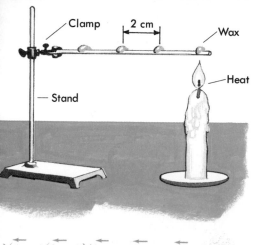

Clamp · 2 cm · Wax · Heat · Stand

Conduction. Place small pieces of candle wax about 2 cm apart on a metal rod. Support one end of the rod and heat the other end. How do you know heat is traveling along the metal rod? In which direction does it travel? Measure the time it takes the heat to travel each distance of 2 cm. How fast, on the average, does the heat travel?

Investigate factors which affect the speed at which heat is conducted. Use a hotter flame to heat the end of the rod. What do you find?

Use a rod with the same dimensions but made of a different substance. What do you find?

Try using a rod made of the same substance but with different dimensions. What do you find?

Collect the results in a table like the one here. Remember, you must keep all factors the same except the one you are investigating.

The Theory of Conduction. All matter is composed of atoms which may combine to form molecules. The atoms and molecules are in constant motion. Even in a solid, metal rod they are vibrating. When the rod is heated at one end, those molecules are caused to vibrate more. They hit neighboring molecules which in turn hit others. In this manner the heat is conducted along the length of the rod.

The molecules of a solid rod are all closely packed. They move back and forth only a short distance. The molecules in a liquid are farther apart than in a solid and move in any direction over longer distances. Molecules in a gas are even less compact and they also move in any direction over even longer distances. Solids tend to conduct heat faster than gases or liquids. Why? Why do you suppose air is a poor conductor of heat?

Convection. Study the temperature of the air around a heat source. Measure the temperature at several locations around the heat source as shown here. Make a diagram of your heat source and indicate the temperature at each location. Where is the air warmer? Cooler?

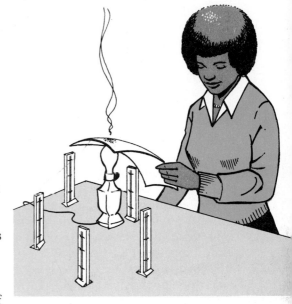

Use smoke to study the motion of air around the heat source. Draw arrows on your diagram to show how the air is moving. What is happening to the air next to your heat source? Where is the warm air going? How does the heat from the source travel to other places?

The Theory of Convection. When gas molecules are heated, the molecules move faster. These molecules hit other molecules, pushing them and causing them to move faster also. In this way the molecules of the gas tend to spread out. This body of gas becomes warmer and less dense than the gas which has not been heated. The warmer, less dense gas is pushed upward as the cooler, more dense gas flows under it. The warmer gas moves up and away from the heat source carrying the heat with it. This manner of transferring heat is called convection.

Most buildings are warmed by convection. In each room a pipe made hot by hot water or hot air runs along the base of an outside wall. The pipe warms the air around it, causing the air to rise to the ceiling, flow along the ceiling, and down the opposite wall. The cooler air on the floor flows toward the pipe and is warmed. Thus, warm air is circulated around the room.

World Wide Convection. Convection also plays an important role in transferring heat around the earth. The warmest part of the earth is the equatorial region. The sun's rays are most intense there. The air, warmed by the earth, rises into the atmosphere and at high elevations flows northward and southward. At about 30° north and south latitudes this air begins to sink. Some of this air flows into the equatorial region. As the air flows along the surface, it forms the trade winds. Some of the sinking air flows northward and southward from the 30° latitude, forming the winds called the westerlies. How is the heat which is absorbed at the equatorial region transferred around the rest of the world?

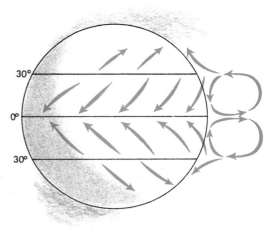

Radiation. Select two identical metal cans. Paint one dull black and leave the other shiny. Cut out a top of cardboard for each can. Cut a slit in each top and suspend a thermometer inside each can. Place each can 30 cm from a bright lamp. Note the temperature of each can at the end of every minute for 10 minutes. After 10 minutes, turn off the lamp and continue noting the temperature of each can at the end of every minute for another 10 minutes. Enter your observations in a table like the one here. From your observations, plot a graph of the changing temperatures compared with time. Which can heated up more quickly? Why? How did the thermometers inside the cans receive energy from the lamp? How was the heat energy from the lamp transferred to the cans?

Once you turned off the lamp, which can lost heat more quickly? At what point would the two cans eventually stop losing heat? If you shined the light at the cans indefinitely, would they get hotter and hotter until they melted? Why? They would get hotter and hotter until what point?

Theory of Radiation. A burner on an electric stove becomes red hot when turned up to "high." You can see that the burner gives off light, and you can feel that it also gives off heat. If you turn off the burner and watch it cool, you notice that it becomes darker red until you cannot see any red at all. But the burner still feels hot. By using special film, it can be shown that the burner still gives off light. The light is called infrared, a "light" to which human eyes are not sensitive.

When light strikes a shiny surface, the light is reflected. But when light strikes a dull, black surface, the light is absorbed and transformed to heat energy. Any object that is warmer than its surroundings will transfer its heat by any or all these methods: conduction, convection, and radiation.

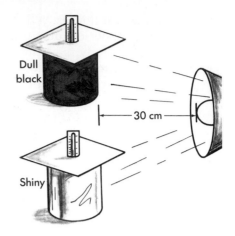

Time in minutes	Black can °C	Shiny can °C
1		
2		
3		
4		
5		
Light on 6		
7		
8		
9		
10		
Light off 11		
12		
13		
14		
15		
16		
17		
18		
19		
20		

SUMMARY QUESTIONS

1. How does heat travel along a metal rod?
2. Name two materials which are good conductors of heat and two which are not.
3. How is convection different than conduction?
4. How can convection be used to warm a room?
5. How are the trade winds formed?

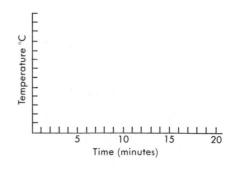

EXPANSION AND CONTRACTION

Most objects become larger or smaller as temperatures change. Sometimes such changes in size are remarkable, as shown in the photograph above. Usually, however, the changes are small and go unnoticed.

Whether large or small, the effects of heating and cooling are often of great importance. Engineers must give special attention to expansion and contraction when constructing roads, bridges, and buildings. The photograph shows what happened to a road one hot day when the concrete expanded more than the builders expected.

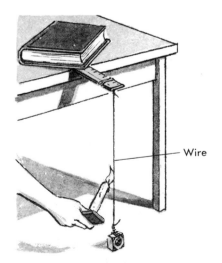

Wire

Expansion of a Wire. Hang a mass from a wire as shown at the left. The wire should be just long enough for the mass to clear the floor by the thickness of a card.

Swing the mass and then heat the swinging wire with a flame. What happens? Suggest an explanation for this effect. What happens after the wire cools? Explain your answer.

Comparing the Expansion of Wires. The apparatus on the next page is used to compare the expansion of different metal wires. Two wooden blocks are clamped to solid tables about 3 meters apart. The wire to be tested is stretched between nails in the blocks.

A double thickness of strong rubber band is stretched between a washer at the end of the wire

Flat washer · Stretched rubber band · Pin · Nail · Soda straw

and a nail in one of the blocks. Note the enlarged view. The rubber band should be stretched almost to its greatest length.

Note also the soda straw which pivots at a point close to the washer. A small movement of the washer moves the opposite end of the soda straw through a large distance. Thus, any change in the length of the wire is greatly magnified.

Hold several burning candles under the wire. Use a ruler to measure the change in position of the tip of the soda straw. Remove the candles and observe what happens to the end of the soda straw. Test other kinds of metal wires in the same way. All the wires must have the same length for accurate comparisons. Do wires made of different metals expand the same amount?

What caused these railroad tracks to bend?

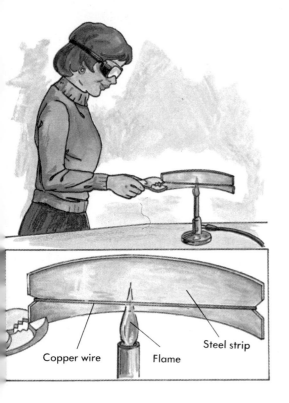

Effect of Unequal Expansion. Cut a 10-centimeter strip of steel from a can. Stretch copper wire between holes or notches in the ends of the strip. The result will be a device resembling the shape of a bow as used with arrows.

Hold this bow in a flame. What happens to it? Let it cool. What happens?

Explain your observations in terms of the unequal expansion of steel and copper.

An Automatic Fire Alarm. Set up the apparatus shown below using the device made in the previous activity. The nail should be less than 0.5 centimeters from the steel strip.

Predict what will happen when the metals are heated. Test your prediction.

How can such a device be used to give a warning in case of fire? What are some other possible uses for this device?

Most types of thermometers, thermostats, and fire alarms depend upon the unequal expansion of two different metals. The two different metals are combined to form a *bimetallic bar*. A bimetallic bar is easily made by riveting together a strip of brass and a strip of steel.

Bimetallic Thermometers. The diagram in the margin shows a type of bimetallic thermometer. This one is used to measure the temperature inside an oven. Where else have you seen bimetallic thermometers in use? Study the diagram and explain how it operates. What would happen if the steel and brass strips were interchanged?

Remove a bimetallic thermometer from its case. Find the parts and observe the effects of heating and cooling the bimetallic coil.

Automatic Temperature Controls. Many heating and cooling devices are controlled automatically by *thermostats*. For example, a thermostat makes a refrigerator operate when air inside becomes too warm, and it shuts off the refrigerator when the air becomes cold again.

What is the purpose of a thermostat in an electric iron? In an air conditioner? In an oven?

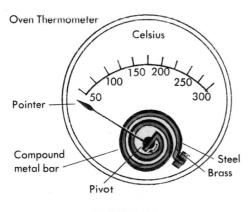

Oven Thermometer

Celsius

150 200
100   250
50   300

Pointer

Compound metal bar

Steel
Brass

Pivot

3mm space   Heat here
Bell

201

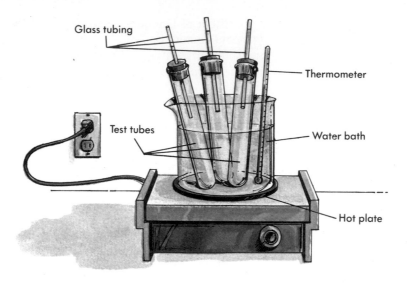

Glass tubing

Thermometer

Test tubes

Water bath

Hot plate

CHANGES IN VOLUME

A metal wire not only expands in length when heated, it also expands in diameter. The result is an increase in volume of the wire. Most other substances, including gases and liquids, increase in volume when heated.

Large changes in volume take place when some liquids cool and change to solids, such as when water freezes. However, the greatest increases in volume take place when liquids are heated and change to gases.

Liquids. Study the effect of heat on the volume of several different liquids. Fill three test tubes, one with water, one with glycerin, and one with alcohol. Using three one-hole stoppers, insert a glass tube in each. Place a thermometer in a beaker. Put each stopper with the glass tube into each test tube. Place the three test tubes in the beaker. Mark the level of liquids in each glass tube. Add water which is nearly boiling to the beaker. Observe the liquids in the tubes as the test tubes warm up and later as they cool off. Which liquid shows the greatest change in volume? Which liquid shows the most constant change in volume with change in temperature?

Theory of Volume Changes. The diagrams on page 203 show models of a liquid at two different temperatures. The atoms which make up the liquid are pictured in the diagram as solid balls for convenience only.

According to one modern theory, atoms in

Atoms

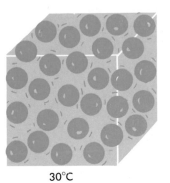

30°C

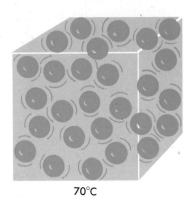

70°C

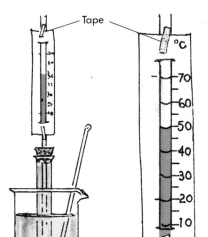

Tape

°C

— 70
— 60
— 50
— 40
— 30
— 20
— 10

Water at 50°C

liquids move freely. The amount of movement depends on the temperature of the liquid.

If heat energy is added to the liquid, the atoms move faster. They push each other farther apart and increase the volume of the liquid. Why does the weight remain the same?

Expansion of Water. Fill a test tube to the top with cold water. Insert a one-hole stopper that contains a length of glass tubing. Attach a piece of paper as shown in the diagram.

Mark the water level in the glass tubing. Heat the water in the test tube. What happens to the water level? Cool the water and notice any change in the water level.

Water is composed of molecules. Explain the volume changes in terms of the spaces between molecules.

A Water Thermometer. Set the device used in the last experiment in a jar of water at 10°C. When the water level stops changing, mark the position of the water level and label the mark, "10."

Set the device in water at 50°C and mark the new level reached by the water. Label this mark. Divide the space on the paper between the two marks into four equal spaces. Label the new marks 20°, 30°, 40°, and 50°. Add marks which divide each ten-degree space in half. Test the accuracy of this thermometer by comparing it with a mercury thermometer.

Add two new marks at equal distances above the 50° mark to represent 60° and 70°. Test the accuracy of your water thermometer at these temperatures.

Mercury Thermometers. Mercury is often used for better-grade thermometers. It expands and contracts uniformly over a wide range of temperatures. Therefore, mercury is suitable for thermometers that must be fairly accurate. In addition, mercury remains a liquid through a great range of temperatures. What are some disadvantages of water as a liquid for thermometers?

Study the construction of a laboratory thermometer. Examine the end of a broken thermometer and note the size of the hole through which the mercury rises. Try to fit a human hair into the hole. Compare the size of this hole with the diameter of the hair.

Why does the column of mercury in this tiny hole look so much larger when viewed from the side?

Clinical Thermometers. Doctors and nurses use a special type of thermometer in which the mercury column can rise but cannot fall. The tube of one of these thermometers is pinched nearly shut at a point just above the bulb. Expanding mercury squeezes past this narrow place but the column breaks when the mercury in the bulb contracts. Thus, the mercury column remains at the highest temperature reached in the mouth.

What is the advantage of this type of thermometer? How does the doctor or nurse return the mercury to the bulb of the thermometer?

Thermometer Scales. Most people (and all scientists) use thermometers marked with the *Celsius* scale. Many English-speaking people use thermometers marked with the Fahrenheit scale. Americans are familiar with both scales.

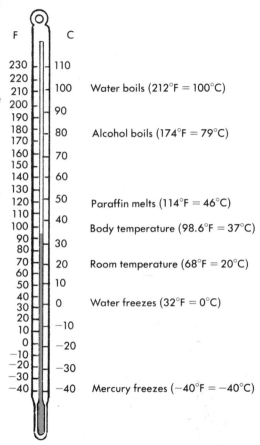

Water boils (212°F = 100°C)

Alcohol boils (174°F = 79°C)

Paraffin melts (114°F = 46°C)

Body temperature (98.6°F = 37°C)

Room temperature (68°F = 20°C)

Water freezes (32°F = 0°C)

Mercury freezes (−40°F = −40°C)

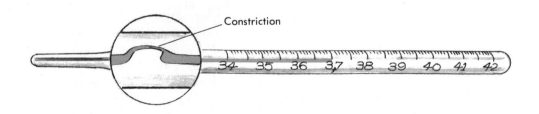

Constriction

To invent a temperature scale for a thermometer, it is necessary to make up a situation in which the temperature of the situation will always be the same. Gabriel Fahrenheit, a German scientist, believed that the normal human body temperature was the same for all persons. He also believed that a mixture of ice and salt made the coldest temperature that could be produced artificially. He marked the level of the liquid of a thermometer in an ice-salt mixture at 0. He marked the normal human body temperature at 100. Then he divided the space between 0 and 100 into one hundred degrees.

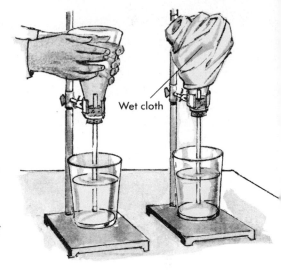

Wet cloth

The Celsius scale sets 0 as the freezing point of water and 100 as the boiling point of water. When does water boil using the Fahrenheit scale?

The thermometer shown on the previous page is marked with both Fahrenheit and Celsius scales. Compare the temperature scales.

Expansion of Air. Fit a flask or metal can with a one-hole stopper and a short length of glass tubing. Hold the flask upside down with the end of the tubing in water.

Warm the air in the flask with your hands as shown at the right. What happens? Cool the flask with a wet cloth. What happens?

Propose an explanation for these observations.

Air Thermometers. Air expands at a uniform rate over a wide range of temperatures. Thus, air is an excellent substance for use in accurate thermometers. Air is especially useful at temperatures too high for liquid thermometers.

Heat a flask with a candle flame until the air stops bubbling out. Let the flask cool to room temperature. Measure the amount of water which enters the flask (this represents the volume of air that was forced out of the flask). Measure the volume of the flask and then calculate the percent of the original air that was forced out. (**CAUTION:** Do not heat with a gas burner.)

The graph at the right describes the behavior of air as it is heated in a flask. More and more air leaves the flask as the temperature rises. Half of the air present at room temperature is forced out at 313°C and two-thirds of the air is forced out at 593°C. Use the graph to determine the highest temperature within the flask while it was being heated by the flame.

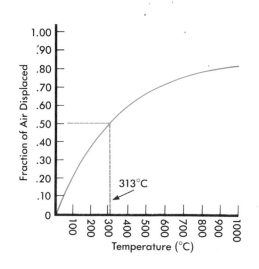

Effects of Freezing. Fill a short, wide can with water. The water level should be even with the top of the can.

Melt some paraffin wax. (**CAUTION:** Do not overheat the paraffin; the gas is explosive.) Pour the liquid paraffin into the same type of can that you used for the water. The level of the paraffin should be even with the top of the can.

Place both cans in a refrigerator to freeze. Compare the changes in volume of the two substances.

Measure the amount of water needed to fill the depression in the solid wax. Compare the volume of this water with the original volume of the liquid wax. Calculate the percent of volume lost as the wax solidified.

Press the top of the can of ice into modeling clay so that the bulge in the ice makes a hollow in the clay. This hollow equals the increase in the volume of the frozen water.

Measure the amount of water needed to fill the hollow in the clay. Calculate the percent increase in volume which resulted from freezing.

Melt the ice and the paraffin in the cans. Do both substances return to their former volumes?

Study the photograph here. Note that solid wax sinks in liquid wax but solid water (ice) floats in liquid water. Explain the photograph in terms of changes in density of the two substances as they freeze.

Effects of Vaporization. Put a single drop of water into one of two test tubes. The other test tube should be dry. Close both test tubes with greased cork stoppers.

Heat the tubes in a flame as shown on the next page. (**CAUTION:** Point the test tubes away from

No water

Greased cork

One drop of water

other people.) Which test tube loses its stopper first? Does the stopper blow from the wet test tube before, during, or after the water boils?

Effects of Condensation. Put a quarter cupful of water into a clean 4-liter can like the one shown in the margin. Heat the open can until steam appears above the opening. Remove the can from the heat, screw the top on tightly, and cool the can. Watch the can as it cools.

Boiling water inside the can produces water vapor. This water vapor forces out much of the air inside the can. When the can is sealed and cooled, the water vapor condenses into drops of water. As a result, a partial vacuum occurs inside the can. Air pressure outside the pan is no longer balanced by the pressure from water vapor inside the can. The can collapses.

SUMMARY QUESTIONS

1. How is the volume of different liquids affected by heat?
2. What is one substance that contracts as it solidifies?
3. What is one substance that expands as it solidifies?
4. What is one theory used to explain volume changes due to heating?
5. Why is mercury used for better-grade thermometers?
6. What conditions are used to mark 0° and 100° on a thermometer using the Celsius scale?
7. How does condensation within a sealed can cause the can to collapse?

Coat hanger wire

MEASURING HEAT ENERGY

Heat energy is commonly measured in units called calories. A calorie is a unit in the metric system of measurement. Physicists and chemists use the small calorie *which is described below. Food energy, however, is expressed in terms of* large calories *(often spelled with a capital) each of which is 1000 times larger than a small calorie.*

There are no instruments for measuring calories directly. Instead, calories are calculated from measurements of the temperature and mass of the substances being studied.

Defining the Calorie. A small calorie is the amount of heat needed to raise the temperature of one gram of water (about 20 drops) one degree Celsius. For example, one calorie of heat energy can raise the temperature of one gram of water from 19°C to 20°C.

Similarly, five calories of heat energy can raise the temperature of one gram of water by five degrees. Five calories can also raise the temperature of five grams of water by one degree.

How many calories are needed to raise the temperature of five grams of water 10 Celsius degrees? 100 Celsius degrees?

Heat Energy from Friction. Add five grams of water at room temperature to a test tube. Record the Celsius temperature of the water.

Pound a large nail into a board. Then pull the nail out quickly and put it in the test tube of water. Record the highest temperature the water reaches. Calculate the increase in temperature and the number of calories which entered the water.

Rapidly bend a piece of coat-hanger wire until the wire breaks. Put the broken ends into five grams

of water. Calculate the number of calories which entered the water by this method.

Heat from Electrical Energy. Put 125 grams of water into a container. Place a thermometer in the water and note the temperature. Place a small heating coil in the water. Plug in the coil and note the time it takes to raise the temperature of the water 5 degrees Celsius. How many calories of heat energy entered the water at this time?

Repeat the experiment with 200 and 300 grams of water, each time heating the water 5 degrees Celsius. How much additional time does it take to heat two times and three times as much water?

Heat Exchange. Mix 250 grams of water (one cupful) at 20°C with 250 grams of water at 30°C. Find the temperature of the mixture.

How many calories did the warm water lose? How many calories did the cool water gain?

Mix 250 grams of water at 35°C with 500 grams of water at 20°C. What is the temperature of the mixture? How many calories did the warm water lose? How many calories did the cool water gain?

Predict the temperature of a mixture of equal quantities of water at different temperatures. Test your prediction. Try other mixtures, making predictions and testing them.

Heat energy passes from substances at high temperatures to substances at low temperatures. Explain what happens when hot and cold water are mixed.

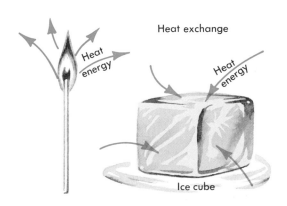

Heat exchange

Ice cube

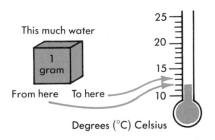

A small calorie is the amount of heat needed to raise the temperature of one gram of water one degree Celsius.

This much water

1 gram

From here   To here

Degrees (°C) Celsius

Heat of Combustion

Substance	Calories per Gram
Alcohol	6 000
Butter	9 000
Charcoal	7 000
Coal	12 000
Egg yolk	8 000
Lignite	9 000
Paraffin (candle)	10 000
Peat	7 000
Rubber	3 000
Wood	4 000

Sand

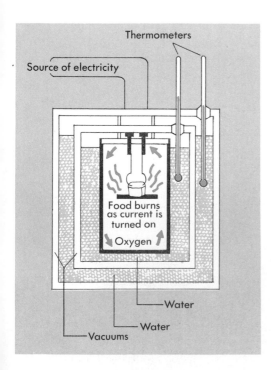

Thermometers

Source of electricity

Food burns as current is turned on

Oxygen

Water

Water

Vacuums

Heat from Chemical Energy. A fuel contains chemical energy which is changed to heat energy during burning. The energy of a sample of fuel is measured by burning the fuel and using the heat to raise the temperature of water.

Make a small cup of aluminum foil and set it in a pan of sand. Place over the cup a small metal dish containing 125 grams of water at 20°C.

Put one gram of alcohol into the cup. Burn the alcohol and determine the change in the temperature of the water. Calculate the calories of heat energy gained by the water.

Test one-gram samples of other fuels. Compare your results with the figures given in the table above. Try to explain why your results are different.

Reducing Measurement Errors. Much heat is lost to the air in the above experiment. Results would be much higher if all the heat produced by a fuel could be passed into the water.

The figures in the above table were obtained by burning fuels inside a device similar to the one at the left. A sample is put inside the strong-walled inner chamber which is then filled with oxygen. The fuel is ignited by an electric spark. Nearly all the heat produced passes into the container and the surrounding water where it can be measured.

Changing Ice to Steam. Add a quarter cupful of crushed ice to a small pan. Heat the pan with a flame or hot plate, stirring the contents slowly.

Record the temperature of the contents every two minutes until nearly all the water has

Crushed ice   Thermometer

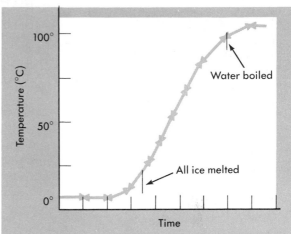

evaporated. (**CAUTION:** Do not hold the thermometer against the bottom of the pan.) Make a record of the time when (1) the last bit of ice melts, (2) the water begins to boil, and (3) the last of the water evaporates.

Make a graph of the time and temperature as shown above. Indicate on the graph the times when the last of the ice melted, the water boiled, and the last of the water evaporated.

Study the graph. What happened to the temperature while the ice was melting? After the ice had melted? While the water boiled?

Explain what happened to the heat that was added to the pan while the ice was melting, after the ice had melted, and while the water was boiling.

Compare the time required to melt the ice, the time required to heat the water to boiling, and the time required to vaporize the boiling water. Which process required the most heat? The least heat?

Cooling Effect of Ice. Heat 125 grams of water to the boiling temperature. Then add enough small pieces of ice to lower the temperature of the water to 20°C. Weigh the mixture and calculate the amount of ice that was added.

Repeat, this time cooling the boiling water with water from crushed ice (but not the ice). Calculate the amount of ice water needed to produce the same cooling effect as the ice. Which has the greater cooling effect, ice or ice water? Discuss the uses of ice for cooling.

Water at boiling temperature

Ice water

Ice water warmed by steam

Ice water warmed by boiling water

Heating Effect of Steam. Put 125 grams of ice water into a beaker. Bubble steam into the ice water as shown until the water temperature rises to 30°C. Weigh the mixture and calculate the amount of condensed steam added.

Repeat, heating the ice water with boiling water instead of steam. How much boiling water must be added to produce the same heating effect as the steam? Which has the greater heating effect, steam or boiling water?

Evaporation. Wrap some moist cotton string around the bulb of one thermometer. Use a second thermometer as a control. Compare the temperature of each thermometer. What happens to the temperature of the water in the cotton as some of the water evaporates?

About 540 calories of heat energy are required to change one gram of boiling water into water vapor at the same temperature. Even more energy is needed to evaporate a gram of water at lower temperatures. The energy taken up by water molecules as they vaporize comes from the heat energy of the liquid. The energy lost by the liquid during evaporation causes its temperature to drop.

Examples of Cooling by Evaporation. The boy at the left below just stepped from the shower. The air was saturated with water vapor, thus none of the water on his damp skin could evaporate. When he stepped into the dry air in the hall, the water on his body began to evaporate. Over 500 calories of heat are lost from his body as each gram of water evaporates.

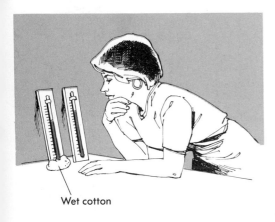

Wet cotton

SUMMARY QUESTIONS

1. What is a calorie?
2. How are calories measured?
3. What are some sources of error in measuring calories?
4. What effect does the process of evaporation have on the temperature of the surroundings?

HEAT CAPACITY

A certain amount of heat will effect the temperature of any substance, but the change in temperature will not be the same for all substances. In other words, each substance has its own heat capacity.

Other substances have different heat capacities, usually much smaller than that of water. Therefore, less heat is needed to raise the temperatures of these substances, and, when they cool, they give off less heat than water.

Heating Rate. Put 20 grams of water into a beaker. Heat the water with a candle flame, taking the temperature every 30 seconds for 5 minutes. Plot temperatures on a graph like that shown below.

Repeat, using half as much water. Repeat again, using twice as much water. Plot each set of measurements on the graph.

How much time was needed to raise the temperature of the first sample 10 Celsius degrees? What was the change in the other two samples during the same length of time?

How does the quantity of water that is in the beaker affect the rate at which the temperature increases?

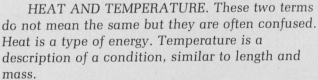

HEAT AND TEMPERATURE. These two terms do not mean the same but they are often confused. Heat is a type of energy. Temperature is a description of a condition, similar to length and mass.

Four liters of boiling water have the same temperature as a liter of boiling water; they are both in the same condition. But the larger quantity requires more heat to reach the boiling temperature; it contains more energy.

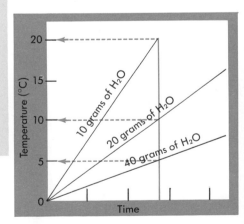

Calculate the number of calories gained by each of the three samples during the same period of heating. Then calculate the calories per minute gained by each sample. How do the heating rates compare?

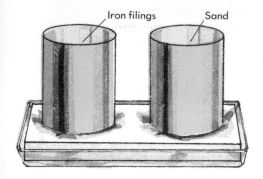

Iron filings   Sand

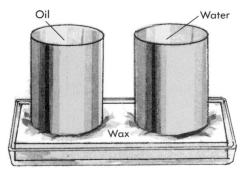

Oil   Water

Wax

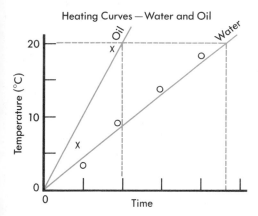

Heating Curves—Water and Oil

Oil

Water

Temperature (°C)

20

10

0

0

Time

Comparing Heat Capacities. Put 125 grams of water into a small can. Put equal masses of other substances, such as oil, sand, iron filings, and lead shot, into other cans of the same size and shape.

Set the cans in boiling water until their contents have been heated to 100°C. Then set the cans on blocks of paraffin.

Compare the depths to which the cans melt the paraffin. Which substance tested has the greatest heat capacity? Which has the least?

Water as a Coolant. Experiments have shown that water has the highest heat capacity of any natural substance. This high heat capacity makes water especially useful for cooling.

Water is excellent for putting out fires because it takes up large amounts of heat from fuels, thus cooling the fuels below their kindling temperatures.

Industries use enormous quantities of water for cooling. The steel industry, for example, uses about 250 000 liters of water for each ton of steel produced.

Heating Rate of Water and Oil. Heat 20 grams of water in a small container over a candle flame. Take the temperature every 30 seconds until it has increased 20°C.

Repeat the measurements using 20 grams of cooking oil. Keep other variables as constant as possible; for example, the size of the candle, the size of the container, and the distance between the flame and container.

Make a graph of the results of the experiment, plotting time and temperature as shown in the margin. Is one calorie of heat needed to raise the temperature of one gram of oil 1°C? Explain your answer.

Water   Cooking oil

Heat Capacity of Oil. Compare the times needed to heat the same mass of water and cooking oil the same number of degrees using the following ratio:

$$\frac{\text{Time needed to heat oil } 20°}{\text{Time needed to heat water } 20°}$$

What does this ratio tell you about the heat capacities of the two substances?

The heat capacity of water is one calorie per gram for each degree change in temperature. What is the heat capacity of cooking oil? Determine the heat capacity of other liquids using the same method described here.

Heat Capacity and Climate. The pictures here were taken about the same distance from the equator, although the climates shown are quite different. Both regions receive about the same amount of heat energy from the sun but the range of their temperatures is quite different.

If equal amounts of heat were given to both regions, which one would increase in temperature more? If equal amounts of heat were lost during the night from each region, which one would decrease in temperature more?

What happens to desert temperatures during the day? During the night? What happens to temperatures in moist regions during the day? During the night?

The graph below shows average monthly temperatures of the two regions pictured at the left. Which region heats up faster during six months of the year? Which region cools quicker half of the year?

The high heat capacity of water is an important factor in determining the climate of a region. To increase the temperature of a gram of water 1°C

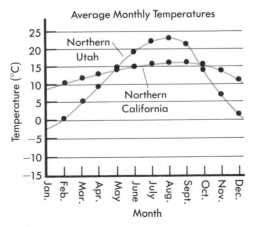

Average Monthly Temperatures

Northern Utah

Northern California

Temperature (°C)

Month

215

requires five times as much heat as increasing the same amount of dry soil by the same temperature. In addition, the evaporation of a gram of water uses as much heat energy that is used in warming 100 grams of soil 30°C.

REVIEW QUESTIONS

1. Why does land heat up faster than water?
2. How many calories must be given to the water in container A below to raise it to the temperature of the water in container C?
3. What would be the temperature of the mixture if the water in containers A and C were mixed?
4. What natural substance has the greatest heat capacity?
5. How are the calories in a sample of food measured?
6. Which requires more heat to raise its temperature one degree, a gram of water or a gram of cooking oil?

THOUGHT QUESTIONS

1. Which of the three containers of water shown below has the most heat? Which has the least heat?
2. What changes would have to be made in an oven thermostat to use it for controlling a refrigerator?
3. Why are orchards near lakes less troubled by spring frosts than orchards farther away from bodies of water?
4. What properties would the ideal coolant have for automobile radiators?
5. Why do people hold tight jar lids in hot water to loosen them?

100g	200g	100g
50°C	50°C	60°C
A	B	C

The Nature of Sound

Much of our information about the world comes to us through our ears. Hearing is especially important for communication with others. The human voice is able to express ideas better than other signs and symbols.

Sounds tell much more than most people realize. A mechanic knows when an engine needs adjustment by listening to it. A biologist identifies many birds, frogs, and insects by their songs and calls. You recognize your friends when you hear their voices. You are able to train yourself to listen and interpret sounds, thus adding to your knowledge of the world about you.

SOUND WAVES

The big drum shown above is a sound radiator. It changes the energy of a moving drumstick into another form of energy which radiates into the surrounding air. This new form of energy produces the sensation of sound when it acts upon our ears.

Sound energy is transmitted from the drum by air molecules moving back and forth regularly with what is called "wave motion." The waves so produced are not at all like the familiar ripples seen on the surface of water. Instead, they are waves of increased and decreased atmospheric pressure.

Energy Transmission by Air. Stretch a piece of thin rubber across each end of a cardboard tube. Fasten the rubber in place with rubber bands.

Push against the rubber at one end of the tube. What happens to the rubber at the other end? Describe the changes in air pressure inside the tube.

Hang a paper clip so that it just touches the rubber at one end of the tube. Snap your finger against the rubber at the other end. Explain what

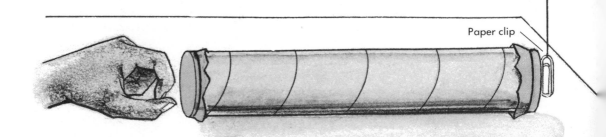

Paper clip

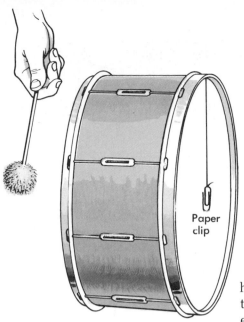

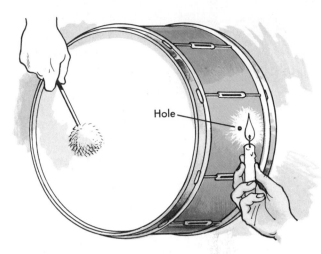

Hole

Paper clip

happens to the paper clip in terms of the energy that is given to it by your finger. Tell how the energy is transmitted to the paper clip.

Behavior of Drumheads. Stand a drum on its side and hang a small mass so that it just touches one drumhead. Strike the other drumhead. Explain what happens in terms of energy transmission.

What is the effect of hitting the drum with a hard blow? With a light blow? Describe the behavior of the drumhead that is struck. Describe the behavior of the other drumhead.

Every drum has a small hole in its side. Hold a small candle flame near this hole as shown above. Strike the drum with a drumstick. Describe the behavior of the flame and explain what happens. What does the flame tell you about pressure conditions within the drum?

Explain why both heads of a drum serve as sound radiators.

Energy of Sound. Lay a drum on its side. Place a light object such as a paper clip on its upper head. Strike notes on a piano or play a stereo loudly. Describe the behavior of the paper clip. What do the movements of the paper clip tell you about the drumhead? Touch the drumhead lightly with your fingertips. What do your fingertips tell you about the motion of the drumhead?

Explain how this experiment provides evidence that the piano or phonograph speaker radiates energy when it produces sound. How is the energy transmitted to the drum? To what form of energy is the sound energy changed when it reaches the drumhead?

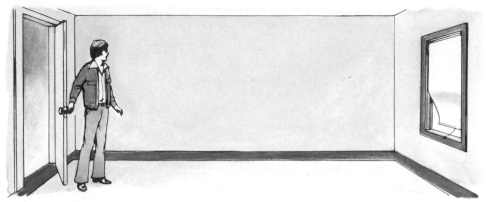

Compression Waves. Open one window of a room and pull down the shade over the opening. If there is no shade, hang a sheet of paper over the opening.

Open a door of the room suddenly. Watch the shade. What happens to it?

The diagrams at the right explain the behavior of the window shade. The rapidly moving door squeezes together the air in front of it before the air can get out of the way. The air is compressed.

This compressed air immediately pushes against the neighboring air which is compressed in its turn. The process continues across the room as a compression wave pushing the shade.

Note that the air next to the moving door does not blow across the room as wind. The air in any one place moves only a short distance and then stops. It is a wave of pressure that travels across the room and moves the shade.

Rarefaction Waves. Repeat the experiment with the door and the window shade, closing the door suddenly instead of opening it. How does the window shade behave?

The swiftly moving door leaves a partial vacuum behind it as shown in the diagrams on the next page. Neighboring air rushes into this space, lowering the pressure in the region from which it came.

Describe the progress of a low pressure wave across the room. Explain why the shade blows inward.

Again, note that the effect is not caused by outdoor air blowing completely across the room to fill in the space behind the door. Instead, a wave of low pressure, called a *rarefaction wave*, moves from the door to the window.

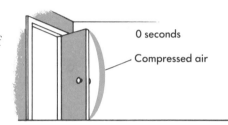

0 seconds

Compressed air

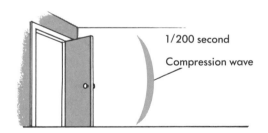

1/200 second

Compression wave

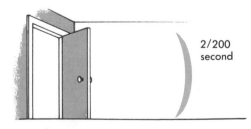

2/200 second

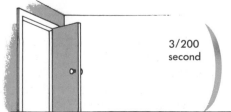

3/200 second

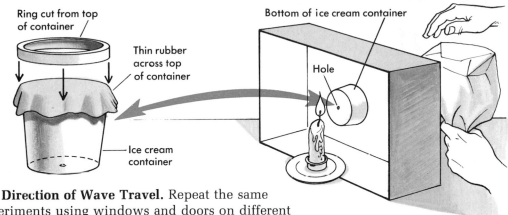

Ring cut from top of container

Thin rubber across top of container

Ice cream container

Bottom of ice cream container

Hole

Direction of Wave Travel. Repeat the same experiments using windows and doors on different sides of a room. Does a compression wave travel only in a straight line or does it spread outward in all directions? Test a rarefaction wave in the same way.

Detecting Compression Waves. Set up the apparatus shown above. The candle is like the ones used for birthday cakes. Its flame must be directly in front of and close to the hole in the container. The large carton protects the flame from drafts.

Burst a paper bag behind the carton. Clap your hands. Stick a pin into an inflated balloon. Hit a large drum. Explain the behavior of the flame in each case.

Shock Waves. Air from a bursting balloon rushes outward, pushing against the surrounding air and setting up a compression wave. This wave travels away from the balloon in all directions. At any one moment, the wave has the shape of a globe.

This compression wave does not travel alone. The air that rushes from the bursting balloon continues to expand until a low pressure results in the region where the balloon had been. A rarefaction wave is produced as air begins to push back into this region of low pressure.

Time for Wave Travel. Send one person away at a distance of 100 or more meters. Ask the person to hit a pail or drum with a downward blow of a stick, raising the stick immediately afterward. A flag on the stick makes it easier to see.

Notice any differences in time between the sight of the stick hitting the pail and the sound of the blow. Test the effect of moving both closer and farther away. What conclusions can you draw about the way sound travels?

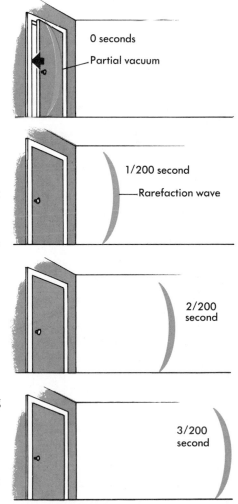

0 seconds

Partial vacuum

1/200 second

Rarefaction wave

2/200 second

3/200 second

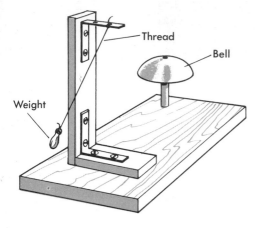

The Speed of Sound. Build the apparatus shown at the left. The swinging mass is a pendulum for measuring time; the thread on which it swings is about 25 centimeters long.

Shorten or lengthen the thread as needed until the pendulum makes exactly 60 full swings per minute. How long does the pendulum need to make one full swing? To swing from one side to the other (a half swing)?

Set the apparatus outdoors on a quiet day. Ask someone to stand behind it and swing his or her arm in time with the pendulum, striking the gong at the end of each full swing.

Move away from the apparatus until you hear the gong at the same time that you see the pendulum at the other end of the swing. Measure your distance from the apparatus.

How far does sound travel while the pendulum swings from one side to the other? How far does sound travel per second?

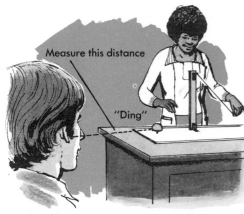

Reflection of Sound. Lay a loudly ticking watch on folded cloth. Place one end of a cardboard tube close to, but not touching, the watch. Place another tube at right angles to the first tube.

Listen at the end of the second tube. Can you hear the watch? If so, note its loudness. Then place a card as shown below at the left. What is the difference in the sound of the watch? Explain what happens to the sound waves.

Echoes. Go to a place where echoes can be heard. Produce echoes by slapping hands, striking a bell, or pounding a board. Produce echoes at different distances, noting changes in time between the production of a sound and the arrival of an echo. Explain these changes.

Ask one person to produce a sound while others stand in a long line on either side. Explain why everyone can hear an echo. Explain why an echo is usually fainter than the original sound.

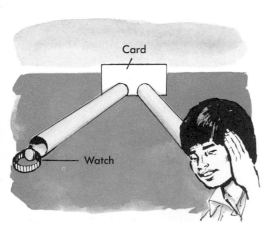

SUMMARY QUESTIONS

1. How are sound waves different from water waves?
2. How can you prove that sound carries energy?
3. What are compression waves? Rarefaction waves?
4. How can you detect compression waves?

TRAINS OF WAVES

We sometimes feel a single pair of compression and rarefaction waves but we cannot hear them. Our ears are sensitive only to three or more waves that arrive close together. A sound as we know it consists of a train of waves that reaches our ears at the rate of at least 16 waves per second.

Making a Train of Waves. Tie a small pebble in a piece of thin rubber. Stretch the rubber over one end of a cardboard tube. Stretch another piece of rubber over the other end.

Use the pebble as a handle to push and pull the rubber as fast as possible. Watch the rubber at the other end. Explain the behavior of this rubber in terms of compression waves and rarefaction waves.

Waves set up in this fashion do not arrive fast enough to produce sound that human ears can hear. Some other means must be used to produce waves arriving fast enough for human ears to hear (at least 16 waves per second).

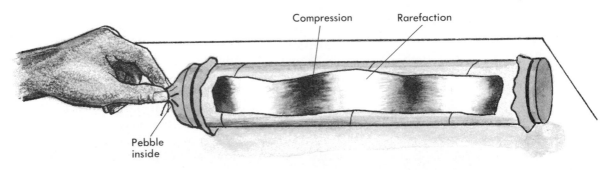

Compression   Rarefaction

Pebble
inside

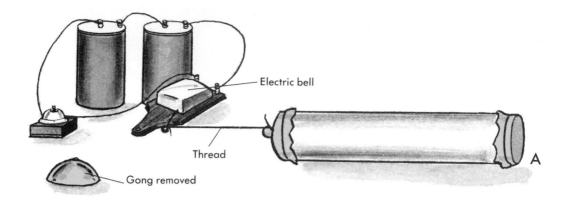

Electric bell

Thread

Gong removed

A

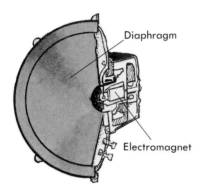

Diaphragm

Electromagnet

A Faster Train of Waves. Set up the apparatus shown above. What do you expect will happen when the push button is pressed? Test your guess.

Watch the rubber at A. Can you see any motion? Touch it lightly with your fingertips. What do you feel? What is the source of energy for the motion of the rubber? Describe the ways in which this energy is transmitted to the rubber.

Hang a small paper clip so that it just touches the rubber at A. Press the push button. What happens? What does this experiment suggest is happening to the air molecules that touch the rubber? What evidence is there that the air molecules are being disturbed?

Loudspeakers. A radio loudspeaker operates somewhat like the apparatus shown above. An electromagnet is connected to a cone-shaped piece of cardboard called a *diaphragm*. The electromagnet moves the diaphragm back and forth, setting up a train of compression and rarefaction waves. This train of waves travels to our ears and gives us the sensation of sound.

A wave detector is described on page 221. Set the apparatus in front of a phonograph speaker. Play a record that has many loud drumbeats. Watch the behavior of the candle flame in the detector. What takes place in the air in front of the loudspeaker?

Trains of Regular Waves. The tuning fork shown on page 225 is sending out a train of compression and rarefaction waves. When a prong of the fork moves to the right, it pushes against the air and sends out a compression wave toward the right. When the prong moves to the left, it leaves a partial vacuum behind and thus sends out a

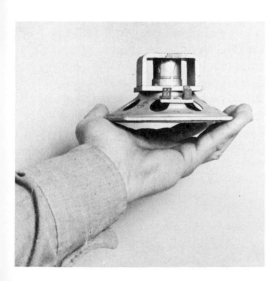

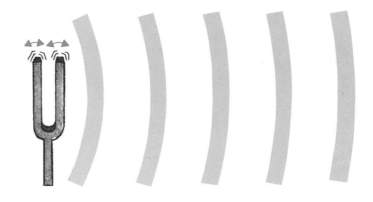

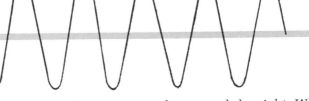

Above normal

Atmospheric
pressure

Below normal

rarefaction wave, also toward the right. What is happening on the other side of the tuning fork?

The prongs of a tuning fork move back and forth steadily. Therefore, the waves follow each other at the same distance.

A very sensitive barometer near the tuning fork would show regular changes in pressure. Atmospheric pressure rises slightly above normal as a compression wave passes by. The pressure falls slightly below normal as a rarefaction wave passes by. Explain the graph below the tuning fork.

Examining Sound Waves. Ordinary barometers are not sensitive enough to show changes in atmospheric pressure caused by sound waves. The study of sound waves is usually carried out with a microphone connected to a device called an *oscilloscope* as shown at the left.

Sound waves vibrate a diaphragm in the microphone just as they vibrate the drumhead studied earlier. The motion of the diaphragm generates a tiny electric current which is then amplified and used to produce an image on a television tube in the oscilloscope. Every response of the diaphragm to changing pressures can be noted.

Study the image on the oscilloscope shown here. Is the train of waves regular or irregular? Is the sound steady or is it fading out? Describe the pressure changes in the air around the violin string.

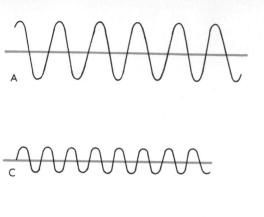

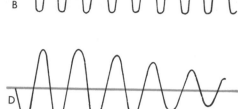

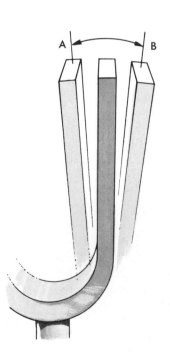

Differences in Sound Waves. Above are shown tracings of images produced on an oscilloscope by six sounds. Compare these images. Which represent loud sounds? Which represent soft sounds? Describe the sounds that produce *d* and *e*.

A train of regular waves produces a pleasant sensation which is called a musical tone. An irregular wave train produces a sensation that is not musical and may be unpleasant. Which of the images above represent musical tones? Which was caused by a breaking dish?

Vibrations. A tuning fork produces sound when its prongs move back and forth rapidly. This type of motion is called *vibrating.*

A single vibration is considered to be the full movement of a prong from one side to the other and back again. Note in the drawing that the prong makes one vibration when it moves from A to B and back to A. The prong makes one vibration while producing one sound wave—a single compression wave followed by a single rarefaction wave.

Frequency. The term *frequency* is used to describe the rate at which a vibration takes place. For example, a tuning fork may have a frequency of 220 vibrations per second. Sound produced by such a tuning fork has a frequency of 220 waves per second.

A complete sound wave is considered to be a compression wave followed by a rarefaction wave as shown on this page. The frequency of sound waves may be thought of as the number of waves produced per second, or the number of waves passing a point per second. Frequency may also be thought of as the number of vibrations per second of air molecules in the path of the sound waves.

The two tuning forks that are shown on this page are sending out wave trains having different frequencies. From which tuning fork do more waves reach your ears each second? Which of the wave trains has the higher frequency?

Pitch. The word *pitch* is used to describe the way that waves of different frequency affect our ears. We say that sounds of high frequency have a high pitch. Generally, the words are used as though they mean the same thing.

Which of the two tuning forks shown here produces a sound of higher pitch?

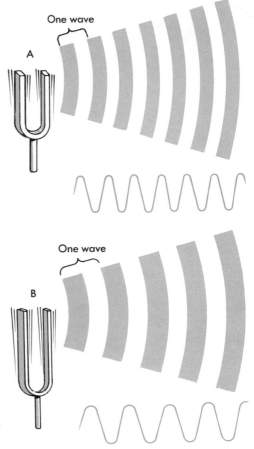

SUMMARY QUESTIONS

1. How many waves must arrive close together in your ear before you hear sound?
2. How do speakers make sound louder?
3. What can be determined about a sound using an oscilloscope?
4. What is a vibration?
5. What does the term *frequency* mean?
6. What does the term *pitch* describe?

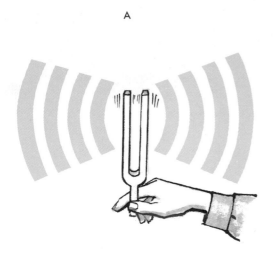

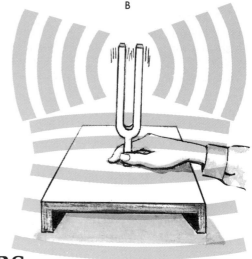

A

B

SOUND RADIATORS AND RESONATORS

Any vibrating object that sets up sound waves is called a sound radiator *because it radiates sound energy. Some sound radiators vibrate chiefly at one frequency, producing nearly pure tones. Most radiators, however, vibrate at several frequencies at the same time. The sounds that they produce may be pleasing or they may clash unpleasantly, depending upon the way the several tones blend within our ears.*

Large sound radiators, such as cymbals and kettle drums, set up waves in large volumes of air and can be heard over long distances. Other radiators, such as violin strings, have only a small area with which to set up waves. Sounds from small radiators must be amplified in order to be heard at a distance.

Small and Large Radiators. Set a tuning fork vibrating. Hold it in your hand and note its loudness. Press the base of the tuning fork against a desk top. What happens to the sound? Touch your fingertips to the desk top at the same time. What do you feel? What is happening to the desk top?

The prongs of a tuning fork are too small to set up strong waves in air. Therefore, a tuning fork is not a good sound radiator. However, when the base of a tuning fork touches the desk top, it sets the whole desk top vibrating. A large area then produces sound waves.

Sounding Boards. Sounds from many weak radiators can be amplified in the manner just described. Thin boards are commonly used for this purpose and so the term *sounding board* has become well known.

Press the base of a vibrating tuning fork against

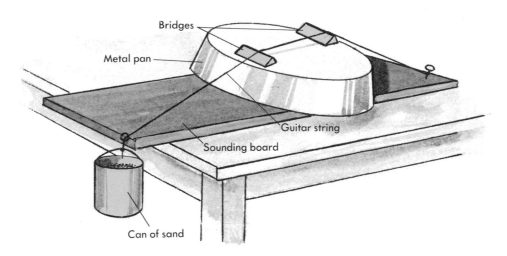

Bridges

Metal pan

Guitar string

Sounding board

Can of sand

thin wood, thick wood, a book, a sponge, a metal pan, and a brick. Which make good sounding boards? What properties do objects need to serve as good sounding boards?

Amplifying Strings. A vibrating string has a very small area with which to produce compression and rarefaction waves in air. The sound of a vibrating string is weak when the string acts alone. It must be *amplified* to be heard very far.

Set up the apparatus shown above. Compare the sound produced with and without the metal pan. Set the bridges on other types of materials such as sponge rubber, a thin board, a thick board, and a brick. Which substances serve best to amplify the sound of the string?

The Violin Family. Musical instruments of the violin family use sounding boards of thin wood. The wood is very carefully selected and shaped to amplify the many frequencies produced by the strings. The most valuable violins have sounding boards that amplify fully all the frequencies of the strings without adding other frequencies that might be unpleasant to hear.

Study the picture here. What part of the violin transmits the energy of the vibrating strings to the upper sounding board of a violin?

Touch a violin while it is being played. Does the top vibrate? Does the bottom vibrate? Look into the openings of the upper sounding board to find out how energy is transmitted to the lower sounding board.

Discuss the purpose of the openings in the upper sounding board.

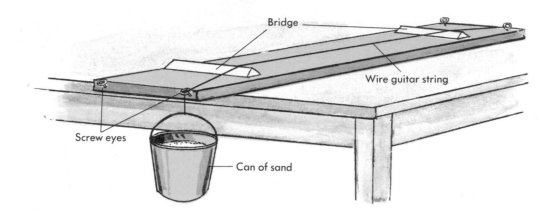

Bridge

Wire guitar string

Screw eyes

Can of sand

String Length and Pitch. Set up the apparatus shown above. Move one of the bridges back and forth. What change raises the pitch of the sound produced by the string? What change lowers the pitch?

Push the bridges far apart and adjust them until the tone of the string matches a low note on a piano. Measure the length of the part of the string that vibrates. Then strike a note one octave higher on the piano and move the bridges until the string produces this tone. What change in length of string raised the pitch of the sound one octave?

Test other octaves in the same way. Is the relation of string length always the same?

Produce the notes of the musical scale. Mark the position of the bridge each time and measure the distances. Note the difference between a whole tone and a half tone.

String Length and Frequency. The keyboard on the next page gives the frequency of each of the notes of a piano. Adjust the bridges of your apparatus until it gives off one of these tones. Find out from the chart how fast the string of the apparatus is vibrating.

Discover the frequency of the string for several other notes in the same way. Measure the length of the vibrating part of the string each time. Then make a graph that shows the relation of the length of a string to the frequency of its vibrations.

Use the graph to predict the frequency of some length not tested. Check your prediction by setting the bridges this distance apart and matching the tone produced with one on the piano.

Octave: The frequency of one note which is double another, or a note on a piano which is 8 notes higher or lower than another.

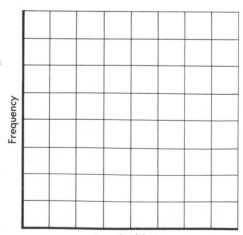

Frequency

Length of String

230

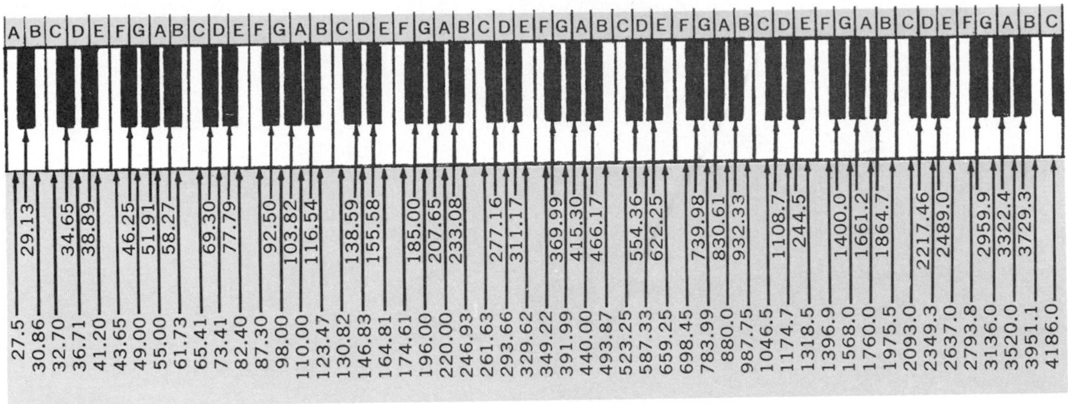
A B C D E F G A B C D E F G A B C D E F G A B C D E F G A B C D E F G A B C D E F G A B C D E F G A B C
29.13 34.65 38.89 46.25 51.91 58.27 69.30 77.79 92.50 103.82 116.54 138.59 155.58 185.00 207.65 233.08 277.16 311.17 369.99 415.30 466.17 554.36 622.25 739.98 830.61 932.33 1108.7 244.5 1400.0 1661.2 1864.7 2217.46 2489.0 2959.9 3322.4 3729.3
27.5 30.86 32.70 36.71 41.20 43.65 49.00 55.00 61.73 65.41 73.41 82.40 87.30 98.00 110.00 123.47 130.82 146.83 164.81 174.61 196.00 220.00 246.93 261.63 293.66 329.62 349.22 391.99 440.00 493.87 523.25 587.33 659.25 698.45 783.99 880.0 987.75 1046.5 1174.7 1318.5 1396.9 1568.0 1760.0 1975.5 2093.0 2349.3 2637.0 2793.8 3136.0 3520.0 3951.1 4186.0

Pianos. Compare a piano with the apparatus on the opposite page. Why are so many strings used? Locate the bridges that support the strings. Find the sounding board that amplifies the sound given off by the strings. What sets the strings vibrating?

The keyboard above gives the rates at which piano strings vibrate. What is the frequency of the string that produces middle C? Compare the length of strings producing sounds of different pitch. How do the frequencies of strings producing sounds one octave apart compare with each other?

Note that some strings are heavier than others. For what frequencies are these heavy strings used?

Large strings radiate more sound energy than small strings because they have more surface in contact with the air. What is done to match the loudness of high-pitched tones with those of low-pitched tones?

Strings that are alike

Unequal amounts of sand

Effect of Tension. The force that pulls a string tight is called *tension*. Set up the apparatus shown above. Use different amounts of sand in the two cans so that there is a different tension acting on each string. Compare the pitch of the two strings. What is the relation between tension and pitch?

Change the tension on the strings. Compare the forces needed to produce two tones one octave apart and two octaves apart.

Adjust the tension in one of the strings until its tone matches a note on the piano. Use the keyboard on the preceding page to discover the frequency of the string. Repeat the process at several different tensions and make a graph of the results.

Use the curve produced on your graph to predict the frequency of the string at a different tension than those tested. Check your prediction by experiment.

Tuning Stringed Instruments. Violins and other stringed instruments are tuned by tightening or loosening the strings. A violin is tuned as shown at the left. Find out how other stringed instruments, including pianos, are tuned. Find out what is used as a standard frequency so that all instruments are tuned the same.

Effect of String Size. Set up the apparatus shown on the next page. Note the effect of the size of the string on pitch. Compare strings of other sizes in the same way.

Most stringed instruments use several strings of different sizes. Actually, the diameter of the string is not important; it is the mass of the string that affects the pitch. To increase the mass of a string without making it clumsy to play, the string is wrapped with wire. This increases the mass without greatly increasing the string's diameter.

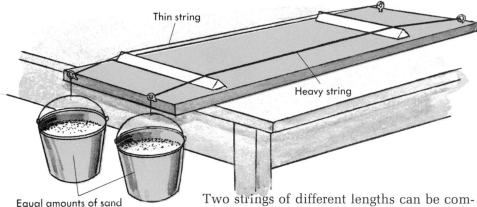

Thin string

Heavy string

Equal amounts of sand

Frequency

Tension

Air column

Air column

Two strings of different lengths can be compared by dividing their masses by their lengths. This gives the mass per unit of length for each string.

Match the tones given off by several strings of different masses with the notes of a piano, keeping the tension and length the same. Then make a graph showing the relation between frequency and mass per unit of length.

Comparing Stringed Instruments. Make a study of several common stringed instruments. How many strings are used on each? What is the frequency of each string when it is properly tuned? How long is the vibrating part of each string? How do the masses of the different strings compare? How does a player change the pitch of the tones he or she produces? How does each instrument amplify the sound given off by the strings?

Amplifying with Echoes. Hold a tuning fork over the top of an empty bottle. Note whether an echo from the bottom of the bottle makes the sound of the tuning fork louder.

Add water slowly to the bottle until the sound suddenly becomes much louder. Adjust the water level until the sound is loudest. Mark the water level.

Test a tuning fork having a different frequency. Test bottles having other shapes and sizes. Compare the results.

Resonance. The prongs of a tuning fork send compression and rarefaction waves into the bottle. These waves echo back from the bottom. If the waves echo back in time to add themselves to waves being sent out in the opposite direction, they make the sound louder. This type of amplification is called *resonance*.

The length of the air column must match the frequency of the tuning fork. A short air column is

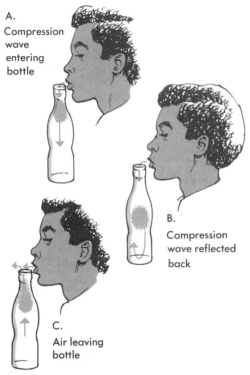

A.
Compression
wave
entering
bottle

B.
Compression
wave reflected
back

C.
Air leaving
bottle

needed for a rapidly vibrating fork so that the waves can return in time to add themselves to the next set of waves. A longer air column is needed for a slowly vibrating fork so that the waves do not return too soon.

The xylophone shown above uses properly tuned tubes to amplify the sound given off by the wooden bars. Explain why a tube is needed for each bar and why each tube has a different length.

Bottle Whistles. Blow across the mouth of an empty pop bottle. Add a little water and blow again. Compare the pitch of the sounds.

Put water into eight pop bottles until they produce the tones of the musical scale. Measure and compare the lengths of the air columns.

Explanation of a Whistle. At first, air enters the bottle, sending a compression wave downward as shown at the left. This wave strikes the bottom, reflects, and returns to the top of the bottle.

The pressure at the mouth of the bottle is now greater than atmospheric pressure. Air rushes from the bottle and starts a compression wave moving outward as shown at C. In the meantime, a rarefaction wave has moved downward and reflected from the bottom. When it returns, pressure at the mouth of the bottle is reduced and air enters once more.

The process is repeated again and again, fast enough to send out a train of waves that gives the sensation of sound. Explain why a short air column produces a sound of higher pitch than a long air column. How does a flute shown at the left

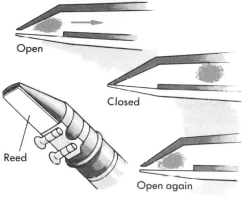

Open

Closed

Reed

Open again

A trumpet Straightened out

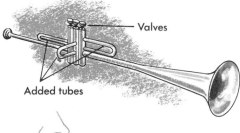

Valves

Added tubes

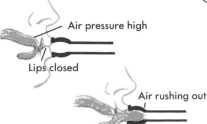

Air pressure high

Lips closed

Air rushing out

Lips open

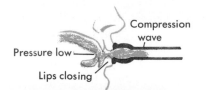

Compression wave

Pressure low

Lips closing

produce musical tones and how is the pitch changed?

Reed Instruments. The sound generator of a clarinet or saxophone is a thin piece of wood called a *reed*. The reed acts as a valve that opens and closes a thin slit in the mouthpiece of the instrument.

A puff of air enters the mouthpiece and sends a compression wave down the tube as shown in the diagrams. However, air pressure drops when air speeds up through a small opening, just as it does when you blow between two strips of paper. Thus the reed is forced against the mouthpiece, stopping the flow of air. A rarefaction wave then travels down the tube.

Pressure in the mouthpiece returns to normal, the reed springs back, and another puff of air enters. The process is repeated many times a second.

The rate at which the reed vibrates depends upon the resonance of the tube. The resonance can be changed by opening and closing holes in the tube, thus changing the frequency at which the reed vibrates.

Ask a musician to demonstrate the action of reed instruments.

The Horn Family. Sounds are generated by the player's lips in all members of the horn family. The diagrams show how waves are made.

The player closes his or her lips tightly and increases the air pressure in the mouth. The lips are finally forced open and a puff of air enters the instrument, sending a compression wave down the tube.

The pressure in the player's mouth drops as air leaves it and his or her lips close again, sending a rarefaction wave down the tube. This process is repeated over and over many times a second.

The tube of a horn resonates to only a few frequencies and the player's lips vibrate at these frequencies only. Thus a bugle player produces only about seven different tones.

The resonance of the tube can be changed by lengthening or shortening it. The resonance of most horns is changed by adding short lengths of tubing with valves so that they can be used as desired. Note the diagram at the left.

How is the resonance of a slide trombone tube changed?

The Human Voice. Our vocal apparatus is made up of (1) the *larynx*, often called the "voice box," (2) the lungs and the muscles that force air in and out, and (3) the mouth, nose, and other cavities that serve as resonators.

Sounds are generated by the *vocal cords* within the larynx. These are not truly cords; they are two elastic membranes that can be drawn across the larynx, stopping the flow of air. The cords are far apart during normal breathing, as shown here.

The cords are brought together for speaking and singing. Air passing through the narrow slit between them causes the edges to flutter, much as a saxophone reed flutters. The fluttering produces a train of compression and rarefaction waves. The three photographs below show how the vocal cords look as they flutter open and shut while producing a sound. The frequency is controlled by tightening or loosening the cords.

The action of the vocal cords can be imitated with a funnel, two pieces of thin rubber, and a string as shown on the next page. Blow through the stem of the funnel and note the behavior of the pieces of rubber.

Male and Female Voices. The chief difference between male and female voices is caused by differences in the vocal cords. Each of the vocal cords of a woman is about two-thirds the length of each of a man's vocal cords and is not so thick. A

 Vocal cords relaxed    Vocal cords closed

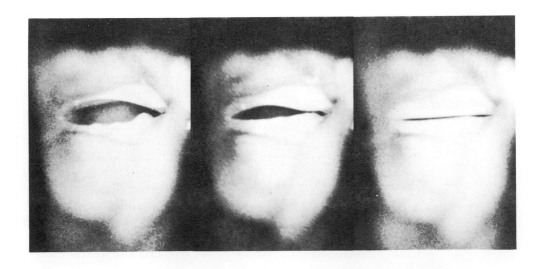

woman can produce sounds that average about an octave higher than those produced by a man.

Change of Voice. A child has a small larynx with short, thin vocal cords. A child's voice is high-pitched. The vocal cords grow longer and thicker as the child becomes older, producing a voice of deeper pitch.

Changes in the vocal cords of a boy may be very sudden when he is about 14 or 15 years old. He may not know how to control his voice at first. It "breaks," sometimes being low-pitched and sometimes being high-pitched. However, within a few weeks or months, he learns to coordinate the muscles of his larynx properly.

Vocal Resonators. The chest, throat, mouth, nose, and sinuses serve as resonators for the larynx. These resonators reinforce the sounds given off by the vocal cords.

Great singers are gifted with unusually fine resonators which change the thin tones of the vocal cords into full, rich musical sounds. The hours of practice put in by singers are planned to develop control of resonance.

The shape and size of some of the vocal resonators can be changed. Repeat the sounds given with the diagrams here. Note the changes you make in the shape of your mouth. Produce other sounds. Study the position of your lips, tongue, jaws, and other organs as you make these sounds.

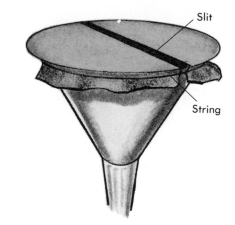

Slit

String

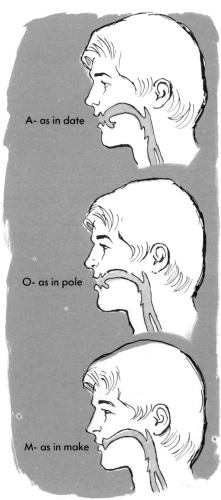

A- as in date

O- as in pole

M- as in make

SUMMARY QUESTIONS

1. What is a sounding board?
2. Why is it necessary to amplify the sound of a vibrating string?
3. How does tension affect the pitch of a vibrating string?
4. How does the mass of a string affect the frequency at which it will vibrate?
5. What is resonance?
6. Give two examples of resonators and explain how they resonate.
7. Voice sounds are produced by vibrating vocal cords. How are these sounds amplified?

REVIEW QUESTIONS

1. What is a sound wave?
2. What are the properties of sound waves that your ears are able to hear?
3. How does a tuning fork make trains of sound waves?
4. Which of the two tuning forks shown here produces the higher pitch? Why do these tuning forks sound louder when they touch the table?
5. How is the energy of the tuning forks transmitted to our ears?
6. In what ways can sound waves be different?
7. How is an echo produced?
8. Why is a child's voice high-pitched?
9. What is the difference between a compression wave and a rarefaction wave?
10. About how many meters per second can sound travel in air?
11. What is the purpose of the reed in a clarinet mouthpiece?
12. Why is the sound produced by blowing over a short test tube different from that produced by a long test tube?
13. What is the purpose of the tubes beneath the bars of a xylophone?
14. What is the difference between a musical tone and a noise?
15. How do we produce sounds to make words?

THOUGHT QUESTIONS

1. If thunder is heard 5 seconds after lightning is seen, how far away is the lightning?
2. Why is the lowest string of a violin often wound with wire?
3. Why cannot a horn be played by merely blowing into the mouthpiece?
4. Why should the timer in the picture at the left start his watch by the flash of the starting gun rather than by the sound of the gun?
5. Study the chart of the velocity of sound through various materials. Propose a hypothesis to explain why sound waves move faster through steel than through air. What properties of a material effect the rate at which sound waves will pass through it?
6. Why do people's voices sound differently?

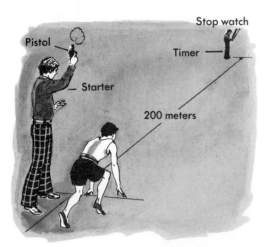

Stop watch
Timer
Pistol
Starter
200 meters

Velocity of Sound

5050 m/s in steel
1500 m/s in water
340 m/s in air
1200 m/s in lead
1240 m/s in helium

3

The Behavior of Light

Light provides us with most of the information about our surroundings. It is not surprising, therefore, that scientists have given much thought to the behavior of light and to methods of controlling it. As a result, we now have thousands of devices invented to produce light in dark places, to improve vision, and to preserve images for later study.

Nevertheless, even though much has been learned about the behavior of light, its true nature is not yet well understood. Theories that have been suggested are not completely satisfactory, either because they do not fit all the facts or because they are very complicated. We are still waiting for a simple answer to the question, "What is light?"

HOW LIGHT TRAVELS

Early scientists suspected that light travels outward from a source at a speed too great for the eye to detect. However, they had no means for proving or disproving this idea. Experiments finally showed that light does travel, and at the same time scientists measured its speed—an enormous 297 600 kilometers per second.

Other experiments have provided much information about the way light travels and about its behavior as it falls on different surfaces. What is not yet known is the method by which light travels, a problem that may not be answered until the nature of light is thoroughly understood.

Outward Spread of Light. According to modern beliefs, light spreads outward in all directions. The light that leaves its source at any one instant may be imagined as an expanding bubble. Other bubbles follow immediately behind, so close together as to be continuous.

The Farther the Fainter. Light spreads over a larger and larger area as it travels outward from its source. The light that falls on one square centimeter of surface near the source will cover many square centimeters farther away.

Hold a light meter (such as used by photographers) near an unshaded electric lamp bulb. Take a series of readings at increasing

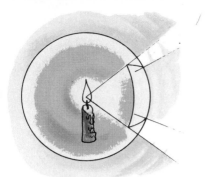

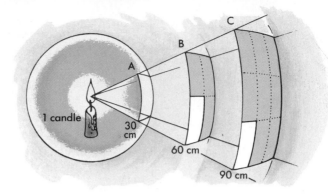

distances from the lamp. What does the meter show about the relation of light intensity to distance from the source?

Comparing Intensities. The apparatus shown below is used to compare the intensity of light at different distances. The smaller card contains a hole one centimeter square. The larger card is marked off in one-centimeter squares.

Remove the reflector from a flashlight and use the bulb as a light source. The bulb and the hole in the small card should be at the same level.

Place the cards in the positions shown below. Count the squares in the lighted area on the large card. How much does the light spread out as it travels from a 30-cm distance to a 60-cm distance? What do you think happens to the intensity of the light at the same time?

Try other combinations of distances. Decide what happens to the intensity of light when the distance it travels is doubled. When the distance is tripled. When the distance is increased four times.

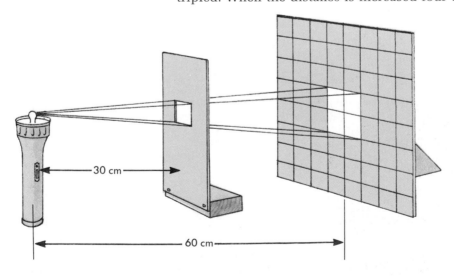

Pinhole

Waxed paper

A Pinhole Viewer. Punch a tiny hole, no larger than a pin, in the bottom of a can. Cover the open end of the can with waxed paper.

Darken the room as much as possible except for one brightly lighted window. Hold the pinhole toward the window and look at the waxed paper. Describe what you see.

Changes in Image Size. Cut a letter from colored cellophane and tape it on the lens of a flashlight. Shine the flashlight toward the pinhole in the can.

Note the position and size of the image on the waxed paper. Move the flashlight farther away. What is the relation of the size of the image to the distance between the pinhole and the flashlight?

Mark the size of the cellophane letter on the waxed paper. Move the flashlight back and forth until the image of the cellophane letter is exactly the same size. Then measure the distance between the pinhole and the flashlight, and between the pinhole and the waxed paper. How do these distances compare?

Calculate the size of the image if the pinhole is twice as far from the flashlight as from the waxed paper. Test your calculations by experiment. Repeat the calculations for other combinations of distances and test the results.

Predict the effect of using a longer can. Test your prediction.

Explaining the Image. Study the diagram below. Trace the paths of the light traveling from different parts of the letter on the flashlight. Why is the image upside down and the same size as the letter?

Make other diagrams showing the flashlight at different distances. Explain changes in the size of the image in terms of the diagrams.

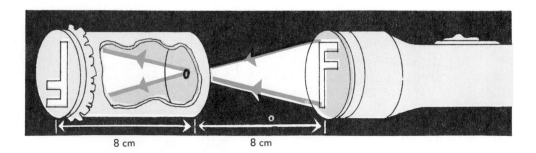

8 cm   8 cm

Light for Seeing. Imagine the scene at the right if the candle were not lighted. No light would be in the room and nothing could be seen. But as soon as the candle is lighted, the teenagers can see both the candle and other objects in the room.

The candle is visible because it gives off light, some of which enters the eyes of the students. Other objects are visible because light from the candle falls on them and reflects toward the eyes of the viewers.

Direct Light. Light from a candle travels outward in all directions. A very small amount of this light enters the eyes of the young woman. A similar small amount enters the eyes of the young man.

Note that the students have different views of the same flame because the light that enters their eyes comes from different parts of the flame. If the young woman wants to see the flame as the young man sees it, she must move into his place.

Reflected Light. Most objects are seen by reflected light. The picture here shows the paths of the light that make the clock visible to both students. Light passes from the flame to the clock and is reflected. Part of this reflected light enters the eyes of both students. Notice again that they have slightly different views of the same object. A third person in the room would see the clock by light that travels along another path.

Absorption. Each time that light falls on a surface, some of its energy is *absorbed*. The energy is taken into the substance and changed into some other form, usually heat. Thus the light that reflects back and forth across a room loses its energy and becomes fainter.

Discuss the absorption of light by mirrors, black felt, white paint, and other materials.

SUMMARY QUESTIONS

1. Why does a light appear fainter and fainter as you go away from it?
2. How does a pinhole produce an image that is upside down and backwards?
3. Why is it possible to see objects which do not give off any light of their own?
4. What happens to the energy of light when it shines on a surface?

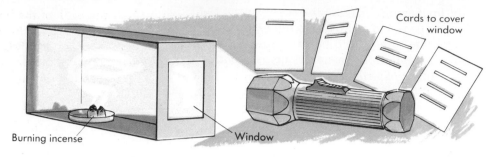
Burning incense   Window   Cards to cover window

REFLECTIONS

Reflected light follows paths that can be calculated beforehand. In other words, the behavior of reflected light can be predicted.

Sometimes people say that reflected light obeys certain scientific laws. This statement is often more puzzling than helpful because scientific laws are not at all like human-made laws. A scientific law is a generalization based upon many experiments which always give the same results when repeated exactly. Many other generalizations in science can do no more than predict average behavior, but the "laws of reflection" have held for every case tested.

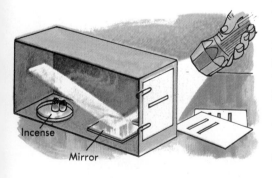
Incense   Mirror

A Light-Study Box. Experiments with light often require a dark place in which to work. The place need not always be large or perfectly dark; a shoe box provides an excellent laboratory for many important experiments. Such a laboratory is easily constructed.

Paint the inside of a shoe box black, and cut a window in one end. Prepare several cards with narrow slits as shown in the diagram. Cut the slits with a single-edged razor blade or a sharp-pointed knife. Fasten the cards over the window of the box with paper clips when needed.

Locating a Beam of Light. Darken the room as much as possible. Place the light-study box on its side and direct a bright light through the window of the box. Fill the box with chalk dust or with smoke from burning incense.

Note how the particles of dust or smoke reflect the light, making the path of the light visible.

Hold a mirror in the beam of light and tilt it up and down. Watch the behavior of the reflected light.

Angles of Reflection. Fasten a card having a single slit over the window of the light-study box. Place a mirror on the bottom of the box as shown on page 245. Ask someone to direct a bright beam of light through the slit to the mirror so that the beam and its reflection appear on the back of the box.

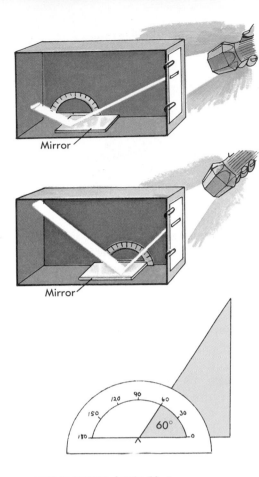

Mirror

Mirror

60°

MEASURING ANGLES. Lay a protractor on an angle with one side of the angle at the zero mark as shown here. Read the number of degrees between the two sides of the angle.

Set a protractor against the back of the box with the center point at the place where the light beam strikes the mirror. Measure the angle at which the light beam strikes the mirror, and the angle at which the reflected beam leaves the mirror. How do the angles compare?

Move the light to change the angles. Measure the angles again. Test several positions of the light source. What general statement can you make about the angles produced between light beams and a mirror?

Position of an Image. Stand a pane of glass in an upright position. Place an unlighted candle behind the glass. Place another candle of the same size and shape, but lighted, in front of the glass. Move this second candle about until the reflection of its flame seems to fall on the unlighted wick of the candle behind the glass.

Measure the distances between the two candles and the glass. How do the distances compare? Repeat the measurements with the unlighted candle in different positions. How do the measurements compare each time?

Describe the path of the light traveling to your eyes from both the lighted and unlighted candles. Where does the reflection of the lighted candle seem to be located?

Without moving the candles, replace the pane of glass with a mirror. Compare the position of the image in each case.

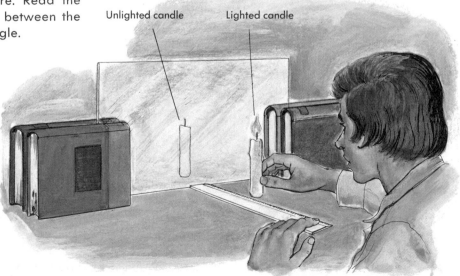

Unlighted candle   Lighted candle

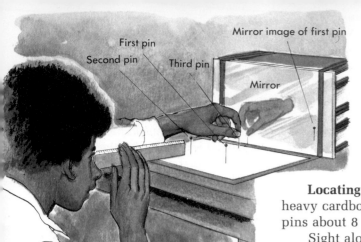

First pin
Second pin
Third pin
Mirror image of first pin
Mirror

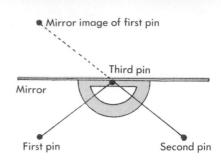

Mirror image of first pin
Third pin
Mirror
First pin
Second pin

Locating an Image. Lay a sheet of paper on heavy cardboard. Stand a mirror on edge. Place two pins about 8 centimeters apart in front of it.

Sight along the second pin at the image of the first pin. Place a third pin next to the mirror in line with the image. Draw lines connecting the pins. Which lines indicate the path of the reflected light?

Remove the mirror. In what direction did the image seem to be? At what distance? Place a fourth pin in the calculated position of the image. Check your calculations by replacing the mirror, sighting at the image as before, and then lifting the mirror without moving your eyes. Where do you see the fourth pin?

Mirror Images. Hold a mirror over a piece of paper, and write the word "mirror" on the paper while looking into the mirror. Try to make the word appear correctly in the mirror. Which letters are easiest and which are most difficult to form? Explain why.

Print a word on a large sheet of paper. View the word in a mirror. Describe the image.

Images that are reversed in this fashion, but not turned upside down, are often called *mirror images*.

Double Reflections. Study the effect of viewing a clock in two mirrors as shown below. Test other arrangements of the mirrors. What happens to an image when light is reflected twice?

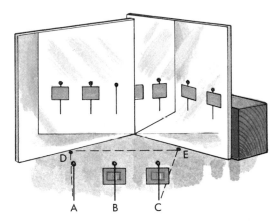

Set up the apparatus shown here. Sight over pin A at the image of flag C in the right mirror. Make a mark D where your line of sight meets the mirror. Draw AD.

Continue to sight along the same line, moving your pencil in front of the left mirror until you find the point where your line of sight meets this mirror. Mark this point E. Draw lines CE and DE. The three lines represent the path of light passing from flag C to pin A.

Find the path of light passing from flag B to pin A by the same method. Use your findings from this experiment to explain why images formed by double reflections appear normal in terms of right and left.

Inverted Images. Reflections of objects in pools of water produce images that are upside down. Such images are said to be *inverted*.

Make a diagram of the paths of light reflected from the surface of water. Explain why the eye sees an inverted image.

Experiment with a mirror to find out when inverted images and reversed images are produced.

Experiment with two mirrors to discover the effect of reflecting light twice from level surfaces. What type of image is produced?

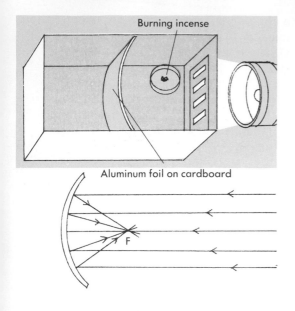

Burning incense

Aluminum foil on cardboard

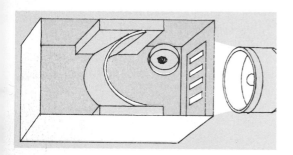

F

Curved Reflectors. Cut a piece of thin cardboard about 3 centimeters longer than the width of the light-study box. Cover the cardboard with shiny aluminum foil, keeping the foil as smooth as possible. Set this reflector in the box as shown at the left. Fill the box with smoke and shine a light into the box through several slits.

Study the ways the beams of light are reflected. Lay a sheet of paper in the box, and trace the paths followed by the beams of light.

Increase the curve of the reflector by placing blocks of wood at the ends of the reflector. What happens to the beams of reflected light? Make diagrams of the beams.

The Focus. The point at which reflected rays come together and cross is called the *focus*. A reflector must have a special curve to have a focus. Try to bend the reflector in the light-study box so that the beams all cross at one point.

Concave and Convex Reflectors. The reflection shown at the top of the page is called a *concave* reflector. This reflector curves away from the light.

Make the reflector curve toward the light as shown below at the left. It then becomes a *convex* reflector. Study the paths of light beams reflected from this convex reflector and make diagrams of them. Does a convex reflector have a focus?

The teenager shown below is looking at his reflection in a sheet of metal that can be bent to make a concave or convex reflector. Where have you seen such mirrors?

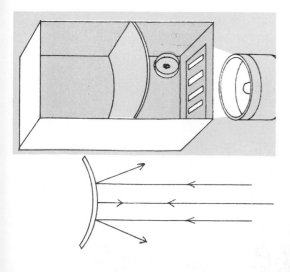

Lamp Reflectors. Flashlights and many other sources of light contain concave reflectors. Remove the reflector from a flashlight and shine the beam of light from the flashlight through smoke or chalk dust in a darkened room. Replace the reflector and test the beam of light again. What effect does the reflector have on the beam of light?

Set up the light-study box with a concave reflector as shown on the opposite page. Find the focus of the reflector. Remove the card covering the window of the box and place a flashlight inside. Move the flashlight until the bulb is at the focus. Notice how the light is reflected. Make a diagram of the light's path.

At the right above is shown a similar experiment using a reflector curved like the reflector in a searchlight. Note that much of the light from a lamp at the focus leaves the searchlight in nearly parallel rays. What is the advantage of this?

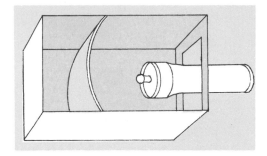

SUMMARY QUESTIONS

1. Along what path is a light beam reflected from a flat mirror?
2. At what distance from the observer does the image in a mirror appear to be?
3. How does the image in a mirror differ from the object being reflected?
4. How are curved reflectors useful?
5. Make a diagram of several light beams being reflected by a concave reflector.
6. Where are convex reflectors found to be useful?
7. What is the focus of a curved reflector?

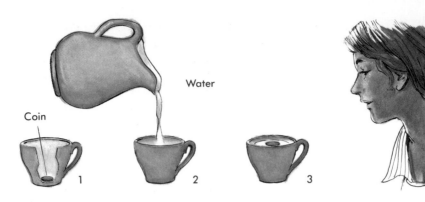

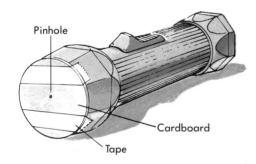

THE BENDING OF LIGHT

We have undoubtedly known for thousands of years that objects in water are not always where they appear to be. Written records show that Arab scientists were studying this subject at least nine centuries ago.

The behavior of light as it passes from water to air or back again could not be explained satisfactorily by early scientists. The explanation had to wait until knowledge was gained about the speed of light passing through different substances.

The Rising Coin Trick. Put a coin in the bottom of a teacup. Push the teacup away from you until the coin is just hidden by the edge of the cup.

Ask someone to pour water into the cup. Describe what you see.

Make two diagrams, one of the cup without water, the second of the cup after water was added. Draw lines on the first diagram to show why light from the coin could not reach your eyes. Draw lines on the second diagram showing the path that light from the coin must have followed to reach your eyes. Locate the image of the coin on the second diagram.

Bending a Beam of Light. Cut a disk of cardboard to cover the lens of a flashlight. Punch a small hole in the center of the disk, and fasten it on the lens with rubber bands or tape.

Fill an aquarium with water to which a few drops of milk have been added. The milk shows the path of light passing through the water. Smoke or chalk dust can be used to show the path of light in the air above the aquarium.

Direct a beam of light from the flashlight toward the surface of the water. Change the angle at which the beam strikes the water. Describe the appearance of the beam. Which way is it bent?

Direct the beam of light toward the glass side of

the aquarium. Compare the appearance of the beam with that obtained earlier.

Refraction. The bending of light as it passes from one substance to another is called refraction. The previous experiment showed the refraction of light as it passes from air into water.

Would you expect light to be refracted as it passes from water into air? Plan an experiment to find out if refraction takes place and, if so, in what direction a beam of light is bent.

Refraction by Different Liquids. For this experiment, use a flat-sided jar or bottle that fits inside the light-study box. Fasten a light source in a position from which it sends a slanting beam against the jar.

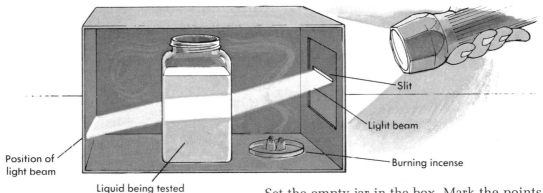

Position of light beam

Liquid being tested

Slit

Light beam

Burning incense

Set the empty jar in the box. Mark the points at which the beam of light enters and leaves the jar.

Fill the jar with water and again mark the points at which the light beam enters and leaves the jar. Repeat using other liquids such as mineral oil, carbon disulfide, and alcohol. Does sugar or salt dissolved in water affect the angle at which light is refracted by the solution?

What is the independent variable in this experiment? What is the dependent variable? How can the other conditions be kept unchanged?

Refraction by Air. Light is known to be refracted slightly as it passes from a vacuum into a gas. The angle through which the light is refracted is too small to be measured in the usual school laboratory.

Effects of this refraction can be seen when light passes through air that is heated by a radiator or other hot object. The light is bent as it passes from the heavy cool air into the lighter hot air and back again. The "twinkling" of stars is caused in the same way.

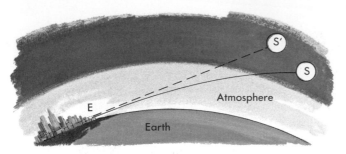

Sunlight is refracted as it enters the earth's atmosphere. The amount of refraction is small at noon when the sun is most nearly overhead. It is greatest at sunrise and sunset when the light strikes the atmosphere at a low angle. Because of this refraction, the sun is seen for several minutes before it rises and after it sets, thus lengthening the day from 4 to 8 minutes.

Refraction by Solids. Obtain triangles and narrow strips of glass about 4 millimeters thick. These can be obtained at hardware stores or building supply stores. Rub the back side of the strips with sandpaper to make beams of light inside the glass show up better.

Darken the room and send a beam of bright light through a slit in the end of the light-study box. Hold a strip of glass in this beam. Trace the path of light as it enters the strip and passes into the air again.

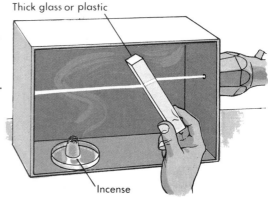

Thick glass or plastic

Incense

Hold a triangle in the beam of light. Trace the direction of light through the triangle. Hold the triangle in different positions, and discover the changes in the path of the light for each position.

Explaining Refraction. Refraction is explained in terms of the speed of light through different substances. Light is known to travel slightly slower through air than through a vacuum, and much slower through water or glass.

If a beam of light falls on the surface of water, the part of the beam that enters the water first slows down first. Therefore, the beam bends until its full width is within the glass. This is shown in the diagram at the right.

When a beam of light enters air from water, the part of the beam that enters first speeds up first, and the beam bends in the opposite direction. Explain what happens when a beam falls straight down on the surface of the water.

Trace around one of the strips of glass or plastic used in the above experiment. Draw lines representing a beam of light entering the strip,

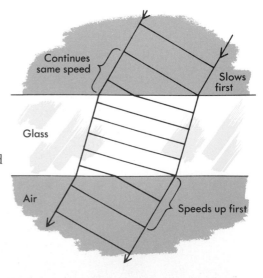

Continues same speed

Slows first

Glass

Air

Speeds up first

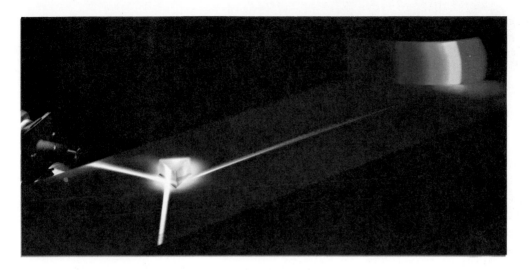

passing through it, and coming out on the other side. Put the strip in the light-study box and check your diagram. Repeat the process with a triangle.

Unequal Refraction. In the photograph above, a beam of white light falls on a triangular glass prism. Part of the light is reflected. The remainder enters the prism and is refracted. Describe the beam of refracted light after it leaves the prism.

The beam entering the prism is made up of light of all colors. These colors would be refracted equally if they all traveled at the same speed through the prism. However, glass slows some colors more than others, and the colors are refracted unequally.

Prop up one end of a light-study box so that a beam of sunlight can enter through a single slit. Hold a glass prism in this beam, and turn the prism until a band of colors is seen. Mark the position of the colors on a sheet of paper.

Which color is refracted most? Which is refracted least? What is the order in which the colors are refracted?

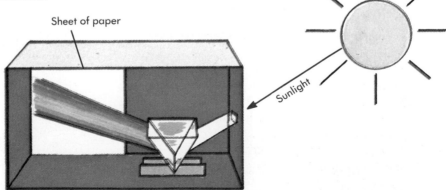

Sheet of paper

Sunlight

Rainbows. Rainbows are produced by the refraction of light through raindrops. Sunlight enters a raindrop and is reflected one or more times. When the light leaves the raindrop, the several colors have been separated by refraction as shown in the diagram below at the left.

Only one color of light can reach your eye from a single raindrop. However, there are many raindrops; red light reaches your eyes from some raindrops, orange light from others, and so on.

The diagram below shows seven raindrops at one instant. However, these raindrops are probably falling. Why does the rainbow remain visible?

Explain why a rainbow seems to move as you move. Explain why two people do not see exactly the same rainbow.

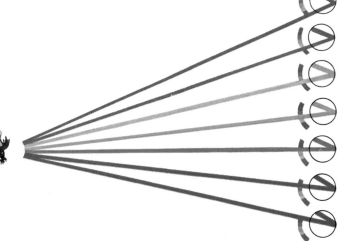

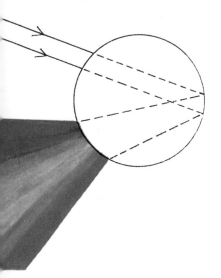

Convergent and Divergent Beams. Place two triangular prisms base to base in the light-study box. Make diagrams of the beams of light which pass through them.

Turn the prisms so that they are point to point. Trace the beams of light that pass through them in this new position.

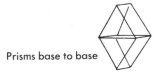

Prisms base to base

Prisms point to point

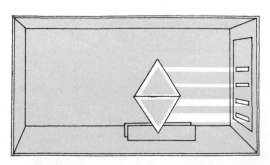

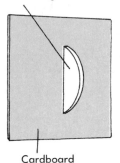

Lens through cardboard

Cardboard

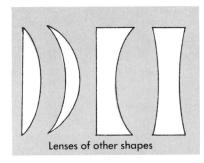

Lenses of other shapes

A. Rays come together after passing through two prisms.

B. Double convex lens acts like two prisms: brings rays together.

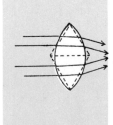

C. Double concave lens acts like two prisms: spreads rays apart.

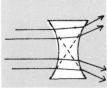

Beams of light that come together and cross are called *convergent*. Beams which spread apart are called *divergent*. In which position did the prisms produce convergent beams? In which position did they produce divergent beams?

Lenses. Lenses are curved pieces of transparent material used to refract light. Most lenses can be classified as either convergent or divergent.

Cut a slot in a piece of cardboard just large enough to hold a lens. Place the cardboard and lens in a light-study box and shine several beams of light through it. Trace the paths of the beams. Is the lens a convergent or a divergent lens?

Test lenses of other shapes in the same manner.

Lens Shape and Behavior. The action of a lens depends chiefly upon its shape. One group of lenses is thicker in the middle than at the edges. The other group is thicker at the edges than in the center. The diagrams at the left show cross sections of the more common types of lenses.

How can you identify convergent lenses without testing them in a light-study box? How can you identify divergent lenses?

Focal Length of Lenses. Parallel beams of light are refracted by a converging lens toward a point called the *focus*. The distance from the focus point to the center of the lens is called the *focal length* of the lens.

Sunlight provides a good source of parallel beams for determining the focal length of lenses. The sun is so far away that the spreading effect of the light is too small to be noticeable.

Hold a lens in sunlight and focus the light on a sheet of paper. The distance from the paper to the lens is very close to the focal length of the lens.

Find the focal length of a lens with the light entering first on one side and then on the other. How do the two focal lengths compare?

Test lenses of different curvature, that is, lenses some of which are curved more than others. What is the relation between the curvature of the lens and the focal length of the lens?

The focal length of a divergent lens cannot be measured by the above method because the beams of light never cross. The focal point is determined by imagining that the diverging beams move backward and cross as shown on page 256. Actual calculation of this point and the focal length is a problem for a more advanced science course.

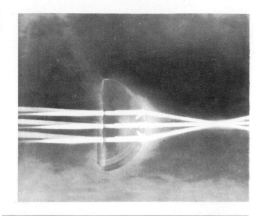

Defects of Lenses. A perfect lens sends all parallel beams of light through one point. However, few lenses are perfect, if any. Usually the beams from one part of a lens pass through a different point than the beams through another part of the lens. The better the lens, the closer together all these points will be. The cost of cameras, microscopes, and telescopes depends to a great extent upon the accuracy with which the lenses are made.

Real and Apparent Images. Hold a convergent lens between a sheet of paper and a distant view as seen through a window. Move the paper back and forth until an image of the scene appears on the paper. Describe the image. Measure the distance between the image and the lens, and compare this distance with the focal length of the lens.

The image formed on the paper is called a *real image* because rays of light are actually focused on the paper. An image you see in a mirror seems to be behind the mirror where no light passes and is called an *apparent image*.

Real images can be formed on a screen where everyone looking at the screen sees the same thing. An image is formed whether there is a screen or not. However, the image is seen only when the screen reflects light to the viewer's eyes.

Replace the convergent lens with a divergent lens. Try to produce a real image on the screen. Explain what happens.

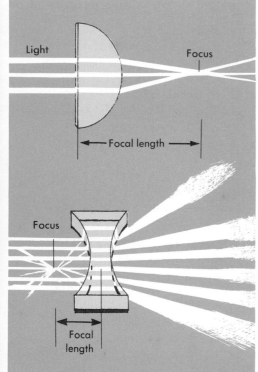

Light

Focus

←— Focal length —→

Focus

Focal
length

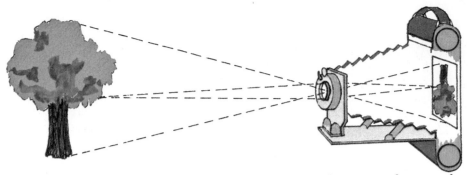

Cameras. The camera lens produces an image of a scene on photographic film, where the light causes chemical changes in the film. Why must the image in a camera be a real image for this purpose? What type of lens must be used in the camera?

The camera shown above is photographing a distant scene. How does the distance between the film and the lens compare with the focal length of the lens?

Image Size and Distance. Cut a disk of aluminum foil to fit over the lens of a flashlight. Make a J-shaped opening in the foil with a sharp pointed knife. Trace the J on a sheet of paper for making comparisons at a later time. Then tape the foil on the flashlight lens.

Direct a beam of light toward a sheet of paper and use a convergent lens to produce an image of the J on the paper. Make the following measurements: (1) the height of the J, (2) the height of the image, (3) the distance between the lens and the J, and (4) the distance between the lens and the image.

What are the distances when the image is equal in size to the J? What are the distances when the image is half the size of the J? When the image is twice the size of the J?

Predict the size of the image when the lens-to-J distance is three times the lens-to-image distance. Test your prediction. Try other distances.

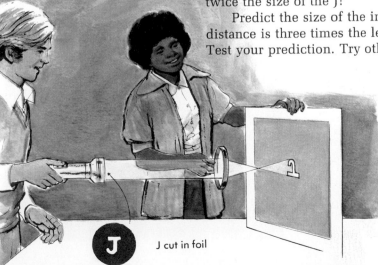

J cut in foil

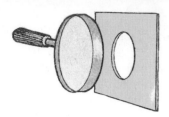

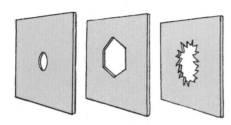

The Lens Opening. Cut several squares of cardboard a little larger than a convergent lens. Make holes of different shapes and sizes in the squares as shown here at the left. Hold each square of cardboard against the lens while producing an image on the screen. Compare the image with the image formed by the lens alone. What is the effect of the size and shape of the lens opening on the image?

Cameras have a device called the *diaphragm* for changing the size of the lens opening. What is the effect of the diaphragm on the image produced in the camera?

Magnifying Lenses. Hold a convergent lens above an object, moving the lens back and forth until you see a magnified image of the object. Measure the distance between the object and the lens. Compare this distance with the focal length of the lens.

Test other lenses in the same way. Decide where an object must be held in order to produce a magnified view of the object. Is the image so produced a real image or an apparent image?

The diagram below shows paths of light passing from an object through a lens into a person's eye. The light enters the eye as though it had come from a larger object located beyond the real object. Black lines represent actual paths of light. Broken lines represent apparent paths. Explain why the image is right side up.

Reducing Lenses. The bottom photograph shows the appearance of an object viewed through a divergent lens. Describe the image. From what kind of an object does the light seem to come? Where does the image seem to be?

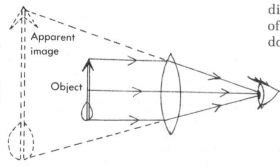

Apparent image

Object

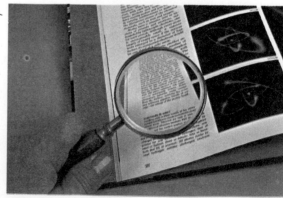

Liquid Lenses. Lenses can be made of any substances that are able to refract light. The sketch at the left shows how you can make liquid lenses. The left side of the sketch shows a round flask containing a little water. Describe the shape of the water. View an object through the water. Is the lens convergent or divergent?

The right side of the sketch shows an empty flask held in a glass dish filled with water. What is the shape of this water lens? Is it convergent or divergent?

Test other liquids such as alcohol and olive oil in place of water. Compare them with water for magnification power.

Magnifying Power. Make a test scale like that shown at the left. Mark heavily one line on a sheet of lined paper and put circles at its ends. This is the zero line. Mark the first, second, fourth, and eighth lines beyond the zero line and number them as shown.

Lay the center of a lens over one of the circles on the zero line. Using one eye only, move the lens toward you until the images are sharp. Line up the image of the zero line with the actual zero line, and compare the size of the image of a space with the actual spaces.

The lens being tested here makes four spaces seem as large as eight actual spaces. Therefore, the lens is a *two-power* lens and the object is magnified two times.

Test the magnification power of several lenses. Does there seem to be any relation between magnifying power and focal length?

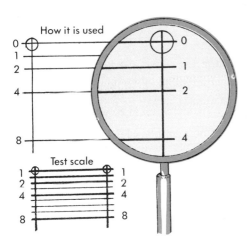

Field of View. Hold a lens over a sheet of squared paper such as graph paper. Make a circle the size of the lens. Count the number of squares in the circle. Then count the number of squares that can be seen through the lens when it is focused for magnifying. What happens to the field of view when a magnifying lens is used?

Repeat the measurements for lenses of different magnifying power. What is the relation of the field of view to the magnifying power of a lens?

Doublet Magnifiers. Hold two magnifying lenses close together and view an object. What happens to the magnifying power when two lenses are used together?

Determine the magnifying power of the lenses when used together and when used separately,

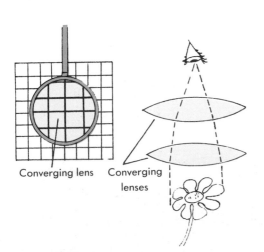

Converging lens   Converging lenses

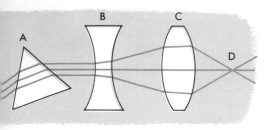

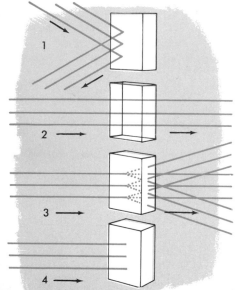

using the scale shown on the previous page. What is the relation of the magnifying power of lenses used together and their separate magnifying powers?

What happens to the field of view when two lenses are used together?

Predict the magnifying power of other lenses when used together. Test your predictions.

REVIEW QUESTIONS

1. What are the names of the three objects shown above?
2. What is the point at D called?
3. Which of the three objects above can be used as a magnifying lens?
4. Which of the three objects above can produce a real image on a screen?
5. How can you see objects that do not give off light?
6. Why does light become fainter as it travels from its source?
7. Why is light refracted when it passes from air into water?
8. How tall is the shadow of a girl 1.6 meters tall when she stands halfway between a small source of light and a screen?
9. Where must a converging lens be held in order to use it as a magnifying lens?
10. What is happening to the light that falls on each of the surfaces numbered 1, 2, 3, and 4 at the left?

THOUGHT QUESTIONS

1. How tall a mirror does a person need to see herself or himself completely at one glance?
2. How can you arrange two mirrors so that light from your automobile headlights shines straight back to you?
3. What must you do to the lens-to-slide distance as you move a slide projector closer to the screen?
4. Why are the images produced by the lenses at the left so different?

Investigating on Your Own

Heat, sound, and light are all related. Each is a form of energy. The knowledge we have about these forms of energy comes from the observations scientists have made throughout the centuries. Some scientists study the sources of these forms of energy, others study how energy is transmitted, while others try to understand the effects of energy on matter.

In this chapter you will use several techniques to study heat, sound, and light. You will set up controlled experiments, make measurements, organize the data, and interpret the results of your investigations.

HEAT

The Effect of Heat on a Gas. What happens to the volume of a gas as heat is removed from it? To investigate this question, you will need to set up an experiment in such a way that you control all the variables except the one you are studying. Which variables will you need to keep constant? Which variable will you want to change?

In 1787, Jacques Charles investigated this question using the apparatus shown at the right. Bend a piece of glass tubing into the shape of the letter "J" and seal the end forming the top of the "J." Partially fill the tube with oil. Heat the closed end with a candle flame to force the air through the oil and out of the tube. Allow the tube to cool. If the air trapped in the tube has a height of more than ten centimeters, drive more air out of the tube by heating the closed end again. The amount of air that is trapped should measure between 6 and 10 centimeters. With the amount of oil in the tube remaining the same throughout these experiments, the pressure on the gas should remain constant.

Prepare a scale marked off in centimeters and millimeters on a piece of cardboard. Attach the tube to the card so that the "J" is upside down as the diagram shows it. Also attach a thermometer calibrated in the Celsius scale. Insert this apparatus into a test tube, and hang the test tube in a beaker of water. The water level in the beaker should be above the "J" tube.

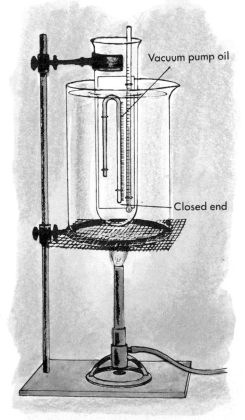

Apparatus for demonstrating Charles' law

Vacuum pump oil

Closed end

Temperature (°C)	Measured Distance in Volume Change (mm)
5	
10	
15	
20	
25	
30	
35	
40	
45	
50	
55	
60	

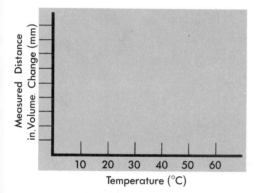

Note the temperature of the water in the beaker and then add several spoonsful of cracked ice. Read and record the change in volume of the trapped air every time the temperature of the water drops one degree. Continue reading and recording until the temperature of the water reaches 5°C. Use a chart such as the one here for recording your observations. What happens to the volume of air as it is cooled?

Gently heat the beaker of water. This time record the change in volume for every degree of temperature the water rises. What happens to the volume of air as it is warmed?

Prepare a graph such as the one shown here, plotting the change in volume of air compared with the change in temperature. How can you predict what the volume of the air will be at 0° or at 100°? What is the limit to which you could cool the air? What would happen to the volume of the air if you cooled it down to −273°?

Do other gases, such as oxygen or carbon dioxide, behave in the same fashion when heated or cooled? You may want to carry this experiment further by testing these other gases.

EXPERIMENTAL RESEARCH

1. Find out how much dry ice (solid carbon dioxide) expands as it becomes a gas. Cut a block of dry ice into a cube, one centimeter on each side. Use the method shown here to collect the gas which comes from the block. Measure the number of milliliters of gas produced. (**CAUTION:** Wear gloves when handling dry ice.)

2. Watch the expansion of water as it freezes. Lay aluminum foil on a block of dry ice. Put a drop of water on the foil. Make drawings to record the behavior of the water.

3. Check the temperatures at which different car thermostats open. Place several thermostats in a container of water. Heat the water and note at what temperature each thermostat opens. Demonstrate your procedure to the class.

4. Use soot to blacken the bulb of an air thermometer like the one constructed earlier in this chapter. Use this thermometer to compare the

Water level

Metric ruler

Water

Salt and ice

Testing flame temperatures

Wire screen

heating effect of an electric heat lamp and a candle flame.

5. Study the behavior of water as it cools toward the freezing point. Set up the apparatus shown at the left. Watch the water level in the tube as the temperature in the flask drops. Note the temperature at which the water level reaches its lowest point.

6. Find out which regions of a flame have the highest temperature. Pull a strand of wire from a piece of wire gauze. Use this wire to study the flame of a bunsen burner. Carefully hold the wire in the flame. Assume that the shorter the time needed to make the wire glow, the greater the temperature. Which is hotter, a yellow flame or a blue flame? Find the hottest part of the blue flame.

7. Make a comparison study of the heating rate of 0.5 liter of water when heated in pans of different shapes, sizes, and materials. In which pan does water heat up the quickest?

OTHER INVESTIGATIONS AND PROJECTS

1. Prepare an illustrated booklet on the history of the steam engine, emphasizing the contributions this machine has made to civilization.

2. Find out what makes spark plugs fire at the proper time in an automobile engine. Prepare a chart or model that shows the mechanism used.

3. Read about the Kelvin temperature scale and report on its use.

4. Prepare a report on the properties of matter near absolute zero.

5. Take apart a discarded electric iron or toaster and find the thermostat. Demonstrate the action of the thermostat to your class.

6. Prepare a report on the heating system of your school, calling special attention to application of the principles studied in this unit.

7. Find out how the muffler of a gasoline engine is made and how it operates. Prepare a chart that illustrates your findings.

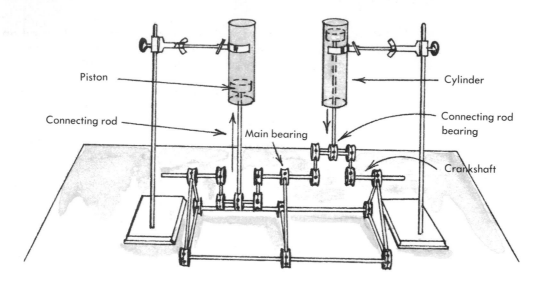

Piston

Connecting rod

Cylinder

Connecting rod bearing

Main bearing

Crankshaft

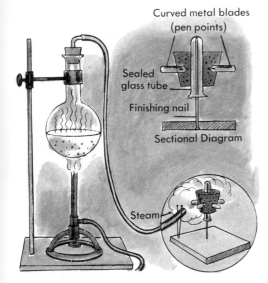

Curved metal blades (pen points)

Sealed glass tube

Finishing nail

Sectional Diagram

Steam

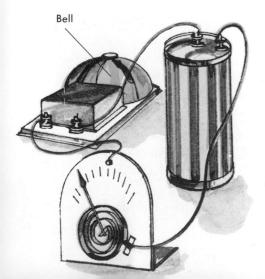

Bell

8.   Make a model of a two-cylinder gasoline engine as shown above. Demonstrate this model to your class and explain how a two-cylinder engine operates.

9.   Make a chart that shows the action of a "heat pump" such as is used in a refrigerator or air conditioner.

10.   Make the model steam turbine shown at the left and demonstrate its action.

11.   Read about the contributions of Benjamin Thompson (Count Rumford) to our present day understanding of heat as energy. Explain to your class how Thompson disproved an earlier theory.

12.   Make a chart that shows how the effects of heat operate a jet engine.

13.   Cut out magazine pictures showing devices which make use of the effects of heat, as studied in this unit. Use the pictures to prepare a bulletin board display.

14.   Make drawings of early steam-powered cars, locomotives, and tractors. What did these steam engines use for fuel? How many years ago did such machines exist?

15.   Bring to class and demonstrate a toy steam engine.

16.   Make a fire alarm like the one shown at the left. Demonstrate its operation.

SOUND

The following investigations will help you to learn about sound on your own. You can measure several characteristics, such as how fast sound travels in air and the wavelengths of various sounds.

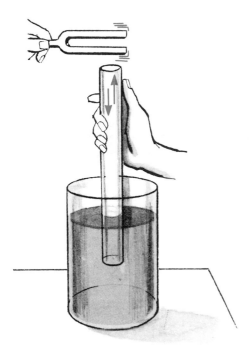

Measuring Wavelength. Measure the distance between two compression waves from a tuning fork, using the apparatus shown at the left. Hold a tuning fork over a glass or metal tube which has one end under water. With the tuning fork vibrating above the tube, move the tube up or down in the water. When you hear sound coming from the tube, you know that the air column in the tube is resonating with the tuning fork. Measure the distance from the water level in the tube to the top of the tube. A compression wave goes down the tube and back for each vibration of the tuning fork. The rarefaction wave also goes down and back. So the distance between waves is equal to 4 times the length of the air column. Measure the wavelength of a tuning fork with a different frequency.

Measuring Speed. The speed of sound can be measured with the same apparatus used above. Each tuning fork has marked upon it the frequency with which it vibrates. For instance, the note A vibrates 220 times per second. When you have measured the length of a column of air that will resonate with a tuning fork of that frequency, you can calculate the speed at which the wave is traveling. Sound travels down the tube and back in the time required for the tuning fork to vibrate once. Test the speed of sound from tuning forks with different frequencies.

Speed of sound = frequency × 2 length of column

EXPERIMENTAL RESEARCH

1. Find out whether the density of air affects the travel of sound waves through it. Put an alarm clock on a sponge rubber pad under a bell jar. Note the sound of the alarm. Pump air from the jar and listen for differences in the sound of the alarm.

2. Tune the string of some instrument, such as a violin, to the same pitch as a tuning fork. Put a small bent piece of paper on the string. Strike the tuning fork and find out if the string resonates to the sound waves sent out by the tuning fork.

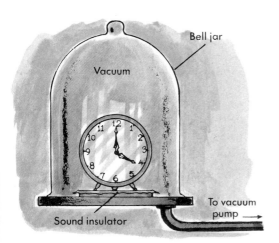

Bell jar

Vacuum

To vacuum pump →

Sound insulator

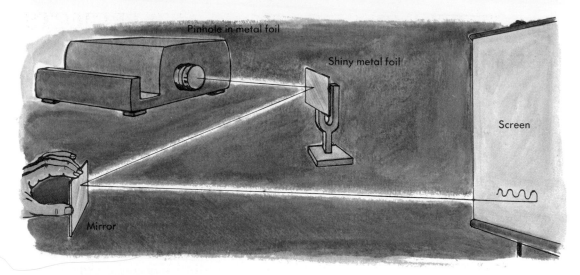

3.   Make a simple oscilloscope as shown above. Fasten a piece of shiny metal foil to one prong of a tuning fork. Cover the lens of a slide projector with foil in which a single pin hole has been made. Shine a beam of light from the projector to the tuning fork. Hold a mirror in the reflected beam and slowly turn the mirror to produce a tracing of the vibrating beam of light on a screen. Test forks of different frequencies.

4.   Study resonance effects of a metal pie plate. Turn the plate upside down and touch different parts of it with the base of a vibrating tuning fork. Locate the points at which the sound is reinforced most strongly.

5.   Find out whether resonance can be produced with a tube open at both ends. Find two cardboard tubes, one of which slides in the other. Hold a vibrating tuning fork over one end and change the length of the tube. If the sound is amplified, compare the length of the open tube with that of a closed tube which resonates at the same frequency.

6.   Hang a small bell in a flask. Boil some water in the flask, remove the flask from heat, and then stopper the flask tightly. Shake the flask and study the sound produced by the bell.

7.   Hang a small bell inside a flask. Shake the bell under water and compare the sound that is transmitted through the water with that transmitted through air alone. Which frequencies travel best through water?

OTHER INVESTIGATIONS AND PROJECTS

1. Make a scrapbook on the instruments of a symphony orchestra showing a picture of each instrument. List the sound generators in each, and describe the method by which the frequency is changed. Include the method by which the sound is amplified if this is done.

2. Take an old auto horn apart and find out how it produces sound.

3. Ask a physics teacher or an electronics expert to demonstrate an oscilloscope connected to a microphone so that it will respond to sound waves.

4. Prepare a report on the types of musical instruments used in ancient times.

5. Find out how sound waves are used to measure the ocean depths. Report to the class, using diagrams to help you make your explanations.

6. Set up the apparatus shown below, and show that compression and rarefaction waves can travel through metal.

7. Plan a visit to a church or other building that contains a pipe organ. Ask an organist to explain how it operates.

8. Read about the invention of the phonograph. Describe the first phonograph to your class, explaining the principle by which it operated.

9. Demonstrate a "howler" by blowing a narrow blast of air over a tube while moving the tube up and down in a container of water.

10. Fill a series of drinking glasses with different amounts of water. Adjust them so they produce the notes of a scale when struck. Play a tune on them.

11. Find out how unwanted echoes are prevented in auditoriums and other large rooms. Explain what happens to the sound energy.

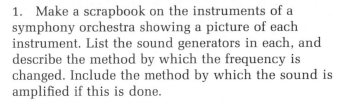

Glass or metal tube

Vacuum cleaner blower tube

"Howler"

Water

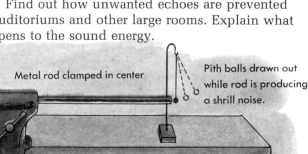

Stroking with damp cloth sets up sound waves in rod.

Metal rod clamped in center

Pith balls drawn out while rod is producing a shrill noise.

LIGHT

The following investigations deal with the nature of light. You will learn how the properties of light can be measured. Some of these suggestions will present you with more questions than they answer, but that frequently happens during scientific investigations.

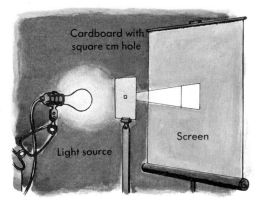

Size of Hole	Distance from Cardboard
2 cm	
3 cm	
4 cm	

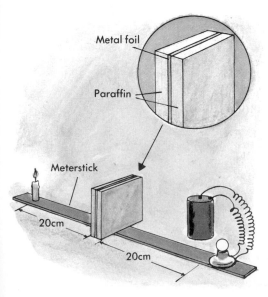

Effect of Distance on Brightness of Light. Investigate this question: Why does a light seem dimmer as you move away from it? Use a flashlight bulb or a high-intensity lamp as a light source. Place the light source at one end of a meterstick. Cut a square centimeter piece of cardboard, and hold it about 10 cm from the light. The cut-out square should be positioned so that the light shines through on to a screen about 10 cm from the square hole. Move the screen toward and away from the hole. What happens to the size of the lighted area on the screen?

To find the relation between the size of the lighted area and the distance of the screen from the hole, you will need to make some measurements. Keep the distance between the light source and the hole the same. Adjust the screen so that the lighted square measures 2 cm on each side. Now measure the distance between the screen and the hole. Record your observations on a chart similar to the one in the margin. Now adjust the screen so that the lighted square measures 3 cm on each side, and again measure the distance between the screen and the cardboard. If you position the screen so that it is twice as far from the light source as from the cardboard, what will be the area of the lighted square on the screen? Why does the size of the lighted square change with distance? Why does the lighted square appear to be dimmer as the screen is moved away from the light source?

Comparing Brightness. Compare the brightness of a light to the brightness of a candle in a darkened room. Use 2 blocks of paraffin about 2 cm thick. Sandwich a piece of aluminum foil between the 2 blocks, and wrap 2 rubber bands around the sandwich to keep it together. Stand the sandwich straight up in the middle of a meterstick. Place a lighted candle exactly 20 cm away. Place the light that is to be tested exactly 20 cm away on the opposite side of the sandwich. Does one of the paraffin blocks appear brighter than the other? If so, which one?

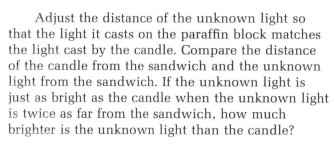

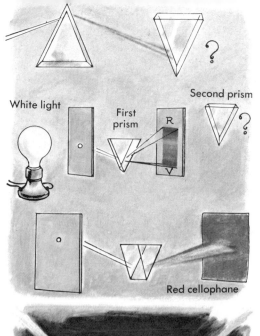

White light
First prism
Second prism
R
V
?
?

Red cellophane

Adjust the distance of the unknown light so that the light it casts on the paraffin block matches the light cast by the candle. Compare the distance of the candle from the sandwich and the unknown light from the sandwich. If the unknown light is just as bright as the candle when the unknown light is twice as far from the sandwich, how much brighter is the unknown light than the candle?

EXPERIMENTAL RESEARCH

1.   Study the light that passes through a prism. Pass the light from one prism through a second prism held as shown above. Use a second slit to select one color of light to be sent through a second prism. Hold colored cellophane in the beam of light entering a prism and then in the beam leaving the prism. Prepare charts of your discoveries.

2.   Use a light meter to measure the light reflected from different types of surfaces. Exhibit the surfaces with proper labels describing their reflective powers.

3.   Study the behavior of light as it falls on water at different angles. Make a cardboard protractor as shown at the left, and tape it to the outside of a flat-sided, transparent vessel. Fill the vessel to the line shown on the protractor. Direct a beam of light at the center point of the protractor. Measure the angles of the refracted and reflected beams. Compare the intensities of the beams as the angles change.

OTHER INVESTIGATIONS AND PROJECTS

1.   Visit someone who makes his or her own reflecting telescopes. Report to your class on the methods of construction.

2.   Prepare an exhibit of materials that are opaque, translucent, and transparent.

3.   Make a periscope and explain its action.

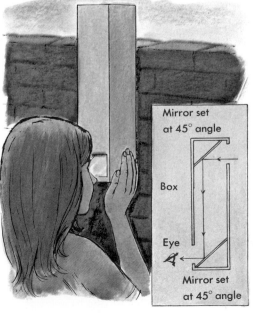

Mirror set at 45° angle

Box

Eye

Mirror set at 45° angle

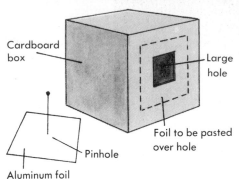

A Pinhole Camera

Cardboard box

Large hole

Foil to be pasted over hole

Pinhole

Aluminum foil

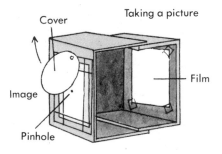

Taking a picture

Cover

Film

Image

Pinhole

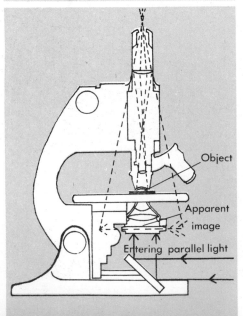

Object

Apparent image

Entering parallel light

4.   Make a pinhole camera like that shown here and take a picture with it.

5.   Investigate the candle power ratings of common sources of light, and give the information to your classmates.

6.   Visit the shop where eyeglass lenses are ground. Watch the process and find out how the desired curvature is obtained.

7.   Take apart the toy known as a kaleidoscope and find out how it operates.

8.   Demonstrate the famous "flame-in-water" magic trick shown below. Place the candle and the drinking glass so that the image of the candle appears to be in the glass. Then pour water in the glass. A screen hides the actual candle from the audience.

9.   Prepare a report on the experiments that Sir Isaac Newton did with a prism and on the conclusions that he drew.

10.   Demonstrate shadows formed by two or more sources of light, as shown at the left. Show what happens when one of the sources is moved. Try the effect of using lamps of different colors.

11.   Read about the operation of binocular microscopes. Explain why enlarged images are formed. Demonstrate the real image by removing the top lens and replacing it with a sheet of waxed paper. Explain how to calculate the magnifying power of the various combinations of lenses.

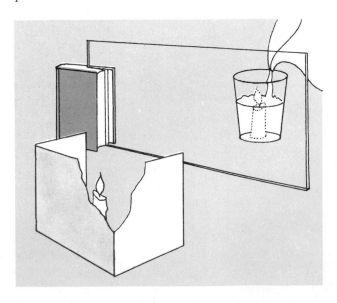

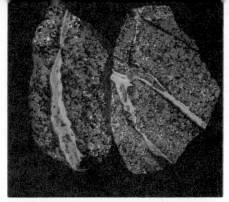

Discoveries About Light

Light is a form of energy. Light can cause some objects to become warmer and in some cases to glow. Plants can use the energy of sunlight to make food, a form of chemical energy. Engineers and scientists have found ways to convert light energy into electrical energy. The energy of sunlight is used to produce electrical energy for satellites and space vehicles.

We understand how most forms of energy travel from one place to another. Simple theories have been accepted for explaining how electrical energy travels as a flow of electrons through a conductor. Other simple theories explain how heat travels through metal, air, water and other materials. But how does light travel through space? This is one of the big unanswered questions in all of science.

This chapter traces ideas about how light travels. You will learn why scientific theories sometimes change, and you will be introduced to current theories about the nature of light.

HOW LIGHT TRAVELS

Some of the oldest records about how we see and hear come from the 2500 year-old writings of the Greek philosophers. Hearing was fairly accurately described as sound coming from a source and being received by the ears. Sight, however, was thought to be "sight rays" sent out by the eyes themselves to explore the objects around them. In support of this thinking, some philosophers pointed out that the ear went inward into the head to receive sound. The philosophers said that the eyes bulged out to send out their "sight rays." Today, this explanation of sight seems to be nonsense. What evidence can you suggest to disprove the sight theory of the ancient Greek philosophers?

Ancient Beliefs. In the 1500's and 1600's scientists developed lenses. From these lenses they fashioned telescopes and microscopes. With these instruments, the scientists were able to see objects previously not seen. With the use of these instruments, scientists started thinking about the nature of light.

Newton's Corpuscular Theory. In 1666, Isaac Newton sent a beam of white light through a glass prism. The prism bent the beam and broke it up into a band of colors resembling a rainbow. As Newton experimented with light, he decided that light must be made up of particles. The path of some of the particles was bent by the glass prism at a greater angle than others; these appeared blue. Those bent the least by the glass prism appeared red. He called the particles of light corpuscles.

Newton based his theory upon several known properties of light. One property is that light travels in straight lines. As you go farther and farther away from a source of light, the light becomes fainter. Imagine that the source is shooting out corpuscles in all directions. When you are close to the light, there are more corpuscles hitting your eyes than when you are far away. How would Newton argue that the light would appear dimmer as you move away from the light?

The reflection of light is another property explained by the corpuscular theory. In Chapter 3,

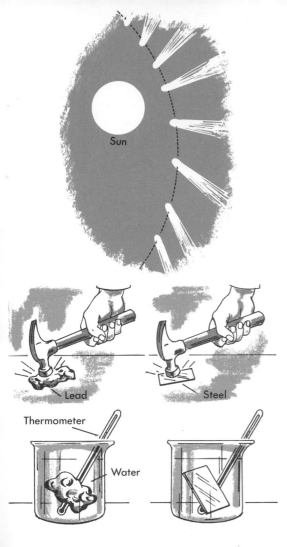

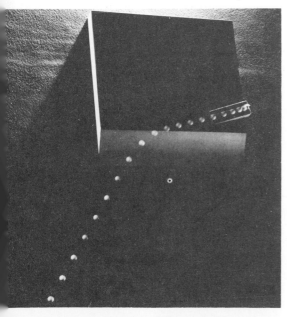

you learned that light is reflected by some surfaces. You found that light bounces from a smooth surface in the same way a ball would bounce. How does this idea help to support Newton's corpuscular theory?

Do corpuscles of light push on the surface upon which they fall? Your experiences with light probably do not demonstrate any push, probably because the push is too small to be noticeable. Very careful experiments with extremely sensitive instruments have found that light does cause pressure. These experiments also show that the pressure of light increases as the light gets brighter. Astronomers believe that a comet's tail always points away from the sun due in part to the pressure of the sunlight on the tail.

It would seem that if particles are striking a surface, they should cause that surface to become warmer. If a surface does not reflect the particles, it should become warmer than a surface that does.

Use a hammer on a piece of lead, striking it 10 times in one minute. Then strike a similar piece of steel the same number of times in one minute and compare the temperature of each material. How can you explain that the lead becomes warmer after the same amount of pounding?

Select two metal cans that are the same size and shape. Leave one shiny and paint the other dull black. Put a thermometer in each can and place them in sunlight for the same length of time. How does the temperature inside each can compare? How does Newton's theory help to explain what is happening to each can?

You might also have experienced the fact that light is bent as it passes through a glass of water. (Refer to the "Rising Coin Trick" in Chapter 3.) Isaac Newton explained this by stating that the corpuscles of light are pulled into the water when they strike the surface. This force makes them travel faster so that they are shifted from the straight path they would normally follow. A model of this bending can be demonstrated by rolling a ball down a tipped board onto a second board tipped at a different angle than the first.

The corpuscular theory seems to explain several observations of light fairly well. Are there some other observations of light which are not well explained by this theory?

What happens when you shine one beam of light across another? Do the two beams interfere with one another? It would seem that if the beams of light are composed of rapidly moving corpuscles, some corpuscles might collide and change the beam. Try shining one light beam across another. Do the beams appear to interact?

Newton proposed that the corpuscles are so tiny and moving so fast that the spaces between them are quite large. The chances of one corpuscle knocking into another are very slim. Whenever a theory needs many changes to account for various conditions, scientists begin to doubt the theory.

Huygen's Wave Theory. The fact that the edge of a shadow is never sharp was difficult to explain by Newton's theory. Christian Huygen demonstrated that an object in the path of a wave did not cast a clearly outlined shadow. He proposed his theory before the French Academy of Science in 1678. The sun, flaming torches, or any light, he said, set up vibrations or waves that traveled outward to the eye of the observer.

Huygen's theory also had problems. Waves such as those on water travel because the material through which they are traveling is moving also. If light traveled as a wave, then how could it travel through a vacuum? Huygen's answer was to point out what had been believed for many years. The ancients had spoken of *ether* which filled the space between the stars and between everything on earth as well. The ether, he claimed, was the material that carried light waves.

Diffraction. However unsatisfactory this explanation of how light travels is, it does offer an explanation of why objects cast shadows with fuzzy edges. Hold an object in the path of direct sunlight or the sharp beam of a flashlight. Note the edges of the shadow cast by the object. If light was transmitted by particles moving in straight lines, the edges of the shadow should be very sharp. Notice what happens to water waves in the photograph as they hit a barrier with a small slit through it. The slit itself acts as a wave generator and waves go out in semicircles from it. This behavior of waves is called *diffraction.*

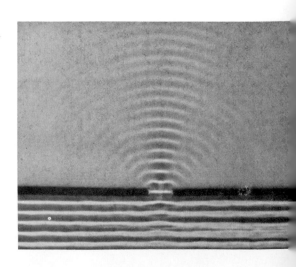

Interference. Water waves can be used to demonstrate still another behavior of waves. If two waves are approaching each other, they seem to

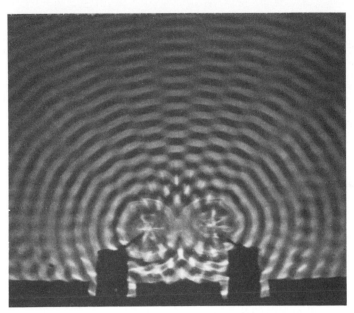

pass through each other and continue undisturbed. A water particle caught in both wave motions will move depending on what part of each wave is acting on it. If the crest of one wave arrives at the same time as the crest of a second wave, the water particle will be pushed up as high as the sum of the heights of the two crests. If the crest of one wave arrives with the trough of the second, the water particle may move little or not at all. The depth of the trough of the first wave subtracts from the motion caused by the height of the crest of the second wave.

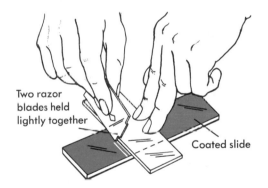

Two razor blades held lightly together

Coated slide

You can demonstrate water waves passing through each other by dipping two fingers into the surface of the water. Move them up and down gently to create waves. The photograph above shows this being done by an electric motor. Notice the pattern made by the waves as they pass through each other. Where in the pattern of the waves is the motion of the waves being canceled? This behavior of waves is called *interference*.

Spray black paint on a glass microscope slide. Using two razor blades held together lightly, scratch two parallel lines through the paint as shown in the drawing. Holding the glass slide up to your eyes, look through the slits at a lighted bulb. The dark and light lines you see are caused by interference. According to the wave theory, what is happening to the dark lines?

Photon = bundle of energy moving at speed of light. The energy of the photon depends on its vibrations, frequency, and wavelength.

More energy = high frequency and short wavelength.

Bright light beam = high density of photons.

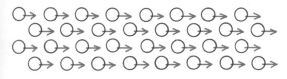

Dim light beam = low density of photons.

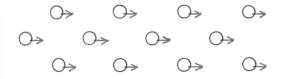

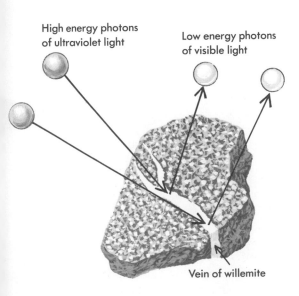

High energy photons of ultraviolet light

Low energy photons of visible light

Vein of willemite

Electromagnetic Waves. In 1860, a noted English physicist, James Clark-Maxwell, developed the electromagnetic theory of light. He stated that light waves were electrical in nature, not mechanical as sound waves are. In addition, he found the speed of certain electromagnetic waves to be the same as the speed of light waves.

Your radio is sensitive to certain electromagnetic radiations. By turning a knob you can select the length of the radio wave that your radio will receive. Your eyes, like the radio, are sensitive to a range of wavelengths of electromagnetic radiation. Of the great range of wavelengths in sunlight, your eyes are sensitive to only a very small range.

Fluorescence. One behavior of light puzzled scientists as late as the 1890's. When ultraviolet light is made to shine on some substances, the substances glow with a light of a different color. This property is called *fluorescence*. Turn to page 271 and examine the photographs of the pair of rocks. Compare the photograph taken in natural light to the one taken in ultraviolet light.

Many organic substances, such as our hair, fingernails, and eyes, emit a bright glow when exposed to ultraviolet light. Some fossils found in sedimentary rocks will fluoresce so brightly that the smallest details of the fossil can be clearly seen.

Modern Theory of Light. The phenomenon of fluorescence remained unexplained until a modern theory of light was introduced. This theory holds that a light beam consists of bundles called *photons*. The brighter the beam the more photons it contains. Each photon contains a certain amount of energy. Those giving our eyes the sensation of blue contain more energy than those we see as red.

Fluorescence may be explained using this theory. An atom may absorb a certain amount of energy if the energy is sufficient to move some of its electrons to orbits of higher energy. That same atom may give up energy when some of the electrons jump back to their previous orbits. The energy is given up in the form of photons. The atoms in the veins of the rocks absorbed energy from the ultraviolet light. Then the atoms gave up some of the energy as photons with less energy than ultraviolet light. In this manner, the light from the veins in the rocks appears as a different color.

SPEED OF LIGHT

If the North Star were to explode, we would not know of it for about 460 years. It takes that long for the light from the North Star to reach us. Light has fascinated us for as long as we have been able to wonder.

Within the short distances we experience on Earth, light appears to be everywhere at once. It does not seem to take any time at all for light to travel even long distances on Earth. Astronomers had no proof that light takes time to travel until they studied the time it takes light to travel across the solar system. Light travels at an incredibly fast rate; in fact, most scientists believe that nothing can travel faster.

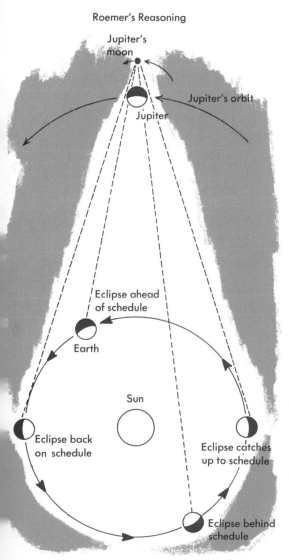

Roemer's Reasoning

Jupiter's moon

Jupiter's orbit

Jupiter

Eclipse ahead of schedule

Earth

Sun

Eclipse back on schedule

Eclipse catches up to schedule

Eclipse behind schedule

Roemer and Jupiter's Moon. In 1675, the French Royal Academy employed a young Danish astronomer to figure out a way for navigators to compare the time of day in one place with the time of day in Paris. Olaus Roemer suggested finding some event among the stars that always occurred at the same time. By noting the time of the event, a navigator could compare the local time with the time when the event was known to occur when viewed from Paris.

The event Roemer selected for study is the eclipse of one of Jupiter's moons. As the moon orbits the planet, it passes behind once every 42½ hours. The moon should follow a timetable of being eclipsed once every 42½ hours. From his careful observations, however, Roemer found that the eclipse lagged farther and farther behind its predicted timetable for six months. Then in the next six months the eclipse occurred more quickly until it was back on schedule. Roemer also noted that the greatest lag occurred when the earth was at its greatest distance from Jupiter. Where in its orbit is the Earth when it is farthest from Jupiter?

Roemer's explanation of the lag in time between predicted and observed eclipses was that it took the light more time to travel over a longer distance. The light took longer and longer to reach the Earth as the two planets moved farther and farther away from each other. The greatest lag of 1320 seconds, Roemer concluded, was the time it took light to travel across the diameter of the Earth's orbit.

Although Roemer's calculations are not very close to modern estimates, he was the first person who tried to measure the speed of light. In fact, he was the first person to demonstrate reasonably that light did travel at a measurable speed. What calculations could have produced the errors in Roemer's estimate of the speed of light?

The Speed of Light Measured on Earth. Once Roemer established that light does travel at a measurable speed, a new task confronted scientists. They had to find a method of measuring the speed by an experiment performed on the earth's surface.

Galileo had suggested that two persons on hills about 2 kilometers apart, each holding a lantern and a timing device, could measure the speed of light. The first person marked the time when the shutter of the lantern was opened so that the second person could see the light. As soon as the second person saw the light, that person noted the time and flashed the second lantern. The two times could then be compared. With the distance between the persons determined, the speed could be calculated. Why do you think this method would not work?

Armand Fizeau, a French physicist, used mirrors to shine a light beam through the spaces between the teeth on the rim of a wheel.

A beam of light was sent from the source to a mirror (B) which reflected the light to another

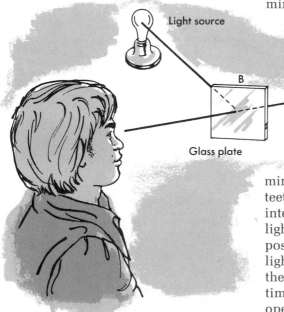

Light source

B

A

Distant mirror

Rotating wheel

Glass plate

mirror (A). Between A and B was a wheel with teeth on its rim. As the wheel turned, the teeth interrupted the light, resulting in short flashes of light. When the wheel was still and in the right position, the observer could see the image of the light source in the space between two teeth. When the wheel was turned faster and faster, there came a time when the light flash passing through one opening returned from mirror A just in time for a tooth to interrupt it. The observer saw nothing. As the spinning became faster, a time came when the observer could see the beam. In the time it took the light to go from B to A to B, the wheel had turned just enough so that the reflected beam passed through the next opening between the teeth.

The wheel had 720 teeth and the light could be seen when the wheel was spinning at 25 revolutions per second. The time needed for the light to travel from mirror B to mirror A and back again was equal to the time it took the wheel to turn from one opening between two teeth to the next opening. That distance is 1/720 of a turn at 1/25 of a second. The light traveled from B to A and back again in 1/720 × 1/25 or 1/18 000 of a second. The distance between the mirrors was 8.67 kilometers. However, the light went back and forth so it traveled a total distance of 17.34 kilometers in 1/18 000 of a second, or about 310 400 kilometers per second.

Since Armand Fizeau's experiments, many refinements have been made and more accurate measurements have been possible. Albert A. Michelson used a revolving mirror instead of a toothed wheel, and measured the distance between the mirrors with greater precision. He determined the speed of light to be 299 730.9 kilometers per second, and the error is calculated to be only plus or minus 3.2 kilometers per second.

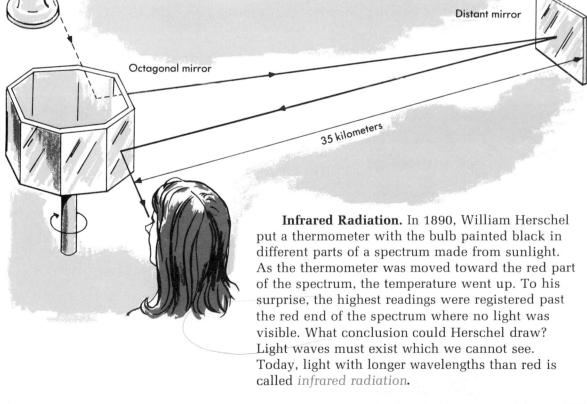

Light source

Octagonal mirror

Distant mirror

35 kilometers

Infrared Radiation. In 1890, William Herschel put a thermometer with the bulb painted black in different parts of a spectrum made from sunlight. As the thermometer was moved toward the red part of the spectrum, the temperature went up. To his surprise, the highest readings were registered past the red end of the spectrum where no light was visible. What conclusion could Herschel draw? Light waves must exist which we cannot see. Today, light with longer wavelengths than red is called *infrared radiation*.

The shore and harbor
at Castine, Maine.

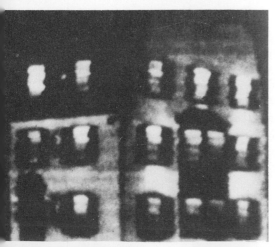

Infrared radiation is given off by all warm objects. Some chemicals change when exposed to infrared rays. Chemists use these chemicals to make photographic paper which can take pictures using infrared and visible light.

Infrared radiation is not scattered by water molecules in the air. Clear pictures of the earth's surface can be taken even through haze and fog. Study the two photographs on this page taken of the same area. The top one was taken with daylight film and the next one with infrared film.

In these days of energy shortages we are particularly concerned about heating. The photograph below taken with infrared film shows where heat is escaping from the building. What step can be taken to conserve the heat in this building?

Ultraviolet Light. If at the red end of the visible spectrum there is radiation that our eyes cannot see, do you suppose there is light at the blue end that we cannot see? It has been found that ultraviolet light exists there which has higher energy than infrared. Sunburns are caused by the ultraviolet rays which penetrate the atmosphere. Fortunately, not much of the ultraviolet light from the sun reaches the surface of the earth. Ultraviolet light is very harmful to living things. Ultraviolet lamps are used in hospital operating rooms to kill whatever germs might be present.

SUMMARY QUESTIONS

1. What did Greek philosophers think about the nature of light?
2. What was Newton's theory about the nature of light?
3. How did Newton explain the formation of the spectrum?
4. What properties of light were not explained by Newton's theory?
5. What did Huygen propose about the nature of light?
6. What is the theory used today to explain the nature of light?
7. How did Roemer establish that light travels at a measurable speed? How did he calculate that speed?
8. How can the speed of light be measured on earth?
9. How are infrared and ultraviolet light useful?

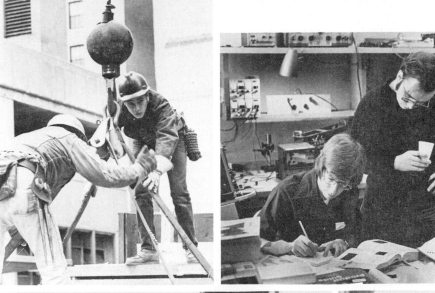

Careers
and Learning Opportunities

The persons photographed on this page are all interested in some aspect of physical science. Note that the persons are of different ages. Each one made important decisions and plans in order to be taking part in his or her particular activity. Some of the people have studied physical science for several years to become expert enough to perform important tasks upon which others can rely. Other people are just beginning to develop their interests. This chapter introduces you to a variety of opportunities available if you are interested in the field of physical science.

IMPORTANCE OF EDUCATION

The wise student plans to get as much schooling as possible. A broad educational background has many advantages. Changing techniques of manufacturing, plus the rapid development of entirely new devices, requires people who can adapt. Usually, persons with more education are the ones who can adapt best to changes. Although an advanced education can be costly, there are various scholarships, grants, and loans which are available. Obtain information about financial aid from the guidance office in your school.

Percent Distribution Rates in the Labor Force of Persons 16 Years Old and Over, by Years of Schooling Completed: 1977

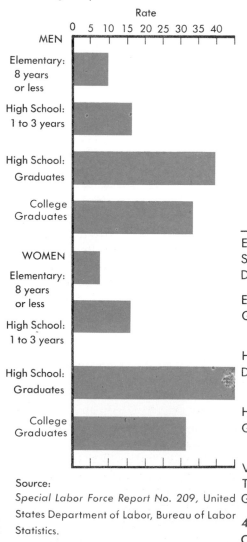

Source:
Special Labor Force Report No. 209, United States Department of Labor, Bureau of Labor Statistics.

Occupational Advantages from Education. The amount of education you receive will influence the amount of money you are likely to earn during your lifetime. American workers who have completed four years of high school outnumber those who did not get their high school diplomas. As a group, skilled workers and supervisors with a high school education look forward to lifetime earnings of about 20% more than people in the same positions who did not finish high school. In addition, the people who didn't finish high school are often the first to be laid off during hard times. Study the graph.

Chart of Lifelong Income Compared with Education

	Lifetime Income	Difference
Elementary School Dropout	$300 000	
Eighth Grade Graduate	$405 000	The 8th grade graduate earns 35% more than the elementary school dropout
High School Dropout	$466 000	The high school dropout earns 15% more than the 8th grade graduate
High School Graduate	$560 000	The high school graduate earns 20% more than the high school dropout
Vocational Training Graduate	$608 000	The vocational training graduate earns 8.5% more than the high school graduate
4-Year College Graduate	$748 000	The 4-year college graduate earns 23% more than the vocational training graduate

Education has a direct relationship on the availability of job openings. Students who finish high school find that they can become employed with less difficulty.

Lifetime Advantages from Education. People who have developed their abilities to reason and to think can lead a richer, fuller, and more satisfying life. Education encourages you to learn, to observe, and to evaluate. A good education means being able to consider and judge the ideas and opinions of others. The experience of getting along with your teachers and classmates will help you in the future to enjoy associations with your employer, co-workers, and neighbors.

OCCUPATIONAL OPPORTUNITIES IN SCIENCE

Junior High School Science Opportunities. Study these photographs of junior high school students experiencing some aspect of physical science. Name the branch of physical science each is experiencing. What are some of the careers suggested? What are some hobbies related to physical science?

Find out about the careers and hobbies your parents or relatives have chosen. When did they decide upon a career? What decisions did they make? What experiences or events influenced them? What training and background do they consider necessary? Survey your classmates for the hobbies and careers of their parents. Prepare a chart showing the number of parents in careers such as science, industry, medicine, sales, services, government, commerce, transportation, and so on. List the hobbies parents have and how many participate in each.

Discuss science related activities in which junior high school students can participate, such as photography, a weather club, repairing and rebuilding old motors, clocks or radios. Prepare a chart of these activities, including some comments about each one. Some comments might include where to obtain material, names of people interested in forming a club, meeting times and places of existing clubs, and persons to contact for more information.

Occupational Opportunities for High School Students. The question of what to do after high school is on the minds of most students. Information about occupations does not always produce satisfactory answers. One approach is to sample an occupation by becoming a helper to someone in an occupation. Even unpaid assistants can experience what a particular occupation requires in training, skill, and time.

Study the photographs on this page. Which show high school students "trying out" a possible future occupation? Which activities provide experiences and skills which might be useful in the future for developing a hobby or for more pleasurable living?

Many schools have separate departments, and some towns and cities have separate schools to prepare students for technical careers. What kinds of science and technical courses are available in your school system?

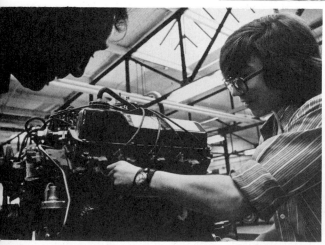

Science Careers for High School Graduates. Millions of workers in the United States provide services for others. These are the people who maintain law and order, work in hospitals, give haircuts and beauty treatments, and see to it that the public is served in hotels, restaurants, and so on.

Other high school graduates may enter directly into semiskilled jobs. Semiskilled workers make up the largest occupational group, and work mostly with their hands. They use, repair, and maintain their own tools, and often keep records of their work. They receive on-the-job training, but do not need to spend much time in learning their jobs. Because their specific tasks may frequently change, the ability to learn new jobs quickly is an important asset to semiskilled workers.

Many jobs require a high school diploma plus special skills or several years of experience. Such jobs are often more interesting and pay more than jobs without the extra requirements. When considering a job, think of the kind of experience it will provide. Will the experience enable you to accept a more challenging and interesting position?

There are other considerations to be made when looking for a job. Read the help-wanted ads on these pages and consider the following questions. How long will the job last? Is the job seasonal? Does it involve building something which, when it is completed, will mean looking for another job? What will the job teach me to do? Will the skills I develop in this job enable me to apply for a better position in the future? Will this occupation be expanding in the future and need more workers? What changes might occur in the job requirements and what new types of jobs might be created?

A few reasons for choosing a job are (1) good pay, (2) working conditions, such as a chance to travel or work outdoors, and (3) a chance to learn important skills. What are some other important reasons?

Bring to class the help-wanted section of a newspaper. The Sunday paper from a large city frequently has many pages of such ads. Cut out several ads which might interest you if you were a high school graduate. Compare the advantages and disadvantages of different jobs. Display the ads along with the advantages of each job.

Careers for Vocational-Technical School Graduates. Many technical occupations are closely related to the professions. People in these occupations work with engineers, scientists, mathematicians, physicians, and other professionals. Among the job titles are engineering aide, programmer, and X-ray technologist. These occupations require a combination of science knowledge and special techniques. People in these jobs receive training in technical institutes, colleges, or on the job.

People in these occupations perform important functions. They help to make the ideas of scientists and engineers available to the public. They operate, repair, and build most of the machines which maintain our standard of living. Their jobs require that they use scientific and mathematical theory to develop practical tools and appliances.

Look at the photographs of technical-vocational school graduates and read the help-wanted ads. Which jobs interest you? Where could you find someone to talk to

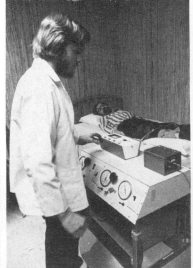

who is doing this kind of work? What promotions are possible in these jobs?

Science Careers for College Graduates. Persons who select a 4-year college will concentrate on one field of interest in the final two years of college. Students who study science, math, or engineering will graduate with a Bachelor of Science degree.

If you have a strong desire to build something that actually works, or design something new, or improve our communication systems, or work with technical things, you may want to consider a career as an engineer. Engineers design and build bridges, superhighways, cars, planes, appliances, office buildings, factories, and so on. Engineers develop new sources of energy and put energy to work.

Look at the pictures of four-year college graduates at work. Read the help-wanted ads on this page. Which type of work interests you? What other science-oriented careers are there? What are some ways you could learn more about careers in science?

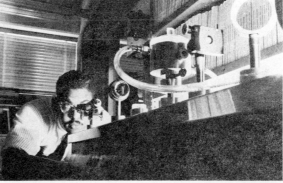

Scientists at Brookhaven Laboratory studying atomic particles.

Astronomer in cage of telescope.

The Future for Careers in Science. The nature and size of major industries have undergone drastic changes. Some industries which employed many workers have all but disappeared, while other industries have grown very rapidly. Aircraft and chemical industries which once grew very rapidly have slowed down, while the electronics industry has grown at an ever-increasing rate. Industries which are just getting started today may become major employers in the future.

Farming is an industry which has had the most dramatic changes in the United States, largely because of technological and biological advances. A century ago, 50% of the workers in the United States were employed on farms. Today, farms employ only 4% of the workers, and they are producing more food than ever. A century ago, one farmer produced enough food for six people. Today, one farmer produces enough food for about 40 people. How has this change become possible? How does machinery help farmers produce more food? What kinds of jobs might the displaced farm workers have taken?

Technological changes refer to industries that become more automated. Jobs which were performed by workers are now performed by machines. Fewer production workers are employed, but more engineers, machinists, maintenance help, and other technicians are needed. In an industry where technological changes are taking place, what is the advantage of an education to the worker?

Most persons face several job changes during their working careers. No longer can you expect just one occupation to last a lifetime. What are some reasons that might cause a person to change jobs? To be able to switch from one specific job to another, a person must have an educational background broad enough to understand the training and retraining necessary for performing new tasks. Young people who have learned a skill and have a good basic education will have a better chance of obtaining interesting work, good wages, and steady employment. The top priority for you today, therefore, should be to acquire as much education, skills, experience, and training as your abilities permit.

SUMMARY QUESTIONS

1. What are three advantages in having as good an educational background as possible?
2. What are three sources of information regarding occupations and careers?
3. Where can you find out about financial assistance for education after high school?
4. Name some industries which have changed in the last 45 years. Propose some explanations for the changes.

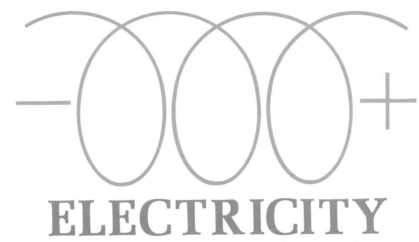

ELECTRICITY

Electric Circuits

A new house today may look similar on the outside to a house built in 1900. Inside, however, there are many differences. One important difference is the number of electrical devices inside the house. Another difference is the number of outlets in the walls and the number of wires inside the walls going to these outlets. A house built in 1920 often had one outlet per room. A house built in 1980 usually has one outlet every 2 meters along each wall!

Electrical devices are used for many purposes. They all are wired similarly. Every electrical device needs a source of electricity such as a battery or a wall outlet. A conductor such as

1

copper wire conducts electricity to the electrical device, such as a fan, toaster, iron or bulb. Because these devices reduce the flow of electricity, they are sometimes called resistances. Then a second conductor conducts the electricity back to its source, the wall outlet or battery. This complete path for electricity is called a circuit.

In this chapter, you will learn about circuits by making them yourself. The source of electricity in these circuits is always 1 or 2 dry cells. You cannot get shocked using dry cells in these circuits.

WIRING A MODEL HOUSE

Models are useful when making a study of electricity in a house. The wiring in real houses is placed within walls and ceilings where it cannot be seen. The wiring in a model house can be left in the open where everyone can trace the circuits.

Building a Model. A model house for the study of electrical wiring need not be large and complicated. A wooden box or a cardboard carton divided into rooms is suitable, or a doll house may be used. On the other hand, ambitious individuals may construct a scale model, complete with foundation, sills, joists, and rafters.

The model shown below represents a slice cut from a house. This type of model requires little lumber and is easy to build. It takes up little space when finished.

The walls, floors, and ceilings are made of lumber about 2 centimeters thick. The back of the model is covered with thin plywood or heavy cardboard.

LEARNING BY DOING. A person can learn much from books and pictures, but he or she usually learns more and understands better when working with actual materials in real situations. That is why this textbook suggests so many experiments, field trips, and models. You will find science much more meaningful when you carry out these activities.

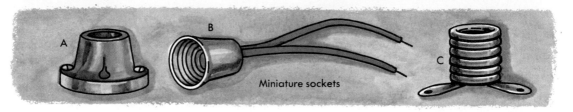

Miniature sockets

Miniature lamp

Socket symbol

Lamp symbol

Same circuit below
using symbols

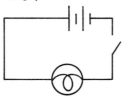

Lamps for the Model House. Flashlight bulbs make good electric lamps for a model house. The kind with screw bases like that shown at the left is the *miniature* size. Buy the type used in two-cell flashlights.

Lamp Sockets. Miniature-size sockets are needed for miniature-size lamps. Three types are shown above; all have screw bases.

Type *A* is especially useful. It has screws for making connections and holes for fastening the sockets to walls or ceilings.

Type *B* can be cut from Christmas tree lights. Connections are made by twisting the ends of wires together as shown below.

Type *C* is the cheapest of the three. However, the wires must be soldered in place because there are no screws for connections.

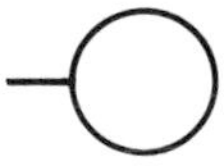

The Energy Supply. Use two dry cells for the source of electrical energy. Connect the cells in series as shown below. With this type of connection, the cells provide electricity having 3 volts of energy.

The path of electricity from its source through a conductor, then a resistance and again through a second conductor back to its source is called a *circuit*. Many circuits contain switches. When the switch is "on" or closed, there is a *closed circuit*. When the switch is "off" or open, there is an *open circuit*.

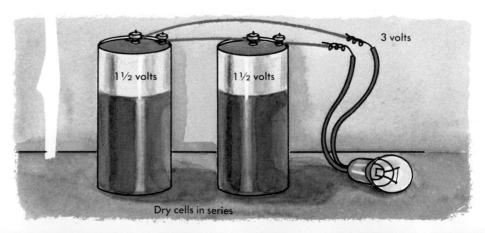

1½ volts   1½ volts   3 volts

Dry cells in series

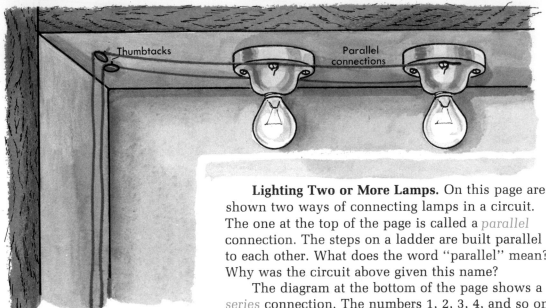

Thumbtacks

Parallel connections

Lighting Two or More Lamps. On this page are shown two ways of connecting lamps in a circuit. The one at the top of the page is called a *parallel* connection. The steps on a ladder are built parallel to each other. What does the word "parallel" mean? Why was the circuit above given this name?

The diagram at the bottom of the page shows a *series* connection. The numbers 1, 2, 3, 4, and so on are an example of a series. What does the word "series" mean? Why is this type of circuit given this name?

Experiments with Connections. Set up a circuit with two lamps in parallel. Unscrew one of the lamps. What happens? Screw the lamp back in again. What happens?

Try three lamps in parallel. Unscrew different bulbs and notice the others.

Set up a circuit with two lamps in series. Compare their brightness with that of lamps in parallel. What happens when you unscrew one lamp? When you screw it back in again? Try three lamps in series.

Connections for the Model House. Do you think that parallel or series connections are used for the lamps in your home? Why?

Are light switches connected in parallel or in series? Why?

Wire each room in the model separately. Use either a series or parallel connection, whichever you think is better. Then test the wiring in each room with two dry cells connected in series as shown on page 293.

Later you will be shown how to connect the different room circuits together.

Diagram of a parallel connection

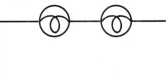

Diagram of a series connection

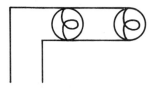

Series connections

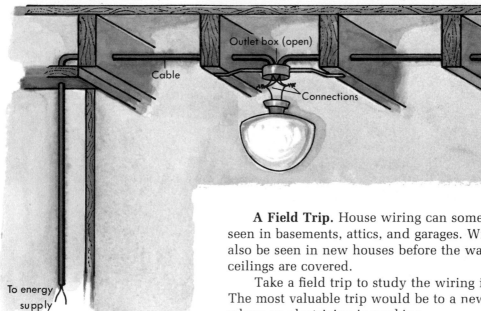

Outlet box (open)

Cable

To another lamp

Connections

To energy supply

Diagram of this wiring

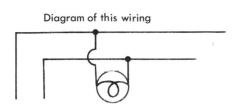

Most electric cable like the one below has a third conductor, called the ground wire, attached to the outlet boxes. Then if either current-carrying wire accidentally touches the box, a complete circuit is formed. How does this circuit protect you?

A Field Trip. House wiring can sometimes be seen in basements, attics, and garages. Wiring can also be seen in new houses before the walls and ceilings are covered.

Take a field trip to study the wiring in a house. The most valuable trip would be to a new house where an electrician is working.

Electric Cable. It may seem that there is only one wire going to some of the lamps. This is because the necessary wires are inside what is called *electric cable*, which strengthens and protects the wires.

Ask for a short piece of electric cable and take it apart. Notice how carefully the wires inside are protected.

The voltage of the electricity in most homes is 120 volts. This electricity has much more energy than that of dry cells. Therefore, the wires must not become wet or touch each other. That is why an electric cable contains so many layers.

Outlet Boxes. All connections are made inside steel *outlet boxes* which are then screwed shut. The boxes protect the connections and keep people from touching the wires. (**CAUTION:** Do not open outlet boxes or try to make connections in the circuits of your homes. The energy of this electricity is great enough to kill you. An electrician knows how to handle electrical wiring properly.)

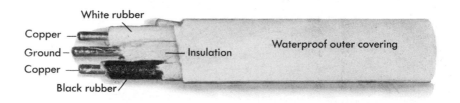

White rubber

Copper

Ground

Copper

Black rubber

Insulation

Waterproof outer covering

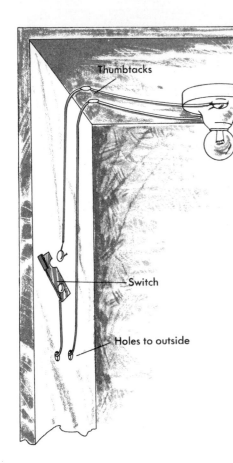

Thumbtacks

Switch

Holes to outside

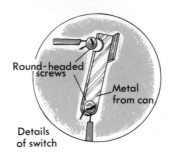

Round-headed screws

Metal from can

Details of switch

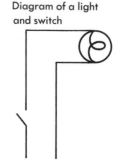

Diagram of a light and switch

Switches for the Model House. Switches are needed to turn lights on and off in the model house. Small switches can be bought in electrical supply stores or they may be taken from old radios. Simple switches can be constructed from round-headed screws and thin metal as shown above.

Note how a switch operates. When the metal strip touches both screws, electricity can flow through it from one wire to the other. When the metal strip is moved away from one screw, electricity can no longer flow. A switch "makes" and "breaks" a circuit.

Build a switch for each room of the house and connect it so that it will control the lights in the room. The diagrams on this page show how to make the connections.

A Circuit Tester. The diagram below shows a useful device called a circuit tester. The bare ends of wires A and B are placed on two electrical conductors. The lamp shows whether or not there is a circuit between the conductors.

Test each switch used in the model house to make sure that it is constructed properly.

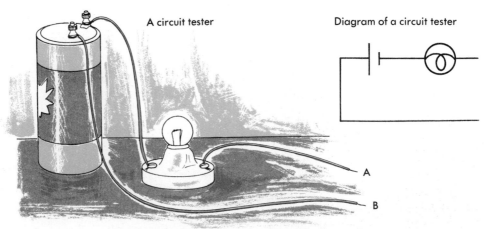

A circuit tester

Diagram of a circuit tester

A

B

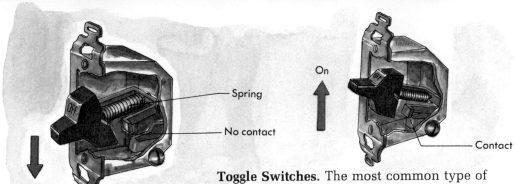

Spring

No contact

Off

On

Contact

Toggle Switches. The most common type of wall switch is called a toggle switch. A toggle switch is shown below together with the steel switch box that protects the wiring.

Study the cut-away views of a toggle switch above. Wires are attached to the screws which lead to metal strips inside the switch. When the knob is down, there is no connection between these strips. When the knob is up, a metal bar is pressed against the strips, carrying electricity from one strip to the other. A spring in the toggle switch keeps the knob from stopping halfway between the "on" and "off" positions.

Other Kinds of Switches. Several kinds of switches can be found in homes. These include push-button switches, pull-chain switches, and turn-key switches. All of these switches operate much like toggle switches; a metal bar is pressed against two strips to make a connection and is moved apart to break the connection.

Collect switches and bring them to school for study. Test them with the circuit tester which is shown on the opposite page. Take some of the switches apart and study the parts inside.

Switch Boxes. Switch boxes are made like outlet boxes. Connect a switch box, a switch, and a piece of cable properly. Test this circuit with the circuit tester. (**CAUTION:** Do not connect the switch circuit to 120-volt electricity.)

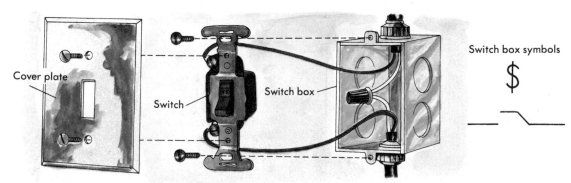

Cover plate

Switch

Switch box

Switch box symbols

$

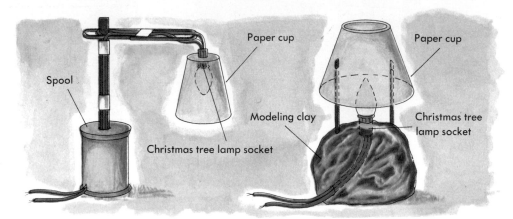

Spool

Paper cup

Paper cup

Christmas tree lamp socket

Modeling clay

Christmas tree lamp socket

Model Lamps. Two types of lamps for the model house are shown above. Other types may be invented as needed.

Sockets from strings of Christmas tree lights are especially useful for making model lamps. Bases for the lamps may be made from wood, spools, and modeling clay.

The model lamps shown above have two wires leading to each although real lamps seem to have only one wire each. Actually, two wires lead to all lamps but the wires may be within a single covering. Take apart some discarded lamps and find out how connections are made. Trace the circuits with a circuit tester.

Floor Plugs for the Model. Below is shown a way to make floor plugs for the model house. Take the covering from 5 centimeters of wire and wind the bare wire around a small nail as shown below at the right. Remove the nail, leaving a coil. Two such coils are needed.

Mount each coil in a hole in the wall of the room. To plug in a lamp, push the end of each lead wire into a coil.

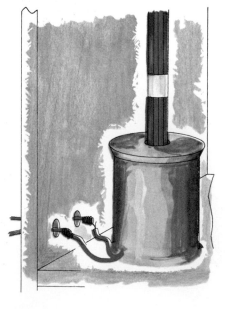

Electric Plugs. Make a study of different kinds of electric plugs. Some types can be taken apart and the wiring traced. The diagram at the right shows how the wires are connected in one common type of plug.

Be sure to notice that there are two wires leading to lamps and other electrical devices. Note the knot in the plug. How does this knot reduce the chances of a short circuit in the plug?

The wire cable used to wire a house is usually a single, solid piece of wire. The wire in lamp cords and extension cords is usually made of many small wires. Why is it more convenient to have extension cords made this way?

Outlets for Plugs. A wall outlet for an electric plug is shown below. When the metal prongs of a plug are pushed into the outlet, they make contact with metal strips which are connected to the electric lines.

Wall outlets are fastened inside steel boxes to protect the connections and the wires. This is the only safe way to have outlets. Some people install other kinds of outlets, but this is dangerous and is forbidden in most cities and towns.

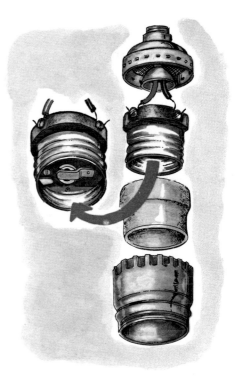

Lamp Sockets. Take apart some lamp sockets and trace the circuits to find out where the electricity flows. Notice the materials that are used to keep the two electrical paths separated. These materials are called "insulators." List the different insulators that you find.

Many lamp sockets have switches inside. You may find sockets with a button to push, a knob to turn, or a chain to pull. Study these to find out how the connections are made and broken. Use a circuit tester to trace the circuits.

Wiring symbol for electrical outlet

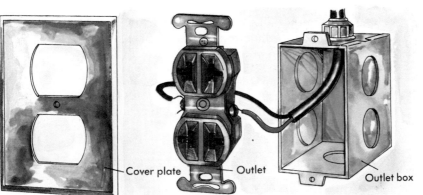
Cover plate   Outlet   Outlet box

To kitchen lights

To basement lights

Master switch

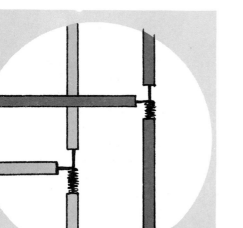

Detail of connections

Installing the Main Line. Directions have been given for wiring each room and testing the circuits. It is now time to connect the rooms to a pair of wires leading from the source of electrical energy.

Stretch two wires up one side of the house, across the top, and down the other side. Connect the bare ends of these wires to round-headed screws. Keep the wires about 3 centimeters apart to make connections easier.

Wherever two wires come from a room, connect them to the main line as shown here. Note that the insulation must be removed from the wires before the connections are made. The connections hold better if they are soldered.

Connect two dry cells in series to the two screws at one end of the main line. All the lamps in the house should now operate.

The Master Switch. Each home has a master switch that can be shut off while repairs are being made to the circuits. Install a master switch in the main line of the model house.

Diagram of main line and master switch

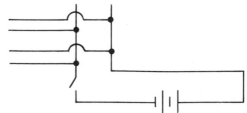

A three-way switch

Diagram of
3-way switch

A Special Circuit. Although the model is now operating, many other circuits can be added. This page describes a circuit that allows a lamp to be turned on and off from two different places. These special switches are called three-way (or double-pole) switches. Are there any circuits like this in your own home? Where?

Three-way switches are needed for this special circuit. A three-way switch makes contact with either of two wires depending upon the direction in which the knob is pushed. The diagram above shows how to make three-way switches with the use of metal strips and round-headed screws.

The colored lines show the wiring. The wires may be placed around the outside of the model for a neater job. For home wiring, a special cable (three wires plus a ground wire) is used in this type of circuit.

When using three-way switches in the model house, always leave the metal strip in contact with one of the screws. Why?

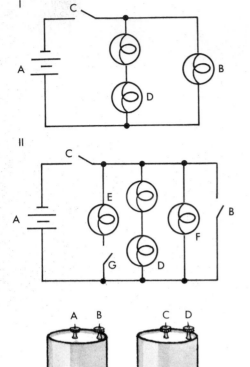

REVIEW QUESTIONS

1. Study circuit I. What does the symbol at A indicate? At B? At C?
2. Compare the brightness of the bulbs when switch C in circuit I is closed.
3. Study circuit II. Compare the brightness of the bulbs when switch C is closed.
4. Compare circuits I and II when switch C is closed in both circuits.
5. Describe the brightness of the bulbs in circuit II when switches B and C are closed.
6. What changes will occur in circuit II if bulb D burns out?
7. What connections should you make so that the two dry cells at the left are in series?
8. What is the voltage of the electricity supplied by one dry cell?
9. What is the voltage of the electricity supplied by two dry cells in series?
10. What is the voltage of the electricity in most homes?
11. Why are the lamps in homes connected in parallel rather than in series?
12. What does an electric switch do?

THOUGHT QUESTIONS

1. Draw a diagram to show how you could turn on a light from three different switches.
2. Most towns require thicker wires going to outlets in kitchens than for wires in other rooms. Propose a possible explanation.
3. Note the two batteries in the photograph. One is labeled 1½ volts and the other is labeled 6 volts, yet each consists of four smaller cells. How is this possible?
4. Describe the brightness of each light in the circuits below. (Assume all bulbs and batteries are the same types.)

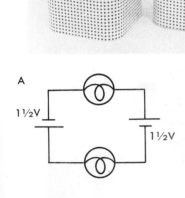

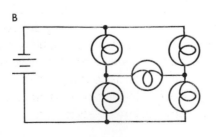

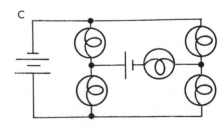

302

Producing Electricity

The study of electricity is often divided into two parts, static electricity and current electricity. In static electricity, electrons in matter are static (do not move). In current electricity, electrons flow, usually through a wire.

The effects of static electricity are usually noticed on materials like paper, rubber, plastic, and clothing. Parts of the surface of these materials may gain or lose electrons. As a result, the surface attracts or repels other nearby objects.

The effects of current electricity are usually noticed on materials like copper wire, through which electrons readily flow. An electric current flowing through a wire always produces some heat and magnetism. These effects have been used to produce hundreds of useful electrical devices. Today, thousands of people work in the electrical industry. Careers include wiring houses, telephone and appliance service, equipment repair, assembling television sets, and developing better ways of producing current electricity. Although current electricity is more useful to us than static electricity, a knowledge of static electricity will help you understand the nature of matter and current electricity.

STATIC ELECTRICITY

Sometimes you feel a shock after sliding across the seat and getting out of a car. Other times you take off a wool sweater and get a shock or hear a crackling sound. These effects are caused by static electricity. This electricity is produced when two different substances are rubbed together.

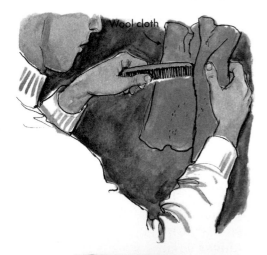

Wool cloth

Small bits of paper

Producing an Electric Charge. Lay a sheet of paper on a desk and rub it with a pencil as shown above. Notice what happens when the paper is picked up. It is now electrically charged.

Place the paper against other surfaces such as the wall, a chalkboard, a door, or window glass. Rub the paper and find out if it can be made to cling to these surfaces.

Rub the paper with other objects such as a comb, a pen, or a book. Make a list of the objects tested, and record whether or not each can produce a charge on paper.

Charging a Comb. Rub a rubber comb briskly against a woolen sweater, shirt, or skirt. See if the comb can pick up tiny bits of paper.

Try charging pairs of materials such as nylon and plastic, or silk and glass. Make a list of the pairs of materials which were rubbed, and record the ability of each material to pick up tiny pieces of paper.

The teen-ager in the photograph on page 305 charged the comb by rubbing it through her hair. What is being attracted to the comb?

Discharging a Comb. When a charged object can no longer attract, it is said to be *discharged*.

Bring a charged comb near a metal object. Then test the comb's ability to pick up the paper. Does the comb still have a charge? Find other ways that it can be discharged.

Charges that Repel. Hang a pair of inflated balloons from short lengths of thread as shown at the right. Rub the balloons with wool. How do the balloons behave? Draw pictures of the balloons before and after being charged.

Find out what happens if only one side of each balloon is charged.

Cut a strip of newspaper about 5 centimeters wide. Hold the center of the strip with one hand and pull it quickly through the fingers of your other hand. How do the strips of paper behave?

SUMMARY QUESTIONS

1. What pairs of items were attracted to each other?
2. What pairs of objects were repelled?
3. Which pairs of objects produced large static charges?
4. Which pairs produced little or no charge?
5. What are some variables which affect static charges?

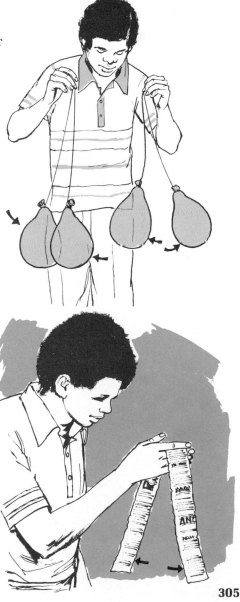

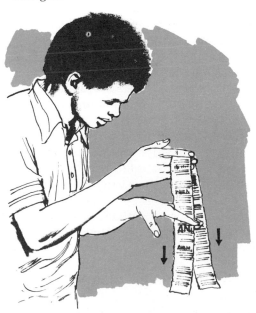

THEORY OF ELECTRIC CHARGES

Scientists have noticed many properties of matter that cannot be explained by observations and experiments. Instead, the idea of atoms, the smallest pieces of matter, have been developed. Then each time matter was observed doing something unexplainable, another property was developed for those atoms to explain the new observations. In this way a theory about the properties of atoms has been developed. These ideas are called a theory because no one has yet seen an atom. The theory is good because it is useful. Thousands of observations of hundreds of different substances can be explained by using the atomic theory. Your observations of statically charged objects also can be explained this way.

HYDROGEN

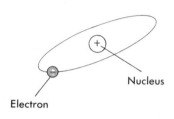

Nucleus

Electron

CARBON

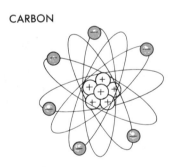

IRON

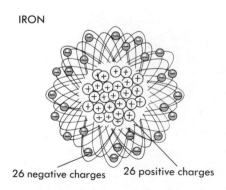

26 negative charges   26 positive charges

Models of Atoms. Models of three atoms are shown at the left. These models were not made to show what atoms look like; no one has yet seen an atom. Instead, the models were made to explain what is known about the parts of atoms, their positions, masses, and electric charges. The models can be and probably will be changed as new information is obtained.

These models describe the atom as having a *nucleus* and one or more *electrons*. The nucleus is believed to be small but dense. Nearly all the mass of the atom is in the nucleus. Electrons are described as being lighter particles traveling in orbits around the nucleus.

The nucleus of an atom is believed to be positively (+) charged. An electron is believed to be negatively (−) charged. Electrons are held in orbits by an attractive force between opposite charges.

Study the models at the left. How many electrons does the hydrogen atom have? How many positive charges are pictured in the nucleus of the hydrogen atom? How does the carbon atom differ from the hydrogen atom? Which atom would have the greatest mass?

Neutral and Charged Atoms. Atoms may have the same number of positive and negative charges. Such atoms are called *neutral atoms*. Sometimes atoms gain or lose electrons. Such atoms are called *charged atoms* or *ions*.

If an atom has more negatively charged electrons than there are positive charges in the nucleus, the atom is *negatively charged*. If an atom has fewer negatively charged electrons than there are positive charges in the nucleus, the atom is *positively charged*.

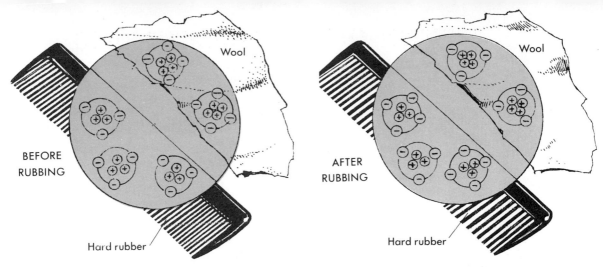

BEFORE RUBBING

Wool

Hard rubber

AFTER RUBBING

Wool

Hard rubber

The effects of static charges are most noticeable on dry days. Charged regions discharge quickly in moist air.

The atoms in the drawings are oversimplified.

Nature of Charged Objects. Models of atoms are shown within the circles on this page. Count the negative and positive charges shown on the comb in the top left diagram. How do the numbers compare? Count the negative and positive charges on the wool. How do the numbers compare?

Count the charges on the comb and the wool in the top right diagram. What changes have taken place? Which substances are charged and what is their charge?

Study the diagrams below. Explain them in terms of charges.

According to modern theory, the charge on an object depends upon whether it has gained or lost electrons. Rubber is believed to gain electrons when rubbed on wool and, therefore, becomes negatively charged. The wool loses electrons and becomes positively charged.

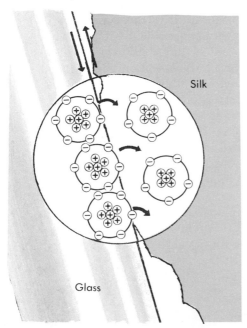

Silk

Glass

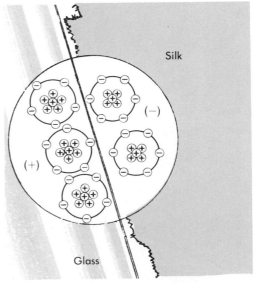

Silk

Glass

307

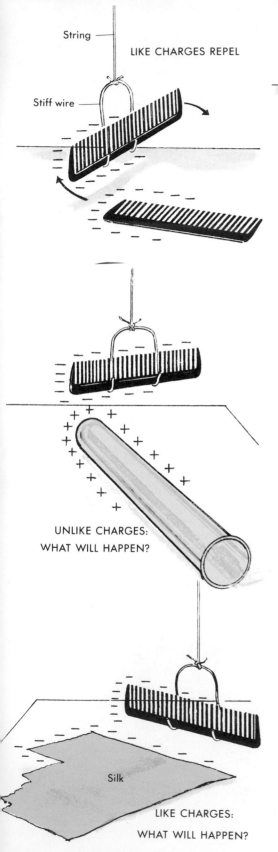

String

LIKE CHARGES REPEL

Stiff wire

UNLIKE CHARGES:
WHAT WILL HAPPEN?

Silk

LIKE CHARGES:

WHAT WILL HAPPEN?

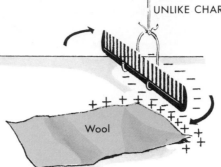

UNLIKE CHARGES ATTRACT

Wool

Predicting Forces. Suppose that two hard rubber combs have been rubbed with wool. According to theory, the combs gained electrons and the wool lost electrons. What is the charge on the combs? What is the charge on the wool?

Predict the nature of the force that will act on the two combs when they are brought near each other. Test your prediction by hanging one of the charged combs in a wire holder as shown above. Bring the other comb nearby and watch the behavior of the hanging comb. How closely do the results of the test agree with your predictions?

Predict the effect of the charged wool on the hanging comb. Test your prediction by the method just described. How good was your prediction?

Predict the effect a test tube rubbed with silk may have on a comb rubbed with wool. What effect may the charged silk have on the comb? Test both predictions.

Use the atomic theory to explain why you may expect any two objects made of the same material to repel each other after being rubbed with the same substance.

Inducing Static Charges. Put an empty metal can on a block of wax or other insulator. Bring a charged balloon very near to the can. Listen carefully as you move your finger to the can and touch it as shown on the next page. Take your finger away first and then take the balloon away. Again move your finger toward the can and listen carefully. What evidence is there that the can was charged?

The negatively charged balloon is shown close to the can in the first picture. More negative charges are shown on the side of the can opposite the balloon than on the near side. Why?

Electrons are shown in the second picture

passing from the can onto the hand. When the hand is removed (third picture), the can is positively charged.

Hold a finger close to the positively charged can. What do you feel (and hear)? Explain what is happening in terms of the movement of electrons.

This method of charging an object is called *induction*. The charge on the can is an *induced* charge.

Attraction by Induction. Set a metal can on an insulated surface. Do not charge the can. Find out whether a charged balloon clings to this neutral can. Recall other examples of a force between a charged object and a neutral object.

The diagram below illustrates an explanation of the force between the balloon and the neutral can. The negatively charged balloon repels electrons away from the part of the can nearest the balloon. Thus, a positive charge is induced in the area near the balloon. The force between these positive charges and the negative charges on the balloon holds the balloon to the can.

Charged objects such as charged balloons also induce weak opposite charges on walls and ceilings nearby. How does this weak opposite charge affect the balloon?

SUMMARY QUESTIONS

1. What observations have you made which support the idea that matter is made up of charged particles?
2. What observations have you made which support the idea that there are at least two different charges in atomic particles?
3. What evidence do you have that similarly charged objects repel each other?

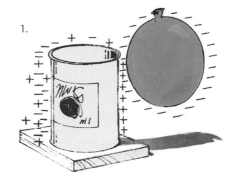

1.

2.

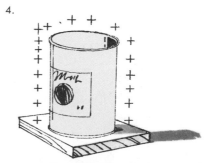

3.

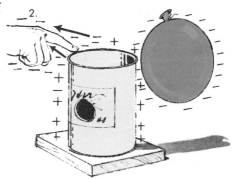

4.

METAL CAN CHARGED BY INDUCTION

ELECTRICITY FROM CHEMICALS

Electricity is a form of energy such as are heat, light, sound, chemicals, and moving objects. Under normal conditions, energy cannot be created; it can be produced only from a different form of energy. Usually, electrical energy is produced from either the energy of moving objects or from the energy of chemicals. In this section, you will learn of the conditions in which chemicals can produce electricity.

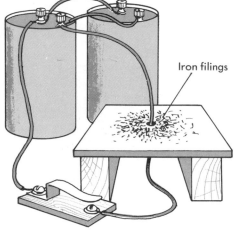

Iron filings

Electric Current and Magnetic Fields. Set up the apparatus shown at the left. Close the switch briefly while lightly tapping the cardboard with your finger. How do the iron filings change when an electric current flows through the wire?

Place several compasses on the cardboard around the wire. Again, close the switch for a few seconds while observing the compass needles. Describe the changes. Switch the wires to the dry cells. Which way do the compass needles point now?

Remove the wire and dry cells from the cardboard. Place a magnet below the cardboard. How do the iron filings and compasses behave?

Magnets exert an attractive force in the region around them. This region is called the *magnetic field*. What evidence do you have for a field around a magnet? What evidence do you have that a wire carrying an electric current also has a magnetic field around it?

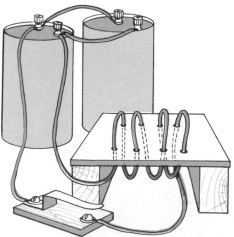

Wire Coils and Magnetic Fields. Make a coil with several large turns, passing the wire through holes in cardboard as shown at the left. Explore the magnetic field around the wires of this coil, using both iron filings and a compass.

Galvanometers. *Galvanometers* are electrical instruments which measure the flow of electricity through a circuit. Galvanometers are especially useful because they can detect even small amounts of electricity.

Wind ten turns of insulated wire around a compass. Turn the compass so that the wires line up with the compass needle. Hold the compass level while touching the two ends of the wire to a flashlight battery as shown here. How does the needle change? Reverse the wires on the battery. How does the needle behave now?

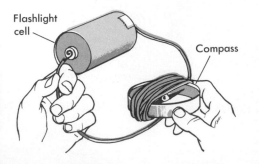

Flashlight cell

Compass

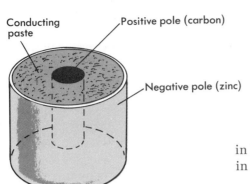

Conducting paste

Positive pole (carbon)

Negative pole (zinc)

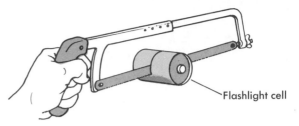

Flashlight cell

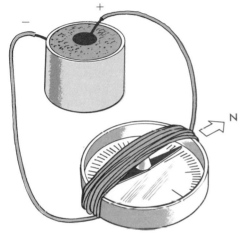

− +

N

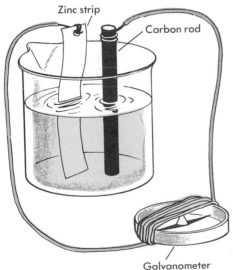

Zinc strip

Carbon rod

Galvanometer

Recipe for sodium sulfate solution. Dissolve 5 grams of sodium sulfate (Na_2SO_4) in 250 ml of water.

Dissecting a Dry Cell. Cut a flashlight dry cell in half as shown above. Note the three parts shown in the diagram at the left.

A dry cell consists of a negative pole where electrons leave when a complete circuit is made. In most dry cells, the negative pole is made of a soft, silvery metal called zinc. In leak-proof cells, the zinc may be covered with thick paper and a thin sheet of steel. The positive pole of flashlight cells is a rod of carbon. In a complete circuit, the carbon rod has too few negatively charged electrons. As a result, this pole is positively charged. Negatively charged electrons are repelled from the negative pole of the dry cell, travel through the circuit, and are attracted back to the positive pole. This flow of electrons is called an *electric current*.

Set up the galvanometer as you did before with the needle lined up with the coil of wire. Attach the two ends of the wire to the two poles of the dry cell. Observe the compass needle. What evidence do you have that electricity is going through the wire coil of the galvanometer?

Testing Wet Cells. The first type of electric cells were called wet cells. *Wet cells* consist of two different metals in a liquid instead of in a paste as you observed with dry cells. Construct a wet cell, using a piece of zinc and a carbon rod from a dry cell. Attach the ends of a galvanometer to the zinc and carbon while holding them in a sodium sulfate solution. Test other liquids such as water, salt water, baking soda solution, and sugar solution. Record how the galvanometer reacts in each case. Which liquids produce the most electricity? Which produce very little electricity?

Not all electric cells use zinc and carbon poles. Test other elements such as iron, aluminum, copper, lead, and tin. Which pairs strongly move the needle of the galvanometer? Which pairs produce little electricity? Study the activity scale of elements on the next page. Predict which pairs of elements would make good electric cells.

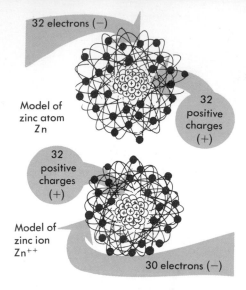

32 electrons (−)

Model of
zinc atom
Zn

32
positive
charges
(+)

32
positive
charges
(+)

Model of
zinc ion
Zn++

30 electrons (−)

Recipe for very dilute hydro-chloric acid. Slowly add 10 milli-liters of concentrated HCl to 110 milliliters of water. (CAUTION: Concentrated HCl should be han-dled only by your teacher or other adult trained in its use.)

When electricity is produced

ACTIVITY SCALE OF SOME COMMON ELEMENTS

Most Active

Magnesium
Aluminum
Zinc
Iron
Nickel
Tin
Lead
Copper
Carbon
Silver
Gold

Least Active

Zinc Atoms and Zinc Ions. The top diagram shows a simplified picture of how chemists describe an atom of zinc. This atom consists of a very dense central nucleus. There are thirty-two positive electric charges in the nucleus of a zinc atom. Orbiting around the nucleus are 32 very light particles called electrons. Each electron has one negative electric charge. Because there are 32 positive and 32 negative charges in the zinc atom, the atom is electrically *neutral*.

Hold a strip of zinc in very dilute hydrochloric acid for two minutes. Remove the zinc and wash away the acid. How has the zinc changed in the acid?

Chemists have noted that many metals, including zinc, slowly disappear when placed in acid. The atoms must be changing. Zinc atoms dissolved in acid are called ions. The next diagram describes a zinc ion. How does a zinc atom differ from a zinc ion? Why is a zinc atom written as Zn, but a zinc ion is written as Zn^{++}?

Ions in Electric Cells. Connect a zinc strip and a copper strip to a galvanometer. Put the metal strips in very dilute hydrochloric acid (HCl). After several minutes, examine the metals. Which metal is being used up? What is happening at the other pole?

According to the theory of atoms, zinc dissolves in the acid, forming dissolved zinc ions which are positively charged. Electrons from the dissolving atoms are left on the remaining zinc pole. These negatively charged electrons tend to repel each other.

When a wire is connected between the two poles (a complete circuit), electrons flow from the zinc, through the conductor, and to the copper pole.

Bubbles of hydrogen collect on the surface of the copper. These bubbles are the result of positive hydrogen ions (H^+) in the acid (HCl) being attracted to the negative electrons on the copper. The negatively charged electrons combine with the positively charged hydrogen ions to form neutral hydrogen, a gas.

Chemists say that a good wet cell is produced when an active element and an inactive element are used for the two poles. Compare this idea with your earlier observations about which pairs of metals deflected the galvanometer the most.

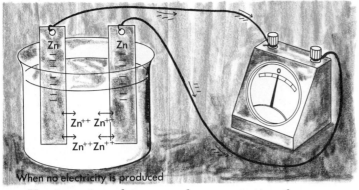

When no electricity is produced

Use two zinc plates as poles, connecting them to the galvanometer as you did earlier. What does the galvanometer tell you about the current?

In this case, some zinc atoms from both pieces go into solution as ions, leaving electrons on the metal. What is the charge on the two zinc strips? Why does no current pass through the galvanometer?

Electric Batteries. Connect three cells as shown here. You now have a *battery* of three cells. Compare the current produced by a three-cell battery with the current produced by a single cell.

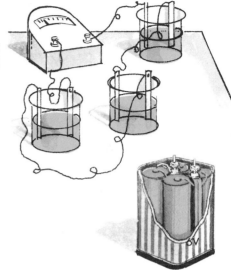

Standard dry cells are labeled 1½ volts. Some are labeled 6 volts; these are batteries of dry cells. Take apart a discarded 6-volt dry cell battery. Count the cells and trace the connections.

Storage Cells. The following electric cell uses sulfuric acid as one of the chemicals. This is a dangerous acid; therefore, the following experiment should be demonstrated by your teacher.

Connect two lead plates to a galvanometer and place them in dilute sulfuric acid. Note whether or not there is a current.

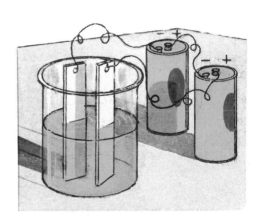

Disconnect the galvanometer and connect the plates to two dry cells in series as shown here. Watch the chemical changes that take place. Hydrogen gas is produced at one electrode. What is the charge at this electrode?

Negative oxygen ions are attracted to the other electrode and unite with the lead to produce lead dioxide. What is the color of this oxide?

After 5 minutes, connect the plates to the galvanometer. Does a current flow? If so, in what direction? Which electrode is positive and which is negative?

How long can this cell ring a bell? Find out if this cell can be recharged again with two batteries. Why is this cell called a *storage cell*?

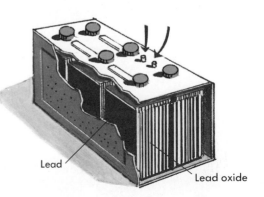

Lead

Lead oxide

Water turbine rotors

ELECTRICAL ENERGY FROM MECHANICAL ENERGY

Although flashlights and transistor radios get energy from dry cells, most electricity we use is made by machines called generators. *Generators convert the energy of motion (mechanical energy) into electricity. The mechanical energy can be supplied in many different ways. For example, moving air can turn the blades of a windmill, or moving water can turn the blades of a water turbine as shown above. Today, however, most electricity is produced in plants which use a fuel to produce steam, and the steam turns a steam turbine.*

Study the diagram below. Note the position of the turbine and generator in this hydroelectric plant. Water moving past the turbine blades causes them to turn. This motion turns the generator. The turning generator gives energy to electrons in the generator's wire, causing the electrons to flow. This flow of electrons is electricity.

Magnetic Induction. Wind a coil of wire and connect it to a galvanometer. Thrust a bar magnet into the coil. What does the galvanometer show?

Which way does the needle move when the N-pole of the magnet is thrust into the coil? When the N-pole is removed quickly from the coil? When the S-pole of the magnet is used?

Try changing the speed of the magnet. Hold the magnet still and move the coil back and forth. Under what conditions can a magnet induce electrons to move in a coil of wire?

Galvanometer
(or use compass and coil of wire)

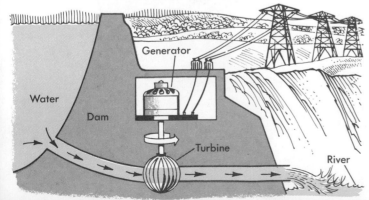

Water

Dam

Generator

Turbine

River

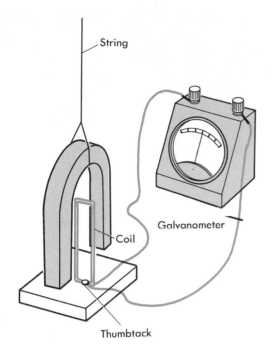

String

Galvanometer

Coil

Thumbtack

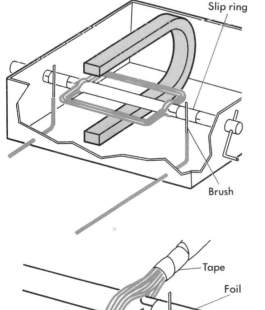

Slip ring

Brush

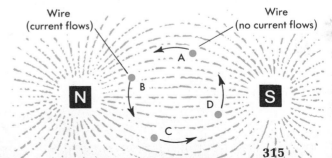

Tape

Foil

Bare wire

Dowel

A Simple Generator. Using No. 24 insulated wire, wind a rectangular coil of 25 turns that can fit easily between the poles of a U-shaped magnet. Mount the coil on a wooden base and connect it to a galvanometer.

Hang the magnet over the coil by a double string. Turn the magnet to twist the string and then release it. Explain the behavior of the galvanometer.

Making a Magneto. The other diagrams show how to make a type of generator called a *magneto*. Use the coil made for the last experiment and mount it on a dowel. Lay the bare ends of the coil along the dowel and wrap a strip of foil around the bare wire and dowel.

Place the strong U-magnet in a box. Make holes in the sides of the box so that the coil turns freely between the poles of the magnet.

Tape two wires to the bottom of the box in such a way that their bare ends touch the rings of foil on the dowel. These rings are called *slip rings*. The wires touching them are called *brushes*.

Connect this magneto to a galvanometer and turn the crank. Trace the flow of electrons. What is the purpose of slip rings and brushes?

Magnetic Fields and Electric Current. The region around a magnet which influences a compass needle is called a *magnetic field*. The effect of this field can be seen if iron filings are sprinkled around the poles of a magnet. The iron filings line up in the field along lines, sometimes called lines of force. Study the diagram below. Note the wire at A moving along the lines of force. Observers have noted that current does not flow through the wire at this time. Note the wire at B moving across the lines of force. Observers tell us that an electric current flows at this time through the wire. Predict what will happen in the wire at C and D.

How else might you get a current to flow in this wire instead of moving the wire?

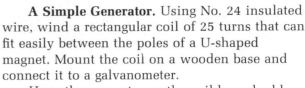

Wire (current flows)

Wire (no current flows)

N   S

A
B
D
C

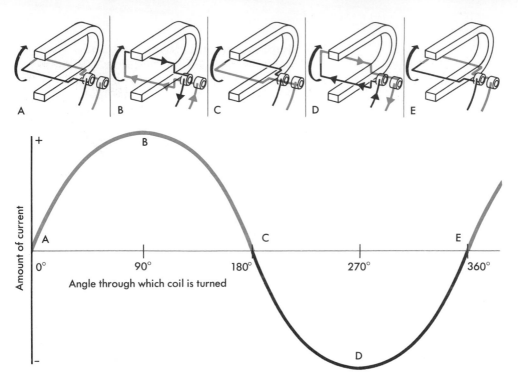

A   B   C   D   E

Amount of current

+

B

A   C   E

0°   90°   180°   270°   360°

Angle through which coil is turned

D

−

Alternating Current. Electrons that flow first one way and then the other way make up an *alternating current* (AC). Describe the current produced by the magneto you made.

The graph above represents the current produced in a coil which is turning at a steady rate. In diagram A above, the coil is moving with the field and not across it; therefore, there is no current, as shown by the graph. As the coil continues to turn, it begins to cut across the field, producing a current. The current reaches a maximum when the coil is cutting straight across the field, as shown in B.

What happens as the coil turns from B to C? Why? Why does the current then reverse itself? Explain the rest of the graph.

Alternators. Alternating current generators are often called *alternators*. The magnetic field is usually produced by electromagnets instead of permanent magnets so the field strength can be changed as needed.

Previous experiments showed that either the coil or the magnetic field may revolve. The automobile alternator shown on the opposite page has field electromagnets in the rotating part, called the *rotor*. The coils in which currents are induced are in the stationary part, called the *stator*.

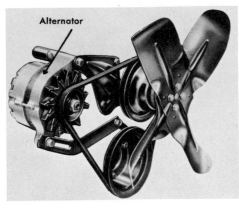

Alternator

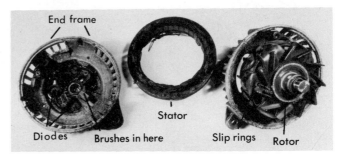

End frame Stator

Diodes Brushes in here Slip rings Rotor

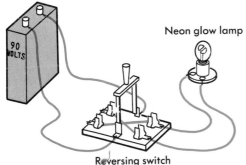

100-watt lamp

Large nail

COFFEE

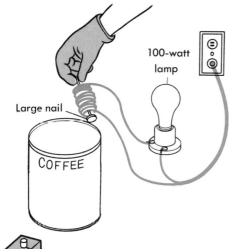

90 VOLTS

Neon glow lamp

Reversing switch

Alternating Current Effects. Wave a pencil in the light of a fluorescent lamp or television set. Note the series of images. Explain them in terms of alternating current.

Wind a coil on a large nail and connect it in series with a 100-watt lamp to an alternating current source. Hold the nail against a can. Explain the vibrations of the can.

Connect a neon glow lamp to a 90-volt battery. Which electrode glows? Reverse the current. Which electrode glows? Connect the lamp to alternating current. Explain why both electrodes glow.

Frequency. The rate at which a current changes direction is called its *frequency*. Frequency is measured in terms of the number of full changes, or cycles, per second. A current that flows in one direction for half a second and then flows in the reverse direction for half a second, has a frequency of one cycle per second.

How long is one cycle of a 25-cycle current? How many times does a 25-cycle current stop and start per second? What is the frequency of a current that stops 120 times per second?

A hand-cranked magneto provides alternating current. Watch a neon lamp in the magneto circuit as the crank is turned with increasing speed. Watch an incandescent lamp in the circuit. Explain your observations.

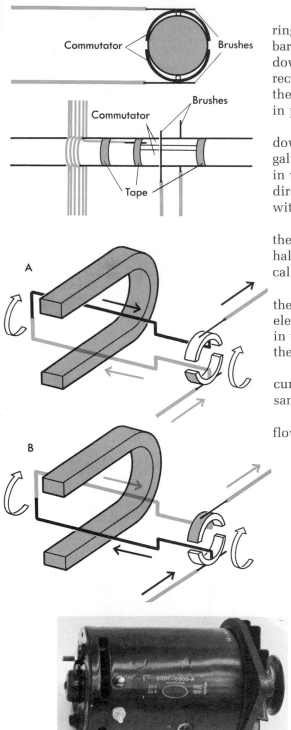

Direct Current Generators. Remove the slip rings from the magneto constructed earlier. Lay the bare ends of the coil along opposite sides of the dowel as shown here. Cover each end with a rectangle of foil that goes almost half way around the dowel; these pieces must not touch. Tape them in place.

Fasten two brushes on opposite sides of the dowel. Connect them to a galvanometer. The galvanometer should show a current which changes in value but which always goes in the same direction. Compare the DC generator shown below with your model.

The Commutator. The two pieces of foil and the brushes automatically reverse the current every half turn. The foil pieces of the reversing switch are called the *commutator*.

Study diagram A. Which way is the blue part of the coil moving in the first magnet? Which way are electrons forced to move? Why does current move in the opposite direction in the black part? Trace the circuit through the generator.

Trace the circuit in diagram B. Why has the current reversed in the coil? Why is the current the same in the brushes as it was originally?

Types of Current. A dry cell produces a steady flow of electrons that move in one direction. This

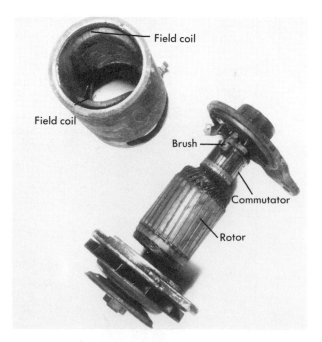

DC generator

flow is called *direct current* (DC). The first graph below shows direct current.

A switch in a dry cell circuit stops and starts a flow of electrons but does not change its direction. The second graph represents *interrupted direct current*.

The current produced by thrusting a magnet in and out of a coil is shown by the third graph. Explain to the class why this is alternating current.

The fourth graph shows the current produced by an alternator. Why are the changes in this current more regular than those shown in the third graph?

The fifth graph represents the current from a direct current generator. This is *pulsating direct current*. How does it differ from the current produced by a dry cell?

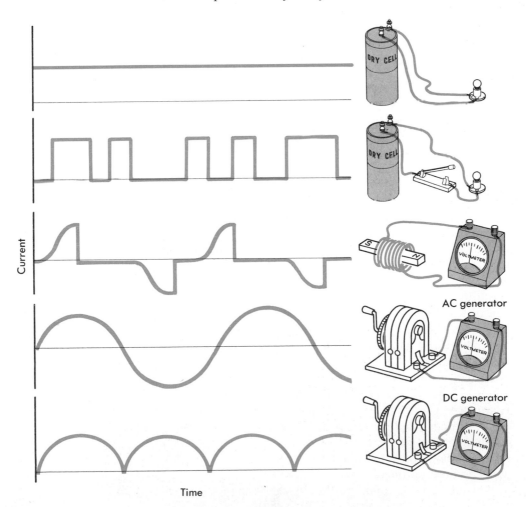

AC generator

DC generator

Current

Time

ELECTRICITY FROM HEAT, LIGHT, AND SOUND

Nearly all the electricity you use daily is produced by generators which are turned by turbines. Occasionally you use electricity from dry cells. There are, however, times when you generate very small amounts of electricity by other means. Although the electricity produced is not great, other characteristics make it useful in our lives.

Magnets Convert Sound to Electricity. Some kinds of microphones make use of currents generated when conductors move in magnetic fields. The type shown in the top diagram has a thin metal ribbon suspended between the poles of a permanent magnet. Sound waves vibrate the ribbon, inducing feeble currents. These signals are very weak and must be amplified before transmitting them to a public address system.

Another type of magnetic microphone has a small coil, which is moved back and forth between the poles of a magnet by sound waves. A current is generated in the coil.

One type of phonograph pickup makes use of magnetically induced currents. The basic parts of such a pickup are shown in the next-to-the-top drawing on this page. The part which holds the needle is kept magnetized by the bar magnet. As the needle vibrates in the grooves of a record, this arm moves back and forth between the poles of the coils and induces currents in the wire coils.

Crystals Convert Sound to Electricity. Certain crystals produce tiny currents when twisted, bent, or squeezed. Electrons are forced from their normal positions as the crystals are deformed. The electrons flow back again when the pressure is released.

The upper part of the diagram here shows how a crystal of Rochelle salt is used in a microphone. Currents are picked up by pieces of foil cemented to the crystal. What happens to the crystal as the diaphragm vibrates?

Obtain an inexpensive crystal microphone cartridge (sold in radio supply stores). Connect it to a galvanometer. Gently touch the diaphragm with a toothpick through a hole in the grill. How does the galvanometer respond?

Some phonograph pickups make use of crystals. The bottom diagram shows one way a crystal is used. What happens to the crystal as the

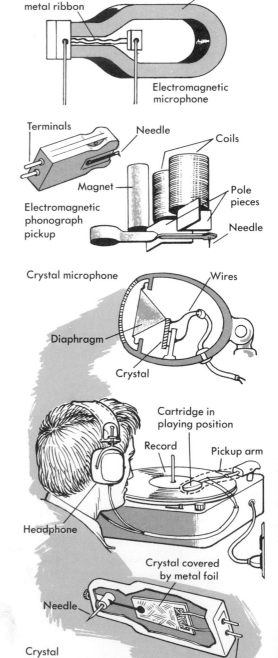

Electromagnetic microphone

Moving metal ribbon — Magnet

Terminals — Needle — Coils

Magnet — Pole pieces — Needle

Electromagnetic phonograph pickup

Crystal microphone — Wires

Diaphragm — Crystal

Cartridge in playing position — Record — Pickup arm

Headphone

Crystal covered by metal foil

Needle

Crystal phonograph pickup — Terminals

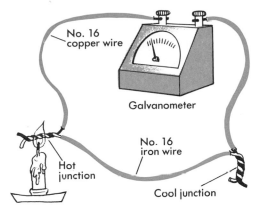

No. 16 copper wire

Galvanometer

No. 16 iron wire

Hot junction

Cool junction

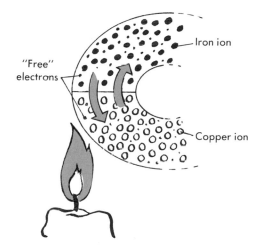

"Free" electrons

Iron ion

Copper ion

Photoelectric cell

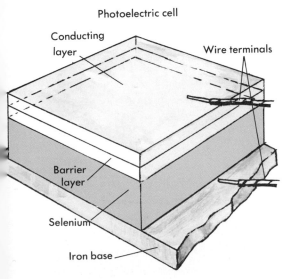

Conducting layer

Wire terminals

Barrier layer

Selenium

Iron base

needle vibrates from side to side in the record grooves? What is the purpose of the foil on each side of the crystal?

Connect headphones to the terminals of a crystal pickup. Rub the needle gently with your fingertip. Explain the sound in the headphones. Place the needle on a revolving record. Explain the sounds produced.

Thermoelectricity. Cut a 1-meter length of No. 16 copper wire and another of iron wire. Scrape both ends of each wire and twist the ends tightly together, as shown at the left.

Cut the copper wire and connect a sensitive galvanometer into the circuit. Heat one junction. Which way do electrons move in the circuit? Try other combinations of metals, such as copper and aluminum.

Heat causes the atoms of the two metals to vibrate. Some of their electrons are dislodged by the motion; their number increasing with the temperature. These free electrons move about at random and some cross the junctions from one metal to the other.

One of these two metals has more free electrons than the second when both metals are at the same temperature. Therefore, electron movement tends to be greater toward the second metal. An electric potential is developed.

A device which produces electricity in this way is called a *thermocouple*. Why must two metals be used in a thermocouple? Why is there no current if both junctions are at the same temperature? Propose some possible uses for thermocouples.

Photoelectricity. A number of substances give up electrons readily when light falls on them. Either selenium or silicon is used in most applications of this method for generating currents.

The bottom diagram shows a selenium photoelectric cell. A thin slice of selenium is mounted on an iron base. The selenium is covered with two transparent layers. One layer is a conducting layer to collect dislodged electrons, the other layer is a barrier which permits electrons to leave the selenium but not to return.

Light energy drives electrons from the atoms in the selenium layer through the barrier layer into the conducting layer. What charge is produced on the conducting layer? On the selenium?

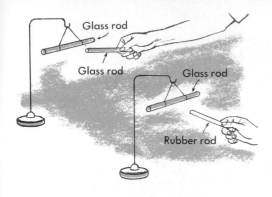

Glass rod

Glass rod

Glass rod

Rubber rod

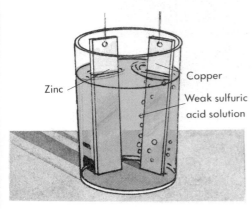

Zinc

Copper

Weak sulfuric acid solution

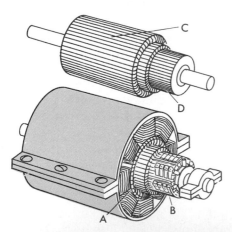

C

D

A

B

Obtain a photoelectric cell (sold in radio supply stores as "solar cells"). Connect the cell to a galvanometer and shine a weak light on it. Increase the brightness of the light and note the effect on the galvanometer.

A light meter contains a photoelectric cell and a galvanometer. The scale of the meter is marked to indicate light intensity.

A battery of photoelectric cells can produce enough power to operate radios and small motors. What are some advantages in using such batteries in artificial satellites?

REVIEW QUESTIONS

1. What is the difference between static electricity and an electric current?
2. What happens when a storage cell is connected to a source of electricity?
3. What happens in the upper diagram if the rods are charged and are moved toward each other?
4. Explain your answer in #3 above.
5. In the wet cell at the left, which metal is positive and which metal is negative?
6. What gas is in the bubbles shown on the copper?
7. Which metal dissolves to form positive ions?
8. Which letters in the generator diagram indicate field coils, commutator, brushes, and rotor?
9. What type of current is produced by this generator?
10. What type of current would this generator produce if the commutator were replaced by slip rings?
11. What characteristic of the electricity is changed by increasing the speed of the generator?
12. What device generates electricity in artificial satellites?

THOUGHT QUESTIONS

1. Why are the effects of static electricity usually more noticeable on poor conductors than on good conductors?
2. Why are many explanations of electricity based upon theories rather than upon direct observations?

Uses of Electricity

Electricity comes into a house as a flow of electrons, usually through thick copper wires. In the house, however, this electrical energy becomes useful as it changes to other forms of energy. A modern record player, for example, uses electrical energy. Some of this energy is converted to mechanical energy that moves the turntable. Some of the electrical energy is changed to sound energy so you can listen to the record. Still some more electrical energy is changed (in some record players) to light energy which shows that the record player is turned "on."

Many helpful devices in a home convert energy from one form to another. Note the lesson on the chalkboard in the photograph. Students are naming devices which convert energy from one form to another. Make a similar table in your class.

There are three parts to this chapter. In the first two parts, you will study the ways in which electrical energy is changed into heat, light, sound, and mechanical energy (motion). In the third part, you will study how electricity can be changed to different forms. For example, electricity is changed in an electric train (or slot car) transformer from a dangerous amount of house current to a much safer form.

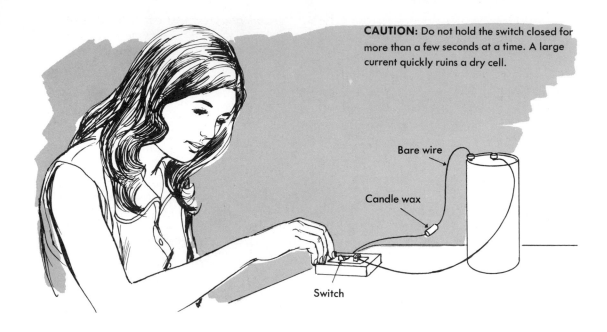

Bare wire

Candle wax

Switch

HEAT, LIGHT, AND SOUND FROM ELECTRICITY

Much of the electrical energy that comes into a home is used to produce heat, light, and sound. Make a list of devices that change electrical energy to each of these forms. How could the heat, light or sound be supplied in each case if there were no electricity? What is the advantage of using electrical devices in each case?

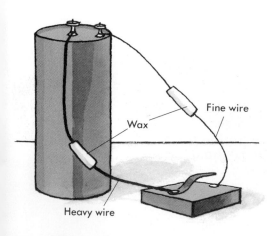

Fine wire

Wax

Heavy wire

Converting Electrical Energy. Soften some candle wax and mold it around a piece of bare copper wire as shown above. Connect the wire to a dry cell and a switch.

Press the switch and note what happens. Touch the wire. How does it feel?

Do not hold the switch closed for more than a few seconds at a time. The large current that flows through a dry cell in this circuit can damage the dry cell quickly.

Effect of Wire Size. Use two pieces of bare copper wire, one having a much larger diameter than the other. Cut equal lengths and connect them in a circuit as shown at the left. Mold equal pieces of candle wax around the wires. Close the switch and discover which piece of wax melts first. Experiment with other sizes of wire. What conclusions can you draw from this experiment?

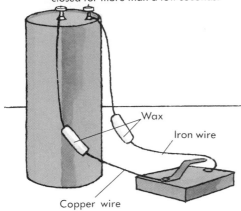

Wax

Iron wire

Copper wire

Effect of Kind of Wire. Cut equal lengths of iron and copper wire having the same diameter. Connect them to a dry cell and a switch as shown here. Mold equal pieces of candle wax around each wire.

Close the switch and note what happens. What conclusion can you draw from this experiment? Test wires of other metals. Why must the pieces have the same length and the same diameter if useful conclusions are to be drawn?

Electric Heaters. The experiment described above helps explain how electric heaters operate. Electricity is made to flow through a wire made of a metal that becomes very hot when it carries an electric current. Such a metal is said to have much *resistance* to the flow of electricity.

Iron has a high electrical resistance compared with copper, but iron cannot be used in toasters or electric irons because it oxidizes rapidly when white hot. A mixture of nickel, iron, and chromium has an even higher resistance and does not oxidize rapidly. Therefore, this mixture, called *nichrome*, is used in most heating devices.

Take apart some discarded electric heaters. Use a circuit tester to trace the circuits. Collect samples of the metals used in the heating wires. Compare these metals with copper in experiments like that shown at the top of the page.

Reducing Energy Losses. The experiments on these pages suggest ways to reduce losses of electrical energy in wires carrying electricity. Such wires should be made of a metal having a low resistance. Why is copper used instead of iron for wires along streets and in homes? Why are large wires used for large currents?

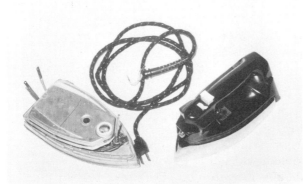

325

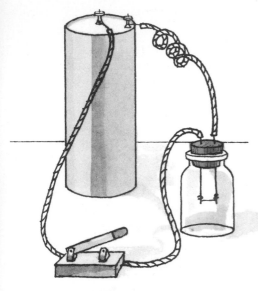

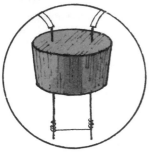

Single strand of picture wire

Light from Electricity. Push the bare ends of two lengths of copper wire through a cork. Connect a short length of very fine steel wire across the ends of the copper wire. A single strand untwisted from a piece of steel picture wire is suitable for this purpose.

Insert the cork in a bottle and connect the wires to a dry cell. Which becomes hotter, the copper wire or the steel wire? Why? When does the steel wire stop glowing? Why?

Incandescent Lamps. Many electric lamps are of the incandescent type. Each contains a wire that is heated white hot by a current of electricity. Anything that glows because it is hot is said to be *incandescent*.

The fine wire in most incandescent lamps cannot be seen because the bulbs are frosted. However, clear glass bulbs are sold in most electrical supply stores. Screw such a bulb in a socket and note the glowing wire.

Make an image of the incandescent lamp in a picture projector. Point the projector toward a screen. Focus the lens until an image of the glowing wire appears on the screen. (With some types of projectors, it is necessary to remove the lens and move it slowly toward the screen until a clear image appears.) Compare the shape of the glowing wire with one in an ordinary incandescent lamp.

The Filament. The fine wire that glows inside an incandescent lamp is called the *filament*. The diagrams on the next page show the location and connections of the filament in a common type of incandescent lamp.

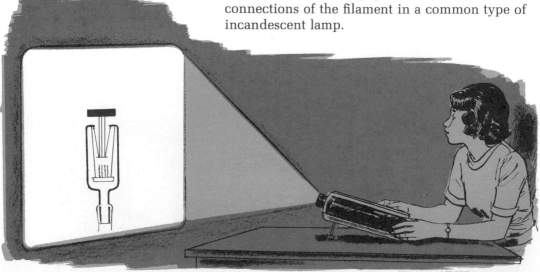

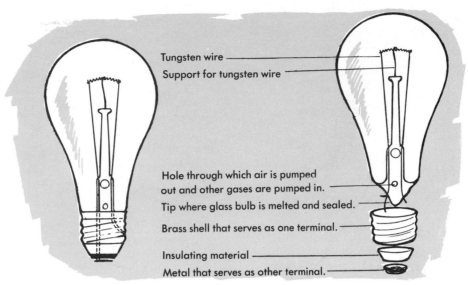

Tungsten wire

Support for tungsten wire

Hole through which air is pumped out and other gases are pumped in.

Tip where glass bulb is melted and sealed.

Brass shell that serves as one terminal.

Insulating material

Metal that serves as other terminal.

Oxidize means to combine chemically with oxygen. Oxidized metal wires are brittle and readily break.

The steel wire used in the demonstration on the opposite page did not last long because it *oxidized* rapidly at high temperatures. Even if all the air had been pumped from the bottle, the steel would have melted or evaporated. Not many metals can be used for the filaments in incandescent lamps.

Modern incandescent lamps contain filaments made of *tungsten*. Tungsten does not melt at white-hot temperatures and it evaporates very slowly. The photograph below shows a close-up of the filament of a light bulb.

The Bulb. The glass bulb of a lamp protects the filament and keeps oxygen away from it. Air is pumped from the bulb and replaced by a mixture of nitrogen and argon which does not react with tungsten. Take apart a discarded lamp and find the parts shown in the diagram above. Use a hand lens to observe the filament. (**CAUTION:** Do not take apart fluorescent tubes. They contain harmful chemicals.)

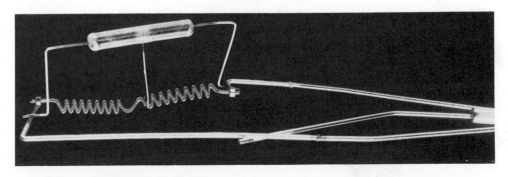

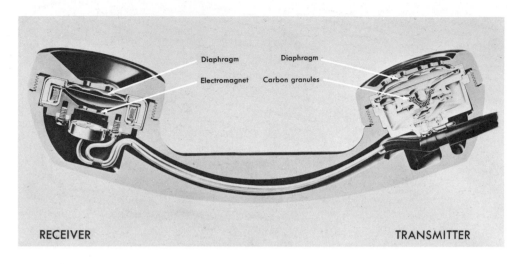

RECEIVER   TRANSMITTER

Diaphragm   Diaphragm
Electromagnet   Carbon granules

The Telephone. The two basic instruments of a telephone system are the transmitter and the receiver. Both instruments are commonly built into one handset as shown above.

Connect a receiver, a transmitter, and a dry cell in series. Ask someone to speak into the transmitter while you listen to the receiver.

Remove the cap from the receiver and find the thin metal plate called the diaphragm. Touch the diaphragm with your fingertips while someone speaks into the transmitter. Explain how the diaphragm sets up sound waves in air.

The Transmitter. A telephone transmitter is a variable resistor controlling the current in the circuit. The resistance is provided by carbon particles.

Attach a wire to each of two metal washers. Lay one washer in a small cardboard box and cover it with carbon particles, made by crushing the carbon rod from a dry cell. Lay the other washer on top of the carbon. Connect the wires to a lamp and two dry cells.

Apply pressure to the upper washer. How does the increased pressure affect the current? How does decreased pressure affect it? Explain the changes in terms of changing electrical resistance.

Study the sound wave diagram on the next page. What does a pressure wave do to the diaphragm of the transmitter? What does a rarefaction wave do?

The graph on page 329 represents the current in a telephone circuit for a brief period. How is this

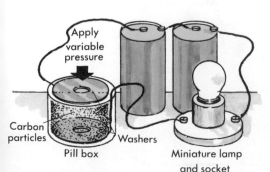

Apply variable pressure

Carbon particles

Pill box   Washers

Miniature lamp and socket

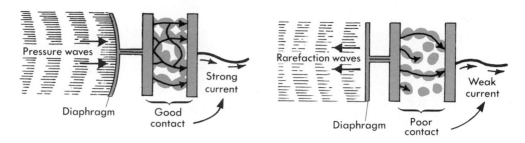

Pressure waves

Diaphragm

Good contact

Strong current

Rarefaction waves

Diaphragm

Poor contact

Weak current

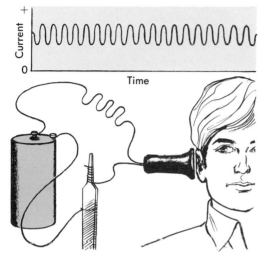

current like alternating current? How is it different? Why does the current never drop to zero?

Suppose middle C (about 260 vibrations per second) of a piano is sounded. Describe the current in the telephone circuit. Describe the change in current if A below middle C (220 vibrations per second) is sounded.

The Receiver. A telephone receiver contains an electromagnet and a steel diaphragm. Unscrew the cap from a receiver and find these parts. How does the strength of the electromagnet change as the current changes? What happens to the diaphragm as the magnetic field changes?

Connect a receiver to a dry cell and a file, as shown here. Touch the file with the loose wire. Explain the click produced by the receiver. Scrape the wire along the file. Explain the changes in the sound as the wire is scraped at different speeds.

Connect a galvanometer in a telephone circuit. What does the meter indicate when someone speaks into the transmitter? Compare the effects of loud and soft sounds.

SUMMARY QUESTIONS

1. Make a list of heat, light, and sound devices in your home which operate by electricity.
2. What are some properties of a wire which would make a good electrical heater?
3. A light bulb consists of a circuit of wire with two ends. Where do these two ends of the wire pass out of the bulb?

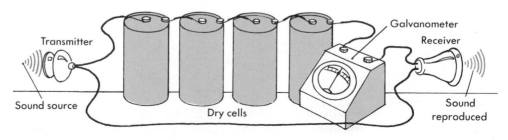

Transmitter

Sound source

Dry cells

Galvanometer

Receiver

Sound reproduced

ELECTRIC MOTORS

Motors use much of the electrical energy produced in this country. The motors in homes, such as refrigerator and washer motors, usually operate on alternating current. The electric motors in automobiles, such as the starter and heater fan motors, operate on direct current.

There are several types of alternating current motors and most of these seem much unlike direct current motors. Nevertheless, all of them operate on the same general principle: like magnetic poles repel and unlike magnetic poles attract.

Forces on a Conductor. Hang a bent piece of wire from a support as shown below. The U-shaped wire should swing freely. Connect this wire to a dry cell and a reversing switch as shown below.

Place a strong bar magnet as shown. Close the switch. What happens? Reverse the current by closing the switch the other way. What does the coil do? Turn the magnet around. What is the effect?

Producing Rotary Motion. Wind a coil of thin, insulated wire on a large nail. Suspend the coil over the poles of a strong U-shaped magnet as shown below at the left. Connect the ends of the coil to a reversing switch and two dry cells.

Close the circuit. Why does the coil turn? Why does it stop? Reverse the current. Explain the motion and the reason that it stops.

Practice reversing the current at the proper moment to keep the coil turning.

Making a DC Motor. The coil just studied needs sliding contacts to prevent the wires from twisting and an automatic reversing switch to keep it turning. The contacts and switch can be provided by adding brushes and a commutator.

The diagrams below show the details of construction. The rotor is a large cork through which a nail has been driven. The nail serves as a core for the coils. Be sure to wind both coils in the same direction. Why?

The bearing of the rotor is a short length of glass tubing which has been closed at one end in a gas flame. The bearing turns on the point of a nail which has been driven up through the base. Smooth the point of the nail with a file to reduce friction, if necessary.

The ends of the coils are made into commutator segments by removing the insulation and taping the ends to opposite sides of the glass tube. Short lengths of bare wire serve as brushes. If needed, adjust the brushes to find the best position for operation.

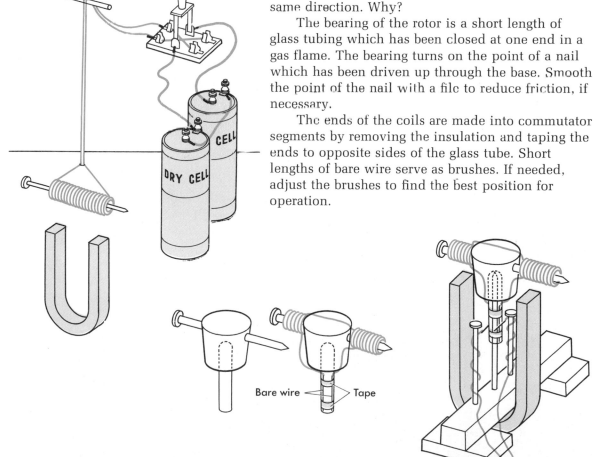

Bare wire — Tape

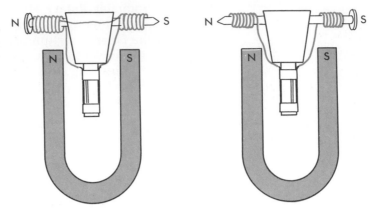

Action of Commutator. Remove the permanent magnet from the newly constructed motor. Hold a compass near one end of the coil while you turn the rotor by hand. How does the magnetism change during one revolution? Test the other end similarly.

The diagrams above explain why a DC motor operates. Each end of the rotor coil has the same magnetic pole as the neighboring end of the field magnet. Therefore, each end of the coil is repelled by the nearby pole and attracted to the more distant pole. What happens as the rotor coil poles approach similar field magnet poles? Why does the rotor keep on turning?

Field Magnets. The rotor of the motor just made turns in the magnetic field provided by a permanent magnet. However, the field of most motors is provided by electromagnets so that the field can be strengthened or weakened as needed.

Add electromagnets to your motor as shown below. Wind the coils of the field magnets so that the top of one has an N-pole and the top of the other has an S-pole. Why? Note the connections in the diagram. Are the rotor coils in series or are they in parallel with the field coils? Explain why.

What happens if the connections to the dry cells are reversed? Explain your answer.

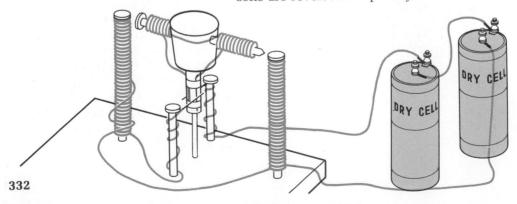

Automobile starter motor

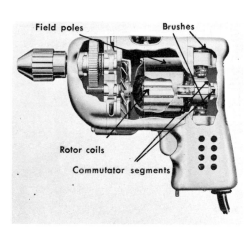

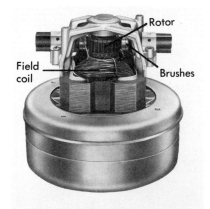

Universal electric motor used in an electric mixer

The Automobile Starter Motor. Compare the pictures of an automobile starter motor with your motor. How many field coils are there? Note that the steel frame connects the poles of the field magnets, making each pair like a horseshoe magnet. What is the advantage of U-shaped cores?

The rotor has more coils than your motor. The sixteen coils form eight electromagnets. How many commutator segments are needed?

This motor has four brushes. How many pairs of rotor coils can be magnetized at one time? Each pair of coils becomes magnetized when near the poles of a field magnet. Each pair of coils is demagnetized just before coming in line with the poles. Explain the reasons for this arrangement.

Universal Motors. One type of motor can be used with either direct or alternating current; these motors are called *universal* motors. They are used in many vacuum cleaners, mixers, and electric drills.

The field coils and the rotor coils of a universal motor are connected in series. Explain the operation of the motor when it is used with direct current.

If a universal motor is connected to an alternating current source, the magnetism of the field coils is changed constantly. Why? How often do the poles change? The motor could not operate if the magnetism of the rotor coils did not change at the same time. Explain how the series connection makes these changes possible.

Synchronous Motors. A synchronous motor is a constant speed motor which operates only on alternating current. It revolves in rhythm with the changes in the alternating magnetic field; in other words, it is *synchronized* with the frequency.

Wind a coil on a large nail. Connect the coil in series with a 100-watt lamp, taping the connections to prevent short circuits.

Set a demonstration compass near one pole of the coil. Snap the needle to set it revolving. The needle should fall in step with the changing field and continue to spin.

What is happening to the magnetic poles of the coil? Explain why first one end and then the other end of the needle is attracted. How fast must the needle spin to keep in step?

Why are synchronous motors used in electric clocks? What determines the accuracy of the clocks?

Making A Synchronous Motor. Cut the rotor from three thicknesses of a tin can and rivet the layers together. The six arms must be at equal intervals. Cement the metal to a large cork. For a bearing, use a short, closed glass tube turning on a sharpened nail.

For the field coils, wind eight layers of insulated wire on headless nails. The tops of the cores must have opposite poles. Connect the bottoms with a strip of iron.

Connect the motor in series with a 250-watt lamp. Start the rotor by giving it a quick twist. Adjustments may be needed; the rotor should spin just below the tops of the poles without hitting them.

At what speed must the rotor turn so that an arm being attracted to a pole arrives at the pole just as the current stops? At what speed must a four-armed rotor spin?

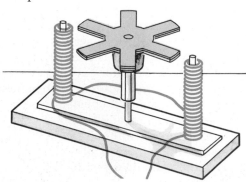

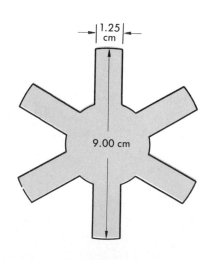

1.25 cm

9.00 cm

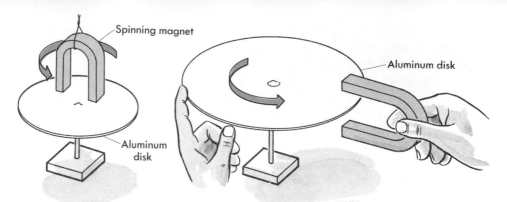

Induction Motors. These motors have no rotor coils, brushes, or slip rings. They depend upon the induction of a current in a metal which is spinning in a magnetic field. The induced current sets up its own magnetic field which interacts with the magnetic field of the field coils.

Cut the rim from an aluminum pie pan. Punch a dimple in the exact center and balance the disc on the point of a sharpened nail. Set the disc spinning. Hold a strong U-shaped magnet with its poles straddling the disc.

Why is a current set up in the disc? Why is a second magnetic field produced? What evidence is there that the new field opposes the first?

Suspend the U-shaped magnet by a length of doubled twine close above the disc. Spin the magnet and explain the behavior of the disc.

The photograph shows the parts of an induction motor. The stator windings set up a changing magnetic field. This field induces currents in the copper bars of the rotor. The induced current produces a magnetic field which interacts with the first field.

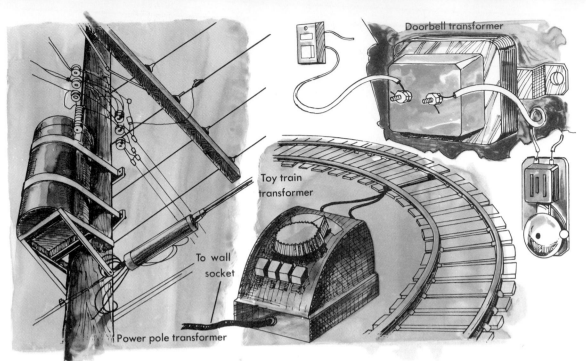

Doorbell transformer

Toy train transformer

To wall socket

Power pole transformer

TRANSFORMING ELECTRICITY

Electrons flow through a wire only if a force acts on them. Thus, every electric current can be described, in part, by the force exerted on the electrons. The units of measurement for the force which moves electrons are called volts. *A dry cell is usually 1½ volts. Most household electricity is 110–120 volts.*

Transformers are used to change or transform the voltage of alternating current. Transformers are important parts of electrical transmission systems and are familiar sights along streets and highways.

Transformers are used to change 120-volt household electricity. Electric door bells, train sets and slot cars include "step-down" transformers to provide a lower, safe voltage. "Step-up" transformers raise the voltage for the operation of neon signs and television sets.

Electromagnetic Induction. Wind two coils of 50 turns each on a large nail. Connect one coil to a dry cell and switch; this is the *primary circuit.* Connect the other coil to a galvanometer; this is the *secondary circuit* as shown on page 337.

Close the switch. When does the galvanometer indicate a current? How long does the current flow in the secondary circuit? Which way does it flow? Open the switch. What happens?

Close and open the switch several times. Describe the current in the primary circuit. Describe the current in the secondary circuit. What happens if the connections to the dry cell are reversed?

Although the two coils are not connected electrically, the primary coil induces a current in the secondary coil by means of an electromagnetic field. Is the current induced when the field is steady or when it is changing?

A Transformer's Changing Field. Replace the galvanometer in the circuit with radio headphones. Close the switch. Open it. When is sound produced? Why? Why is no sound produced when the switch remains closed?

Replace the dry cell with a doorbell transformer. Connect the transformer to an alternating current source. Describe the sound produced. Why is it unnecessary to open and close the switch to produce sounds?

As current builds up in the primary coil, the magnetic field passes through the secondary coil causing electrons to flow in the secondary coil. As the current in the primary coil slows down, the magnetic field collapses. How does this change in the field affect electrons in the secondary coil?

Magnetic Coupling. Two coils are said to be coupled if the magnetic field produced by one coil lies within the other. Coupling would be perfect if all the field lay in the second coil, but this condition never exists.

Wind two coils of 50 turns each. Connect one coil to a doorbell transformer. Connect the other coil to radio headphones. Hold the two coils closely together and separate them slowly. How is the sound affected? Explain why coupling is weakened in terms of the magnetic field.

Hold the two coils in a parallel position and turn one coil until they are at right angles to each other. How is coupling affected?

Test the effect of a straight iron core on coupling. Try a U-shaped core. What happens if a soft iron bar is placed across the poles of the U-shaped core?

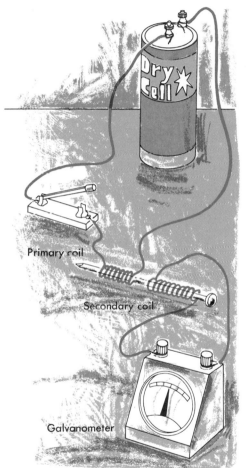

Primary coil

Secondary coil

Galvanometer

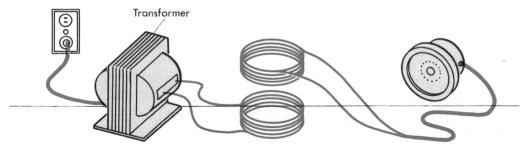

Transformer

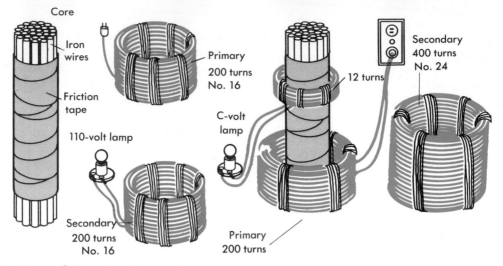

Core
Iron wires
Friction tape
110-volt lamp
Primary 200 turns No. 16
Secondary 200 turns No. 16
C-volt lamp
12 turns
Secondary 400 turns No. 24
Primary 200 turns

Stepping Up and Stepping Down. Tape 30-centimeter lengths of iron wire in a bundle to make a core 5 cm in diameter. Wind on it a primary coil of 200 turns of No. 16 insulated wire. Connect the coil to a lead cord.

Wind a secondary coil of 200 turns of No. 16 wire. Connect this coil to a 110-volt lamp. Place the coils around the core and plug in the lead cord. Explain why the lamp lights. Replace the secondary coil with another coil of 12 turns. Does the 110-volt lamp light? Try a 6-volt lamp.

Estimate the number of turns needed to light a 1½-volt lamp. Make a coil and test your prediction.

Wind another secondary coil having 400 turns of No. 24 wire. What voltage do you expect? Connect the coil to two 110-volt lamps in series.

Under what conditions does a transformer raise voltage? Lower voltage?

Voltage Ratios. Assuming perfect coupling, the output voltage of a transformer equals the input voltage when the primary and secondary coils have the same number of turns. A secondary coil with twice as many turns doubles the voltage because its turns are in series. What happens to the voltage if the secondary coil has half as many turns as the primary coil?

How is voltage stepped up from 110 volts to 2200 volts? What is the approximate ratio of turns in a toy transformer having a 15-volt output?

Electrical Transmission Systems. A small current at high voltage can deliver as much energy as a large current at low voltage. Since small currents can be carried by small wires, high voltage is preferred for long distance transmission systems.

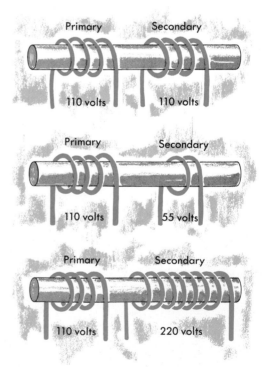

Primary — Secondary
110 volts — 110 volts

Primary — Secondary
110 volts — 55 volts

Primary — Secondary
110 volts — 220 volts

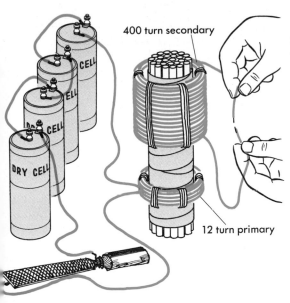

400 turn secondary

12 turn primary

Induction Coils. An induction coil is able to raise the voltage of direct current electricity. Like a transformer, an induction coil has a primary coil and a secondary coil on the same core. An induction coil also needs a device for rapidly interrupting the primary circuit. Why?

Use the apparatus designed for the last experiment to demonstrate the action of an induction coil. Put the 400-turn coil and the 12-turn coil on the core. Connect the 12-turn coil to a file and four dry cells in series. Scraping the file serves to interrupt the primary circuit.

Touch the bare ends of the secondary coil. A shock indicates a high voltage in the secondary coil.

Automobile Ignition Systems. An induction coil is a basic part of the ignition system of gasoline engines. Several thousand volts are needed to produce a spark across the gap of a spark plug. The other parts of the electrical system, such as lights, horn, and storage battery, operate best at low voltages. Therefore, a car generator produces low voltages (commonly 12 volts). This voltage is stepped up by the induction coil, also called an *ignition coil*, to the high voltage needed for the spark plugs.

Obtain a car ignition coil. Connect the top of a spark plug to the top of the coil with a spark plug wire. Wrap wire around the base of the plug and connect it to the negative terminal of the coil. Connect four dry cells in series with a file in the primary circuit. What part is represented by the file? Explain why a spark is formed at the spark plug.

Examine the ignition system of an automobile. Find the ignition coil. Find the parts of the primary (low-voltage) circuit and trace the connections. Find the parts of the secondary (high-voltage) circuit and trace the connections. Remove one spark plug and lay it on the head of the engine. Watch for sparks at the electrodes as the engine is turned over.

Locate with the help of your classmates the part of the system where a switch opens and closes. Why is such a switch necessary?

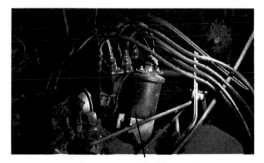

Ignition coil

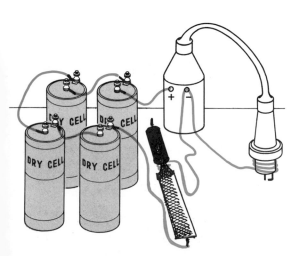

REVIEW QUESTIONS

1. What kind of wire is usually used in electric toasters?
2. What kind of wire is usually used in

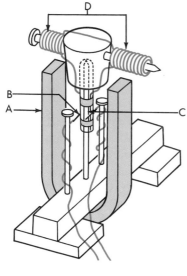

incandescent lamps?

3. What kind of wire is usually used in home wiring?
4. Why does the wire in an electric lamp give off light?
5. What are some factors which affect the temperature of wire carrying an electric current?
6. Slight changes in air pressure at a telephone transmitter are changed into slight changes in an electric current. How is this done?
7. Slight changes in an electric current at a telephone receiver are changed into slight changes in air pressure (air vibrations). How is this done?
8. Identify the following parts of an electric motor: rotor, field magnet, commutator, brushes.
9. Why must insulated wire be used in making a motor such as the one shown here?
10. Does the motor at the left operate on AC or DC? How can you tell?
11. What is the approximate ratio of primary to secondary turns in a transformer that steps down the voltage from 2200 volts to 110 volts?
12. Why must a DC motor have a commutator?
13. What is a synchronous motor?
14. How does an iron core affect the coupling of a transformer?
15. What is the difference between DC, synchronous, induction, and universal motors?

THOUGHT QUESTIONS

1. Why are insulators on high voltage lines larger than those on low voltage lines?
2. The diagram of a bulb at the left shows the path of electricity. Why does only the top part of the wire circuit give off light?
3. Why are some appliances (stoves, dryers) connected to 220-volt electricity, while others, such as lights, clocks, and radios, use 110 volts?
4. Why does the small mass of an electron make rapidly changing currents possible?
5. What will happen to the right coil in the apparatus shown in the bottom diagram if the left coil is made to swing?

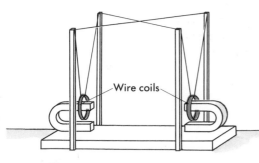

Wire coils

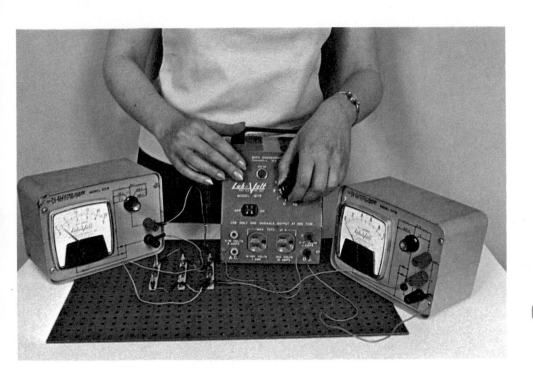

Measuring Electricity

Discoveries in science and the understandings which follow the discoveries are usually linked by measurements. For example, primitive people probably recognized the difference between hot objects and cold ones. However, a much better understanding of the nature of heat occurred only after the thermometer was invented. Thermometers can measure the same temperature many different times with less error than earlier methods. Thermometers also can detect smaller differences in temperature than could earlier methods.

Our present knowledge of electricity depends on measurements at least as much as our knowledge of heat depends on thermometers. Unfortunately for students, there are many different quantities to measure in order to describe the flow of electricity. These units, their importance, and how to measure them are the subject of this chapter.

ELECTRICAL MEASUREMENTS

Two measurements, current and voltage, provide most of the information needed to analyze simple circuits. Other characteristics can be calculated if current and voltage are known.

Many units used here have been named to honor famous physicists. For example, the volt was named for Alessandro Volta, who contributed much to our knowledge of electricity. Unfortunately, the names do not suggest the nature of the units; they must be understood to be used properly.

The Unit of Quantity. It is difficult to measure quantities of electricity; therefore, the unit for quantity is not often used. It is introduced here to give meaning to other measurements.

The unit of quantity is the *coulomb* (named for Charles Coulomb). This unit was defined before the electron theory was established; in today's terms it represents the charge on approximately 6 000 000 000 000 000 000 (6×10^{18}) electrons. This number is enormous but electrons are very small. A penny contains enough free electrons to equal several thousand coulombs.

Electric Current. An electric current is a flow of electrons. If a coulomb of electricity flows each second through a conductor, the electron flow is described as one *ampere* (frequently shortened to one amp). The ampere was named after André Ampére, another early investigator of electricity. An ampere represents the flow of 6×10^{18} electrons every second.

Current is determined by an *ammeter*. As you probably recall, electricity passing through the ammeter coil sets up a magnetic field. This magnetic field pulls the needle. The larger the

1 ampere = 1 000 **milliamperes**

1 ampere =
     1 000 000 **microamperes**

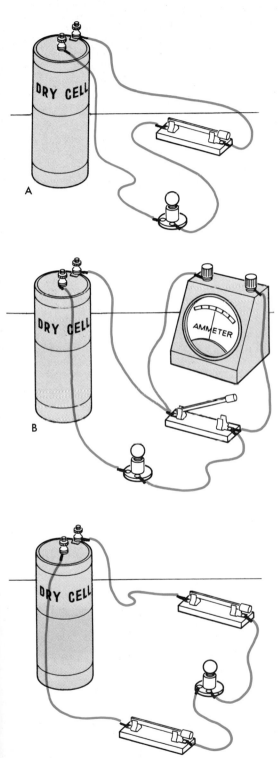

current passing through the ammeter, the stronger the field and the more the needle moves.

The needle of a direct-current ammeter, by swinging right or left, indicates the direction of the current as well as the rate. Most alternating current reverses too fast for a needle to show the direction electrons flow. Thus, an AC ammeter is designed to show only the rate of flow.

What is happening when an ammeter reads 2 amperes? How many electrons pass through the ammeter in 5 seconds?

Many electrical circuits are designed for currents much smaller than 1 ampere. A smaller unit is used in describing the electric current. One milliampere is a current of one thousandth of an ampere. One microampere is a current of one millionth of an ampere. What would you call a meter which measures electric current in milliamps?

Protecting an Ammeter. All the current in a circuit passes through an ammeter. If the current becomes large, the ammeter may be damaged.

Diagrams A and B show how to protect an ammeter from mistakes in wiring. The ammeter is not connected directly into a circuit; its place is taken by a switch which bypasses the ammeter. The circuit is then checked and errors are corrected.

The current is measured by clipping the lead wires of an ammeter to the switch, opening the switch, and reading the ammeter. The switch is then closed and the ammeter wires detached.

Measuring Current. Connect a two-cell flashlight lamp in series with a bypass switch and a dry cell. Measure the current. Explain the reading in terms of the number of electrons passing through the wire.

Replace the lamp with a three-cell lamp. What is the current? Compare its brightness with that of the other lamp. Try a four-cell (6-volt) lamp. What is the current?

Exploring a Circuit. Compare the current on both sides of a lamp as shown here. Take a meter reading at one bypass switch and then at the other switch. What difference, if any, do you find in the current on each side of the lamp? Explain your findings in terms of the flow of electrons.

Predict the current on each side of two lamps in series. Set up the circuit, using three bypass switches, and test your predictions.

Rock

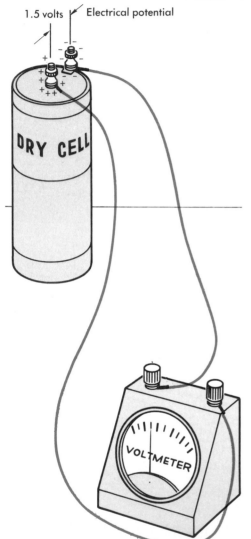

1.5 volts   Electrical potential

DRY CELL

VOLTMETER

Examples of Potential Energy.
One important trait that we possess is the ability to construct and use devices which do work for us. In order for machines to do work, energy must be used. Three examples of stored or *potential energy* are shown above: a waterfall, a rock on a hill, and a dry cell. How can each be made to do work? How might each be changed to have more energy, thus being able to do more work?

Electric Potential.
Electrons at the negative terminal of a dry cell have energy given to them by chemicals which forced them from the carbon electrode to the zinc. The electrons can do work as they return through a circuit to the carbon.

The energy of electrons at the negative terminal is called their *electric potential*. The unit of potential is the *volt*. A volt represents the ability of a coulomb of electricity to do a certain amount of work. At a potential of one volt, 6×10^{18} electrons can lift a kilogram mass about 10 centimeters, or heat a gram of water about 0.75 degree Celsius.

Measuring Potential.
A voltmeter measures the energy of a small sample of electricity. This sample does work by moving the needle of the voltmeter. The work done is assumed to be the potential.

Measure the potential of some dry cells, including flashlight cells. Touch the positive lead of the voltmeter to the positive terminal and the negative lead to the negative terminal.

Determine the potential of two dry cells in series, of two dry cells opposed to each other, and of two dry cells connected in parallel. Compare the potential of good and "dead" cells.

Potential Drop.
The diagram on page 345 shows

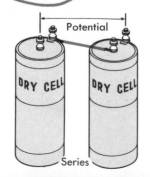

Potential

Series

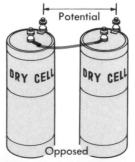

Potential

Opposed

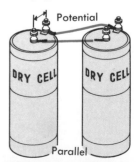

Potential

Parallel

344

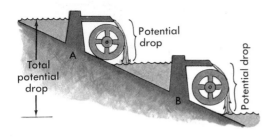

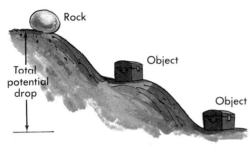

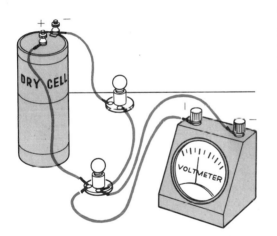

dams and water wheels on a river. The potential energy of the water at each dam allows the water wheel to turn and do work. The loss in energy of the water from having moved from the top of the dam to the bottom of the wheel is called the *potential drop*. If the potential drop at each water wheel is added, the *total potential drop* for water on this part of the river is determined.

Imagine that a higher dam were built at *B* so that the lake was the same height as at *A*. What would happen to the potential drop at *B*? What would happen to the total potential drop in this part of the river? Describe the potential drop and total potential drop in the second diagram.

Electrons give up part of their energy as they pass through each part of a circuit. A voltmeter measures this loss of energy or potential drop.

Connect two 2-cell lamps in series with a dry cell. Touch the voltmeter leads to the screws of one lamp socket. What is the drop in potential in this lamp? What is the potential drop in the other lamp? Compare and explain these two potential drops with the potential of the dry cell.

Replace one lamp with a 3-cell lamp. Measure the potential drop in each lamp in the circuit. Explain the brightness of the lamps in terms of energy conversion. What is the total energy drop? Test a 4-cell lamp (6 volts) in the same circuit. What is the total energy drop?

Connect two 2-cell lamps in parallel. Measure the potential drop in each lamp. Explain why lamps are brighter when they are connected in parallel.

UNCERTAINTIES IN MEASUREMENT. *Few measurements are exact. An instrument may be out of adjustment. Its scale may not be correct. Some instruments are less sensitive in the low and high parts of their range than in the middle. A reading that falls between two divisions on a scale must be estimated.*

Reading errors are common. One number may be mistaken for another number. A scale division that represents two units may be read as one unit. Viewing a needle at an angle, instead of from directly above, introduces parallax.

Instruments may change conditions being measured. Ammeters add potential drop into circuits. Voltmeters increase current.

Discuss these possible errors in terms of their size, their effect on the results, and the methods for reducing their influence.

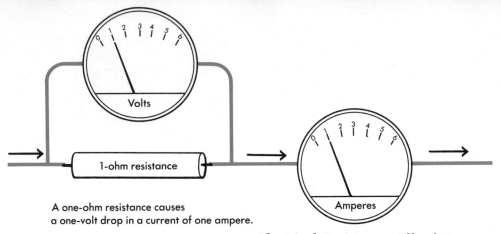

A one-ohm resistance causes
a one-volt drop in a current of one ampere.

$$\frac{\text{Potential Drop}}{\text{Resistance}} = \text{Current}$$

$$\frac{\text{Voltage}}{\text{Ohms}} = \text{Amps}$$

$$\text{Volts} = \text{Amps} \times \text{Ohms}$$

$$\frac{\text{Volts}}{\text{Amps}} = \text{Ohms}$$

Electrical Resistance. All substances present some resistance to the flow of electrons. Most metals have a very low electrical resistance. Insulators have a very high resistance.

Resistance causes a potential drop in a device. The greater the resistance, the greater the energy loss from each coulomb passing through the device.

Measuring Resistance. The unit for measuring resistance is the *ohm* (named for Georg Ohm). A resistance of one ohm causes a one-volt potential drop in a current of one ampere.

Resistance is usually calculated from current and potential drop. If the potential drop in a device is 6 volts and the current is 2 amperes, the resistance is 3 volts per ampere or 3 ohms. What is the resistance if the potential drop is 100 volts and the current is 2 amperes?

Calculate the resistance of flashlight lamps and other devices. Enter the results in a table like that shown below.

Conductance. Conductance is the reverse of resistance. Metals have high conductance; insulators have low conductance.

The unit for measuring conductance is the *mho* (*ohm* spelled backward). One mho of conductance permits a flow of one ampere for each volt of potential drop. If the potential drop in a 2-ampere current is 6 volts, the conductance is 1/3 ampere

Device	2-cell lamp		
Potential drop	2.5 volts		
Current	0.50 amp		
Resistance	5 ohms		
Conductance	0.20 mho		

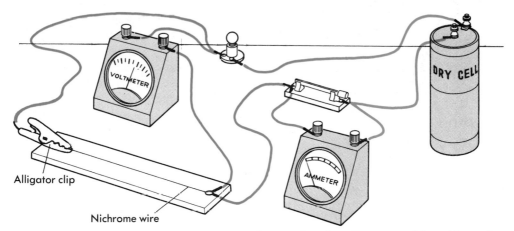

Alligator clip

Nichrome wire

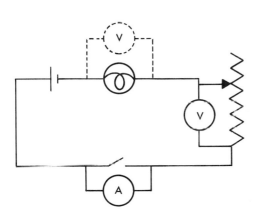

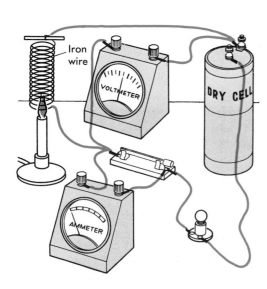

Iron
wire

per volt, or 1/3 mho. Compare this with resistance.

Calculate the conductance of the devices whose resistances were determined in the previous experiment. Add these values to the table.

A Variable Resistance. Attach 1 meter of nichrome wire (or other high resistance wire) to screws in a long board as shown above. Connect one end of the wire to a lamp, a bypass switch, and a dry cell. Use a clip as a sliding contact for the other connection to the nichrome wire.

Move the clip back and forth, noting changes in the light of the lamp. At the same time, note changes in the current as indicated by an ammeter.

Measure the potential drop in the lamp and in the nichrome wire for different positions of the clip. What happens to the potential drop in the lamp as the potential drop in the wire increases? Explain the changes in the light of the lamp in terms of both current and potential drop.

Temperature and Resistance. Wind a 1-meter length of iron wire on a dowel to make a coil. Slip the coil from the dowel and hang it from an overhead support.

Connect the coil in series with a dry cell, a lamp, and a bypass switch. Measure the current and potential drop through the coil. Calculate the resistance of the coil.

Heat the coil with a flame and note any changes in the lamp. Determine the current and potential drop while the coil is hot. Calculate the resistance again and compare it with the resistance of the cold wire.

Test other kinds of wire in the same way. What generalization can you make? To what substances should your generalization be limited?

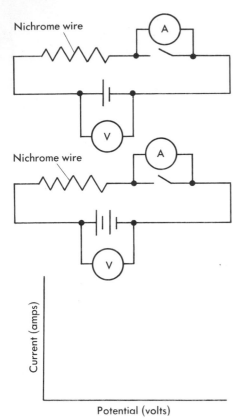

Current (amps)

Potential (volts)

Relation of Current to Voltage. Connect a nichrome heating element from a toaster or iron to a dry cell and a bypass switch. Measure the potential of the cell and the current in the circuit.

Repeat, using two dry cells in series to increase the potential. Repeat again, using three dry cells in series. Plot the results on a graph. What relationship is shown between current and voltage?

Replace the nichrome heating element with a 2-cell flashlight lamp. Measure the current at three different potentials as you did before and plot the results. Explain the differences between the two graphs.

Resistances in Series and Parallel. Determine the resistance of two 2-cell flashlight lamps separately. Connect the lamps in series and determine the resistance of the two lamps together. What happens to the resistance of a circuit when two devices are connected in series? What happens to the conductance under the same conditions?

Connect the same two lamps in parallel. Determine the resistance of the circuit. What happens to the resistance and the conductance of a circuit when devices are connected in parallel? Test combinations of other lamps in the same way.

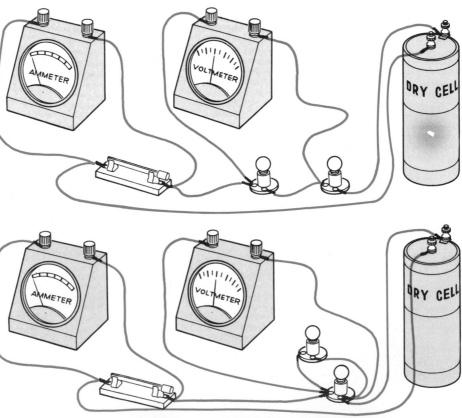

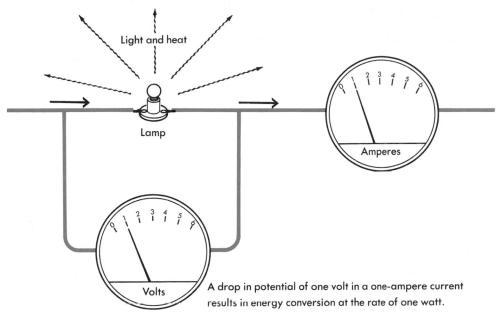

A drop in potential of one volt in a one-ampere current results in energy conversion at the rate of one watt.

Rate of Energy Exchange. A lamp converts electrical energy to light and heat. A motor converts electrical energy to mechanical energy and heat. The rate of conversion depends upon the current and potential drop in a device. The greater the current, the faster energy is being delivered; the greater the potential drop, the more energy that is given up by each coulomb.

The unit for measuring the rate of energy conversion is the *watt* (named for James Watt). A 1-volt drop in potential for a 1-ampere current results in energy conversion at the rate of one watt. What is the rate of conversion if the potential drop in a 2-ampere current is three volts?

Calculate the rate of energy conversion in the lamps you have tested.

Total Energy Exchange. The amount of energy converted by a device is measured in *watt-seconds*. A 1-watt device operating for one second converts a total of one watt-second of electrical energy to some other form.

The watt-second is a small unit. The electrical energy used by your house is too great to be measured conveniently in watt-seconds. A larger unit, the *kilowatt-hour*, is used. How many watt-seconds equal one kilowatt-hour?

How much energy would be used by a 100-watt bulb in 10 hours? A 200-watt television set in 20 hours?

Lamp type		
Potential drop		
Current		
Rate of Conversion		

Watts = Volts × Amps

349

A. Charged area of a balloon

C.

D. Flowmeter

B.

E.

F.

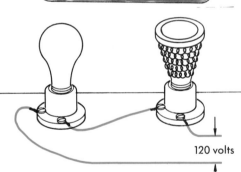

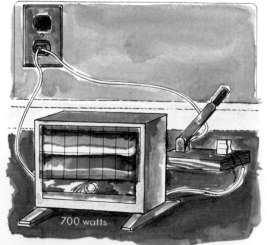

120 volts

700 watts

REVIEW QUESTIONS

1. The diagrams here show a balloon with an area of static charge as well as a circuit through which negatively charged particles flow. Select from the list of words below the word which best describes each letter here.
    voltage
    ampere
    ohm
    watt
    watt-hour
    coulomb
    ohm-meter
    ammeter
    resistance
2. What is the potential drop in the lamp at the left if the potential drop in the heater is 30 volts and the voltage of the source is 120 volts?
3. What is the resistance of the same lamp and heater if the current is one ampere?
4. At what rate is the lamp converting electrical energy to heat and light?

THOUGHT QUESTIONS

1. Many houses have 100-amp service leading to them from the power pole at the street. The appliances in the houses, however, are rated at a total of several hundred amps. Explain this difference.
2. Most houses are provided electricity through three wires. One wire has no voltage. The other two are 110 volts AC with the first wire. How is it possible for these three wires to provide the house with both 110 and 220 volts? (Hint: No transformers are involved.)
3. A 25-watt incandescent bulb provides much less light than a 25-watt fluorescent lamp. Explain why.
4. How many kilowatt-hours of energy would be consumed in the circuit at the left in 8 hours?

Investigating on Your Own

  The field of electricity has many other topics which students your age might study and understand. Few areas of study in school have as many different career or hobby possibilities. In this section you will study some topics in electricity on your own. Some of the projects in this chapter allow you to plan and carry out investigations as an electrical engineer. Other projects allow you to use your hands and develop the skills of a technician. In other projects, you can learn ideas and teach them to others.

  Many of the activities suggested in this section require more time than one class period. You may also need to locate special materials on your own. For these reasons, identify several activities which interest you from these pages. Carry out the activities on your own, outside of regular class time. Ask your teacher, however, for class time to help you plan and schedule your work. Plan to share your findings with others during later class periods.

CONTROLLED EXPERIMENTS

The physical sciences have made the greatest progress in the last two hundred years. Most people would agree that electricity and electronics have changed our daily lives as much as any part of the physical sciences. Biologists and geologists often state jokingly that more progress has been made in the physical sciences because the problems have been easier. Another way of stating this idea is that electricity and electronics can be readily studied and understood by the use of simple controlled experiments.

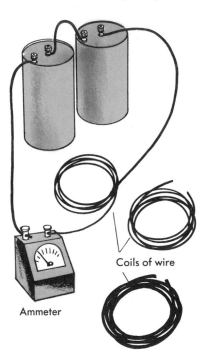

Coils of wire

Ammeter

How To Do a Controlled Experiment.
Physicists studying how the diameter of a wire affects the current flowing through the wire would probably do a controlled experiment. They would obtain wires of different diameter and measure the current flowing through each with an ammeter.

Experiments are set up so that one factor or *variable* is changed at a time. The variable that the physicists changed in the above experiment was the diameter of the wire. This variable is called the *independent* (or manipulated) *variable*. What is the independent variable in each experiment below?

Usually experimenters observe the effect the manipulated variable has on one other variable. This other variable in the above experiment is the current in the wire. This variable is called the *dependent* (or responding) *variable*. What is the dependent variable in each experiment below?

Experimenters plan so that all other possible variables will act the same way on the dependent variable during the experiment. When this is done, the investigation can be called a *controlled experiment*. During the above experiment, the ammeter, source of voltage, length, type, and temperature of the wire should all be kept the same. What variables should be controlled in the experiments below?

Problem	Independent Variable	Dependent Variable	Variables to be Controlled
How does current affect temperature of wire?	Current	Temperature	Type, diameter, length of wire
How does humidity affect static charge?		Time balloon stays charged	
How does time of year affect electricity used in my house?	Month of year		

ELECTRICITY EXPERIMENTS

Investigations like those listed below can give you experience in doing controlled experiments. Carry out one or more experiments on your own. If possible, carry out a controlled experiment which you thought up on your own. When reporting your findings, list the independent, dependent, and controlled variables.

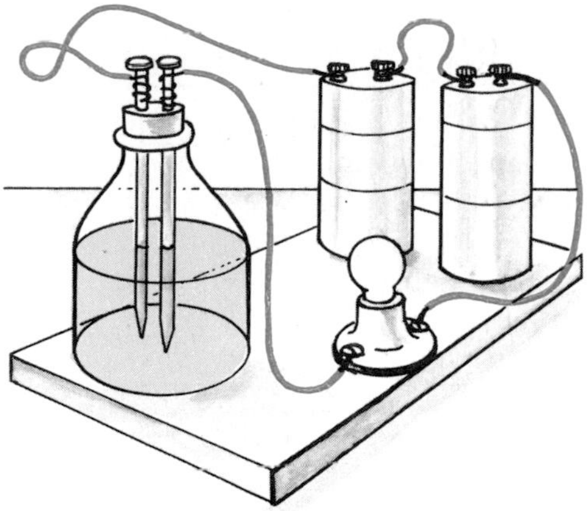

1. Using the tester shown at the left, find out whether any liquids can conduct low-voltage electricity. Try fresh water, salt water, syrup, orange juice, milk, and so on. If the lamp does not glow, can you say that the liquid is a nonconductor?

2. Test iron and copper to see which conducts electricity better. Use wires of equal lengths having the same diameter. Put one wire and then the other in a lamp circuit. Note any difference in the way the lamp glows.

3. Experiment with the effect of the length of a piece of wire in a circuit. Compare 50 centimeters of wire with 100 cm of wire. Try other lengths.

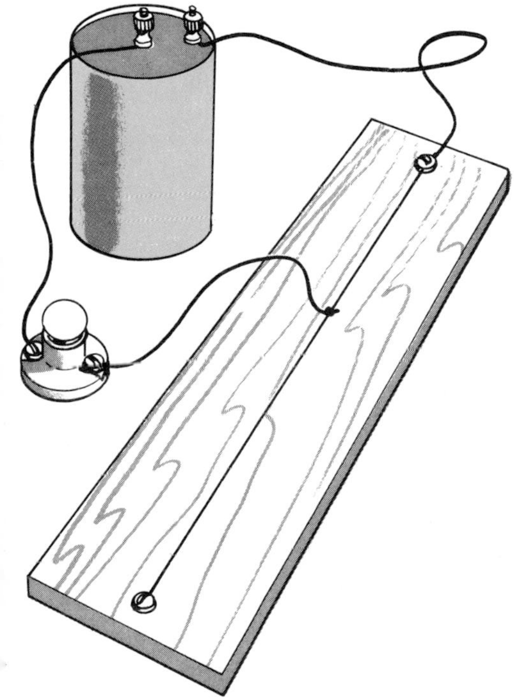

4. Fasten 1 meter of nichrome wire to two screws at the ends of a long board. Set up the circuit shown at the left. Then slice the loose end of the copper wire along the nichrome wire. What happens? How can this device be used?

5. Connect two dry cells in series and light a lamp. Find out what happens if the connections to one of the dry cells is changed.

6. Compare the heat given off by electric lamps of different wattages. Put 2 cups of water in a can and heat it for 20 minutes with one lamp. Note the rise in temperature. Repeat, using other lamps. What factors have you controlled? What factors are not controlled?

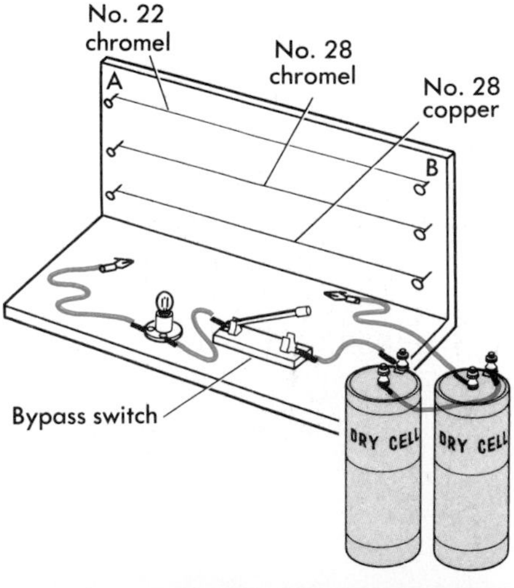

7. Find out whether a dry cell can operate if truly dry. Cut an opening in a flashlight cell, dry it in a warm oven, and test it with a galvanometer. Moisten the contents and test it again.

8. Determine the resistance of different kinds and sizes of wire. Use lengths of 1 meter or more to obtain measurable potential.

No. 22 chromel
No. 28 chromel
No. 28 copper
A
B
Bypass switch
DRY CELL DRY CELL

ASSEMBLING AND TESTING CIRCUITS

Electricity and electronics play such important parts in our lives that it is not surprising that there are numerous career opportunities. Some examples include house wiring electricians and stereo set assemblers, power company lineworkers and television set repairers, designers of new equipment such as computers and calculators, and hobbyists who may combine the skills of assembler, repairer, and designer.

In this section are projects similar to the tasks described above. Carry out at least one project from this list or a similar project of your own choosing.

1. Using the tester shown at the left, find out which pipes in a building are connected.

2. Drive two pipes into the ground a few meters apart. Using the tester shown at the left, find out if the soil will conduct low-voltage electricity. Compare the conductivity of wet and dry soil.

3. Work out a plan for wiring dry cells, a bell, and two push buttons in such a way that the bell can be sounded from two different places.

4. Design a burglar alarm that will ring a bell when a door is opened. (Hint: Make a switch that stays open as long as the door is closed.)

5. Make a photoelectric cell. Clean two copper plates with fine steel wool. Set the plates in a dilute copper sulfate solution for several days to coat them with a film of copper oxide. Then connect the plates to a galvanometer. Shine a strong light on first one plate and then the other.

6. Study the effect of alternating current on a neon glow lamp with a stroboscope. Cut a 15-cm disk of cardboard and make a slot near the edge. Mount the disk on a stove bolt with two washers, and fasten it in a hand drill. Watch the lamp through the slot as you gradually increase the speed of the drill.

7. Determine the speed of an electric fan with the stroboscope described above. It will be necessary to count the turns of the crank per minute, while viewing the fan at a speed which makes the blades seem to stand still. The speed of the disk can then be calculated from the velocity ratio of the drill.

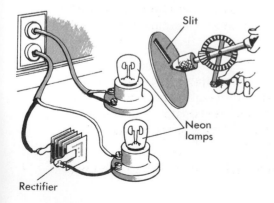

Slit

Neon lamps

Rectifier

8. Connect one neon lamp directly to an AC source. Connect another lamp to the same source

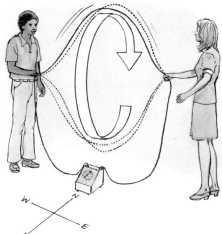

Phonograph record

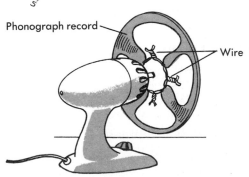

Wire

Copper

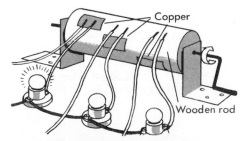

Wooden rod

with a rectifier. Observe both lamps at the same time through a slit in a disk of cardboard held by a hand drill. Speed up the drill slowly and study the synchronization of the flashes.

9. Make and study the waterdrop electrostatic generator shown at the left. Four No. 2 cans are attached to wallboard. The upper cans have no bottoms; the lower cans have small holes for draining off the water. Adjust the flow to about two drops per second.

Determine the charges on the cans with an electroscope. Connect an electroscope to a can and watch its behavior as the drops fall. Connect a neon lamp between the bottom cans.

10. Take apart a broken electrical appliance such as an iron or toaster. Trace the circuit with a battery operated circuit tester as described in the first chapter of this unit. Display and explain the circuit to your class.

11. Find out if the earth's magnetic field can induce an electric current. Connect the ends of a 10-meter wire to a galvanometer. Swing the wire like a skipping rope. Try standing in different directions.

12. Construct an electrical circuit as described in a book from your school library.

13. Read about and set up a circuit to demonstrate electroplating.

14. Set up and demonstrate a circuit containing a doorbell.

15. Set up and demonstrate a circuit containing a telegraph.

16. Wire a phonograph record to an electric fan. Explore the region around the rotating record with a compass. Charge the record with a wool cloth and find out if moving free electrons set up a magnetic field.

17. Construct a device to operate the three lights for a traffic signal as shown at the left. As the wooden rod (broom handle) is turned, the strips of copper make complete circuits touching the brushes, one at a time.

18. Rectifiers change alternating current to direct

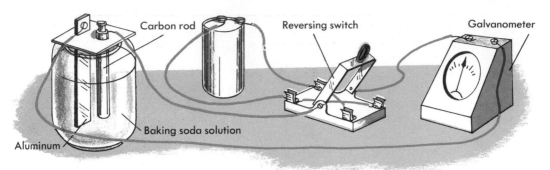

Carbon rod · Reversing switch · Galvanometer · Baking soda solution · Aluminum

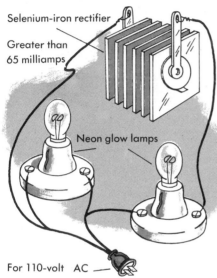

Selenium-iron rectifier

Greater than 65 milliamps

Neon glow lamps

For 110-volt AC

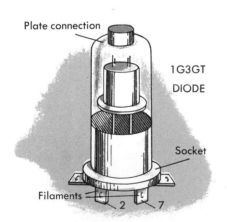

Plate connection

1G3GT DIODE

Socket

Filaments

2   7

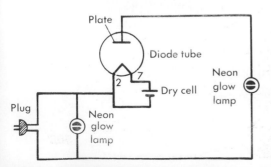

Plate

Diode tube

Neon glow lamp

7

2

Dry cell

Plug

Neon glow lamp

current. Construct a rectifier as shown above. Dissolve 15 milliliters of baking soda ($NaHCO_3$) in 1 liter of water. Trace the path of electrons from the negative terminal of the battery through the baking soda solution when the switch is in either position. In which direction does current flow through the baking soda solution?

19.   Demonstrate a solid-state rectifer using a selenium-iron rectifer with a rating above 65 milliamperes. Construct the circuit at the left. Explain your observations when the circuit is plugged in.

20.   Construct and demonstrate a vacuum tube rectifier. Obtain a diode tube of Type 1G3GT (or a substitute recommended by a radio expert). Connect terminals 2 and 7 to two dry cells as shown here. Note the tube filament heat up. Construct the rest of the circuit as shown below. Explain your observations of the glow lamps when the circuit is plugged in.

21.   Demonstrate the effect of a power diode. Connect the diode to a 6-volt battery and lamp. Reverse the connections on the battery. Describe and explain your observations.

22.   The noise-maker diagramed on page 357 will attract attention. The core of the coil is a 5-cm iron bolt with a 6-millimeter diameter. Cover it with insulating tape. Wind on 150 turns of No. 30 insulated wire, bring out a loop for the center tap, then wind 150 more turns. The 0.05 microfarad capacitor and the 62 and 220 kilohm resistors may be changed as desired to produce other effects.

23.   Construct the blinker shown on page 357. The transformer is a universal output type. The transistor

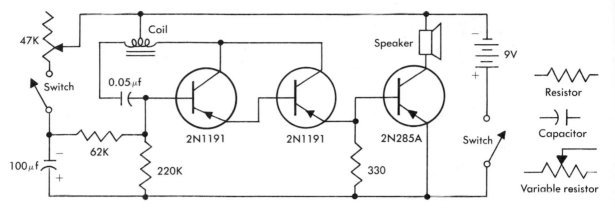

is for medium power. The potentiometer (volume control) controls the rate of flashing.

24. Study the action of capacitors. Construct the next-to-the-bottom circuit on this page. Close the switch. Which way do electrons flow? How long? Why do they stop?

Open the switch. Touch an insulated wire across the two ends of the capacitor. Explain the results.

25. Construct a flasher circuit as shown at the lower right. Observe and explain what happens. Try capacitors and resistors of other ratings. How do they affect the circuit? Propose an explanation for your observations.

26. Assemble a display of the parts of an automobile ignition system, including a battery, induction coil, distributor, and spark plug(s).

27. Design and construct an electric answer board which lights a bulb or rings a bell when a wire is touched to the correct answer.

28. Locate a thermistor (sold in radio and hobby stores). Set up the circuit shown below. Study the effect of the temperature of the thermistor on current flow. Then use the thermistor to measure the temperature of some unknown environments.

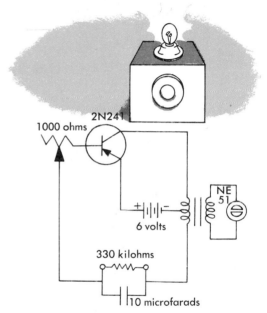

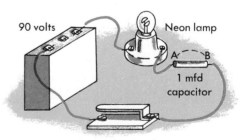

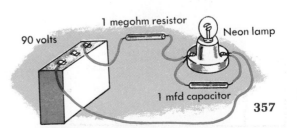

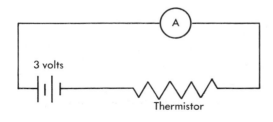

ADDITIONAL INVESTIGATIONS AND PROJECTS

Smoke stack — electrostatic precipitators off

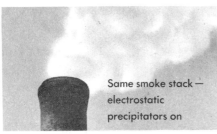

Same smoke stack — electrostatic precipitators on

Six diodes

Alternator

1.   The following people have played important roles in the development of electricity. Prepare a report on one of them: Volta, Ampére, Galvani, Faraday, Ohm, Hertz, Franklin, Steinmetz, Edison, Bell, Morse, Watt.

2.   Find out how electric charges are used to remove dust particles from smoke. Make a chart that explains the process.

3.   Collect objects that have been electroplated and display them with a chart that describes the electroplating process.

4.   Prepare a report on lightning, describing it, its effects, and telling what is known about its causes.

5.   Paint a mural that shows some of the highlights in the history of electricity.

6.   Make a chart that shows the parts of an automobile storage battery.

7.   Read about electric eels and find out how they produce electric charges.

8.   Take photographs of lightning, setting the shutter for a time exposure and closing it after a flash appears in front of the camera.

9.   Find out how the sound track of a moving picture film produces sound as it runs through a projector.

10.   Make a chart of the symbols used for electronic devices to help you in reading wiring diagrams.

11.   Find out how semiconductor diodes are used to provide direct current from the alternators in modern automobiles.

12.   Write an illustrated report on the invention and development of vacuum tubes.

13.   Make a large scale model that shows the construction of a transistor or triode tube.

14.   Prepare an exhibit of different kinds of capacitors, taking some of them apart to show the nature of the insulation between the plates.

15. To help classmates interested in electronics projects, prepare a list of books available in local libraries and for sale in bookstores and radio supply stores.

16. Take apart a discarded microphone and study its construction. Prepare and give a report on its operation.

17. Find out how a tape recorder is able to record and reproduce sounds. Experiment with recording sounds at one speed and reproducing them at another speed.

18. Write a short history on the development of electric motors and generators.

19. Visit a steam-electric plant or a hydroelectric plant. Find out the voltage at which electricity is generated and the voltage to which it is raised for transmission. Find out also the maximum power, in kilowatts, which the plant is capable of producing.

20. Use topographic maps to locate the transmission lines, transformer stations, and generating plants that service your community.

21. Make the shocking coil shown at the left. Wind two layers of No. 16 wire on a bolt for the primary coil. Fill the remaining space with windings of No. 32 wire for the secondary coil. Scrape one of the primary wires along the nails to produce an interrupted current. Ask someone to hold the secondary wires to feel the shock.

22. Using doorbell transformers, set up a model of a high-voltage transmission system as shown below. The first transformer represents a generator and may be concealed in a box that looks like a steam-electric plant.

23. Read and report about the methods being used to convert nuclear energy to electrical energy and

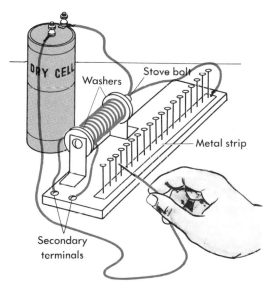

DRY CELL

Washers

Stove bolt

Metal strip

Secondary terminals

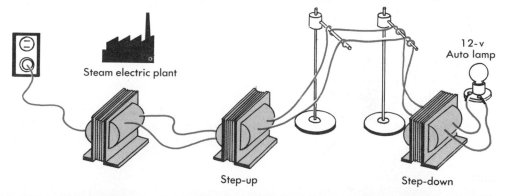

Steam electric plant

Step-up

Step-down

12-v Auto lamp

Transformer station

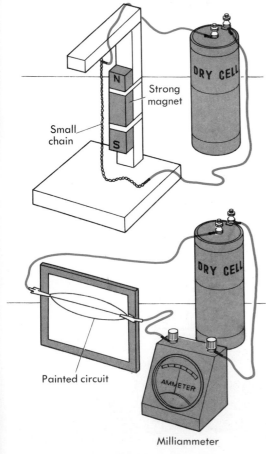

Strong magnet

Small chain

N

S

DRY CELL

Painted circuit

DRY CELL

AMMETER

Milliammeter

the advantages of nuclear power sources over conventional sources.

24.   Obtain discarded alternating current motors. Take them apart and locate the parts. Ask a physics teacher or an electronics technician to help you understand their operation.

25.   Visit a transformer station and ask to be shown the transformers, voltage regulators, lightning protectors, and switches. Take photographs and prepare a report on the system.

26.   Make a chart which shows how the magnetic poles of a DC motor reverse every half revolution.

27.   Take apart an automobile ignition coil and exhibit the parts, together with labels which explain the function of each.

28.   Find out how ammeters and voltmeters operate and make charts that explain their construction.

29.   Ask an electrical engineer to explain "three-phase" alternating current systems that use three-wire transmission lines, such as those seen throughout the countryside.

30.   Obtain a discarded automobile generator, take it apart, and identify the parts. Explain its operation to your class.

31.   Make the device shown at the left and demonstrate the force on a conductor which carries a current in a magnetic field.

32.   Investigate the ignition system of an outboard motor or a power mower to find out how high-voltage electricity is supplied to the spark plugs.

33.   Find out whether a small DC motor can operate as a DC generator when provided with mechanical energy.

34.   Demonstrate printed circuits with the materials shown at the left. Use salt water tinted with dye for the conducting ink. Moisten the paper around each clip with a watercolor brush, then draw a line between them. The milliammeter should show a small current. Other lines may be drawn to provide parallel circuits.

Getting Electricity to Your Home

6

The most common form of energy used today is electricity. In the home, electricity powers radios, televisions, and telephones. Electricity runs motors to wash and dry clothes and dishes, heat and cool homes, or freeze foods for storage. Electricity also provides light as well as high temperatures for cooking.

Outside the home, electricity provides the energy for many devices. Highways are lighted and traffic is controlled by electricity. Electric motors run elevators, buses, and subways to transport people and materials. Computers, billing machines, typewriters, and duplicators all rely on electricity. Electricity is also used for separating metals from their ores and for welding steel to make metal structures. Most farmers, repair shops, and manufacturers depend on electricity to run machines.

Electricity is used in so many situations that without it our lives would change dramatically. Where does electricity come from? How is it produced? How does electricity get from where it is produced to where it is used? These are some of the questions you will examine in this chapter.

TURNING THE GENERATORS

For several hundred years, much of our manufacturing has depended on rotary (turning) motion. The photograph above shows a mill originally built in 1635 in Plymouth, Massachusetts. What force acts on the water wheel to produce rotary motion?

The photograph below shows a Chicago factory 100 years ago which made farm tools and machinery. Estimate the number of machines in this picture which depend on rotary motion. Suggest why such factories were usually built along fast-moving rivers.

Imagine this factory 40 years later when the owners were able to install electric wiring and small electric motors to operate each machine. How would the picture change? What would be some advantages of electrically powered machines over the earlier versions shown here?

Turbines. Records do not tell who the first person was to devise a simple engine to turn an electric generator. Probably this involved joining the armature shaft on a generator to the axle of a water wheel as shown at the left. Soon, early experimenters realized that water, wind or steam could be used to turn electrical generators. The experimenters also learned that if the blades were curved, the engine would turn faster, producing more electrical energy. Such engines with curved blades are called *turbines*.

Water Turbines. Water wheels mark the beginning of producing large amounts of rotary motion without using the energy of animals or people. Perhaps this is one reason we still enjoy visiting old, restored water-powered mills. Many people do not realize that water wheels are still used today, producing more power than ever. These present day grandchildren of the early water wheels are now called *water turbines*. Unfortunately, water turbines are hidden from view, usually located at the bottom of dams such as shown below. Note the different way this turbine will turn compared to the old style water wheels. When this turbine and generator are installed, all that can be seen of the turbine, even from within the power station, will be a large steel cover.

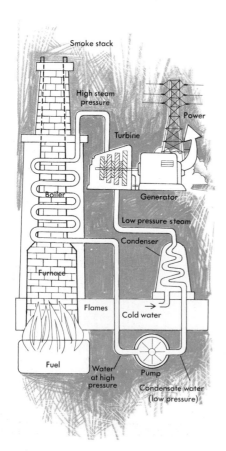

Smoke stack

High steam pressure

Power

Turbine

Boiler

Generator

Low pressure steam

Condenser

Furnace

Flames

Cold water

Fuel

Water at high pressure

Pump

Condensate water (low pressure)

The Steam Turbine. Today, over 90% of the electrical generators receive their rotary motion from steam turbines. As in the case of water turbines, these engines are rarely seen by most people. Only the occasional visitor to an electric generating station ever sees the outside cover. Only the rare visitor (or worker) at a steam turbine factory sees the precision parts which make up this important engine.

Turbine designers and builders have one goal. They build turbines which give the greatest amount of rotary motion from steam so that the greatest amount of electricity can be produced.

The photograph shows a steam turbine being assembled. When completed, it will be able to turn a generator large enough to supply the electricity for the houses, schools, streets and businesses for a city of 550 000 people.

Fossil Fuel Generating Plants. Over 90% of the energy to make steam for turbines comes from fuels such as coal, lignite, oil, or natural gas. These fuels have been formed from the matter of animals and plants that lived hundreds of thousands of years ago. These fuels are called *fossil fuels*.

Chemical energy in a fossil fuel is turned into heat energy by burning. The heat energy is used to boil water which expands enormously as it changes to steam. The steam is piped under great pressure to a turbine. When the turbine blades are hit by the steam, they spin rapidly, converting the steam's energy into mechanical motion. The turbine's shaft is also the shaft of the generator, thus the mechanical energy of the turning turbine converts mechanical energy into electrical energy.

The steam which leaves the turbine has lost most of its energy. It is much cooler when it enters the condenser. The condenser further cools the steam, turning it into water. The water is pumped back to the boiler to be heated again.

Nuclear Powered Generating Stations. Fossil fuels are becoming scarce so scientists and engineers are searching for other energy sources. One source is nuclear energy. The nuclear furnace, called a *reactor*, uses radioactive elements to heat water to steam. Nuclear fuel is usually uranium, thorium, or plutonium. The radioactive atoms in nuclear fuels release neutrons. These neutrons in turn hit the nuclei of other atoms, causing them to release more neutrons. All this action heats up the reactor.

The neutron activity and heating can be slowed down by placing rods of material into the reactor which absorb (stop) neutrons. As these control rods are raised out of the reactor, it heats up; as the rods are lowered, the reactor cools down. One advantage of a nuclear reactor is that the control rods can regulate the heat produced as the need for electricity changes.

Water is circulated through pipes in the hot reactor. The heated water is pumped through a heat exchanger where it cools somewhat and then flows back to the reactor where it is reheated. The water is used over and over again, never leaving the primary loop.

Water in the secondary loop flows around the pipe containing the hot water from the reactor. This water in the secondary loop is heated to form steam. The steam drives the turbine.

Many people are opposed to nuclear power plants because they fear a power station may heat up out of control, causing a nuclear accident. Such an accident might involve spilling radioactive materials into the air, contaminating many square hectares of land, or killing people. Another problem with nuclear plants is the disposal of the "ashes" of radioactive fuels. These "ashes" remain dangerously radioactive for years.

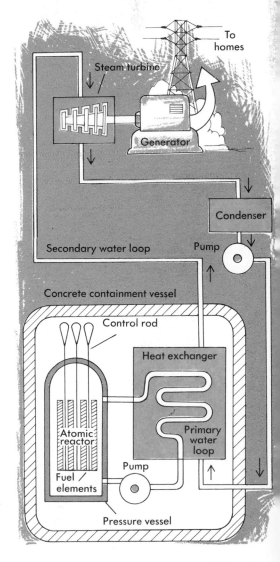

GENERATING AND TRANSMITTING ELECTRICITY

Electricity is one form of energy that cannot be stored in large amounts. There are storage batteries in automobiles, but these batteries store very little electricity considering their size. In addition, batteries will not store electricity for very long. The next millionaire inventor may be the person who discovers how to store large amounts of electricity for long periods.

As electricity use changes, the ability of a power generating station to produce electricity must also change. Every evening at dinner time, if 30 million electric stoves are turned on, more electricity must be produced and transmitted instantly.

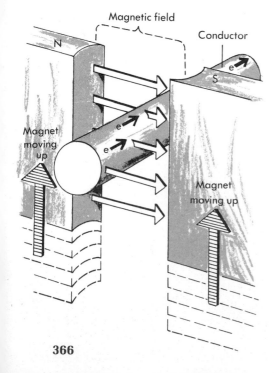

Reviewing Electricity Generators. The drawing shows a wire moving through the lines of force in a magnetic field. In such a conducting wire, electrons tend to move toward one end of the wire and away from the end. If a return path is set up, an electric current will result. The device which makes use of this idea is called an electric generator. Generators are of two general types: those which produce a steady or direct current (DC), and those which produce a current in one direction and then in the other direction (AC). These two types of generators were described on pages 314–319.

AC or DC Generators: An Early Decision. In the early 20th century, decisions were being made about how to set up electrical systems for cities and towns. The engineers faced the difficult decision about whether to generate direct current (DC) or alternating current (AC).

The early planners predicted a great demand for electricity. To meet this demand, there appeared to be two choices for generating and transmitting the electricity. One choice was to generate and send electricity at household voltage and at very high current over a transmission line to the houses which wanted electricity. Unfortunately, high current in wires produces heat, and most of the electricity is wasted. For this system to work, every house would have to be within 3 kilometers of a generating station and the wires would have to be very thick (and expensive).

The second choice was to transmit electricity at very high voltages and low current. Much less energy (heat) is lost this way. Fewer generating stations would be needed and thinner wires could be used. Before the electricity entered houses, however, the voltage would have to be reduced with a transformer to a safer level. DC transformers are much more expensive and less reliable than AC transformers, which have no moving parts. Consequently, nearly all electricity produced today is generated as AC and is transmitted at high voltages.

Transmission Lines. Transmission lines leading from generating stations generally carry very high voltages of more than 60 000 volts. The lines usually are strung from poles or towers. Insulators keep the lines from coming in contact with the towers. The lines must be strung tightly enough to prevent sagging and swinging. In addition, the towers must be strong enough to support ice that may form on the lines in winter. As ice forms on the lines, they will be blown more by the wind.

Some people feel that overhead transmission lines are an eyesore. Many cities, towns, and villages have refused to allow them. In these cases the lines are buried underground. Generally, solid insulation is used for the cables and they are protected further by being placed in iron pipes. Sometimes the pipe is filled with oil kept under pressure by pumping stations along the line. The oil is a very effective insulator for lines carrying very high voltages. The cables are pushed through ducts running underground.

Underground transmission lines have several problems, however. They are expensive to set up and maintain, about ten times more expensive than

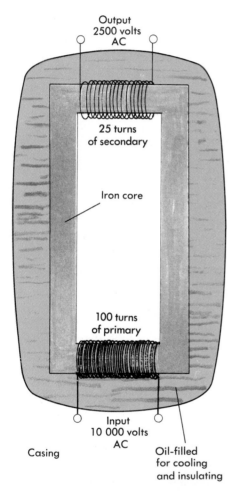

Output
2500 volts
AC

25 turns
of secondary

Iron core

100 turns
of primary

Input
10 000 volts
AC

Casing

Oil-filled
for cooling
and insulating

overhead lines. When breaks occur, they are difficult to find and require more time to repair. Customers depending on a broken underground line are likely to be without electricity for a long time.

Engineers have designed more appealing towers to carry the lines. Compare the towers on page 367. Which is least attractive? Most attractive?

Transformers. An AC transformer operates at very high efficiency, that is, the power given off is nearly equal to the power put into it. This remarkable efficiency is one of the major reasons that AC current is used throughout most of the world today.

The AC supplied today is a 60-cycle current which means that it alternates 120 times a second. Therefore, a magnetic field around a coil would change its direction 120 times a second. An electric current will start flowing in another coil of conducting wire cutting across the changing magnetic field. The number of coils of the primary wire compared to the number of coils in the secondary wire determines the voltage induced in the secondary coil. The drawing shows how to step down 10 000 volts to 2500 volts. The primary wire is wrapped around an iron core. The secondary wire is wrapped around the primary wire (without electrical contact) with one-fourth as many turns. Although the input is 10 000 volts, the output is 2500 volts. Note that the amperage or amount of current entering is one-fourth that of the current being put out. The transformer is very efficient, but it loses some energy because it heats up. Oil is used to cool the transformer.

Distribution Systems. A distribution system begins with a substation connected to a transmission line. The voltage is reduced to a level considered safe for populated areas. (This distribution is called the primary distribution system. The secondary system further reduces the voltage to the level at which electricity is used by the consumer.) Primary systems may range up to 13 800 volts. Systems leading to houses and apartments usually operate at 2400 to 4800 volts. (Secondary circuits are fed from transformers that reduce the voltage of the primary circuit to 110 or 220 volts for supplying family units.)

Distribution Circuits. A line leaving the substation may have several lines running off it to

serve various side branches. This arrangement is
called a *radial circuit*. It has the drawback that if a
short circuit occurs in the main line, the whole line
goes dead and many customers have no electricity.

Another design for a circuit has several lines
running from the substation to supply electricity to
the primary system. Such a circuit is referred to as
a *network*. If one line is short circuited, there may
be no interruption of service because other lines
can carry the electricity.

Power Sharing. Electric power, as you already
know, cannot be stored in any large amounts, so it
has to be used immediately. Suppose a generator is
supplying a steady load of power to a town. Then
as the sun sets, all the streetlights are suddenly
turned on and people begin coming home and
cooking their dinners. What happens to the
generator? As the load of electric current jumps
up, the generator starts to slow down, just as any
machine does when it becomes overloaded. This
slowdown is noted by a sensitive speed control that
automatically feeds just enough extra steam to the
turbine to speed up the generator again. Drawing
more steam reduces the temperature in the boiler.
This temperature change is sensed by a device
which then feeds more fuel to the generator's
furnace or moves a control rod in a reactor. The
extra electric power needed at dusk comes from
using extra fuel.

Brownout. What happens if the turbine is using
all the steam it can handle and the generator is
turning out all the current it possibly can, and there
is demand for more power? If the overloaded power
plant is part of a regional network, the overload
signal is sent to the nearest underloaded power
plant which takes up the load. If still more power is
needed and no more is available, then the power

Radial circuit

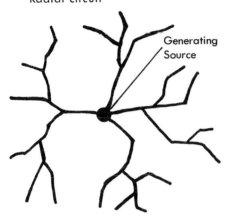

Generating
Source

Network circuit

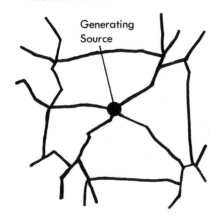

Generating
Source

369

company must decrease the voltage on the lines. When this happens it is called a *brownout*.

Blackout. Almost all electrical equipment is designed to run at a single voltage. Many appliances run at 110 volts. Heavier appliances, such as air conditioners and electric clothes dryers, run at 220 volts. (Changes in power are made by changing the current.) Most appliances have a built-in safety margin so that they will operate on as much as 3% less voltage. A 3% drop-off is hardly noticeable. But sometimes 3% is not enough and the power company has to cut the voltage by 5%. This causes a few problems because some electric motors will not start and lightbulbs are noticeably dimmer. If the power has to be cut still further, an 8% cut will be used. With normal voltage drops as occur regularly in a house or office when automatic devices start up, many motors simply will not start. If an 8% brownout is not enough, then the power must be cut off completely to some customers. This is called a *blackout* to those affected. The power company has the difficult task of deciding which sections will be least affected by a blackout.

Protective Devices. Various safety features have been designed for use in distribution systems. Some protect against sudden surges caused by lightning. Some protect against short circuits. Others offer protection from overloading.

Protection from Lightning. Since transmission lines usually run through open country and often over mountain tops, they are vulnerable to lightning. Should a lightning bolt hit an overhead cable, the lightning will cause a surge of extremely high voltage along the cable. The surge causes heavy damage to the insulators and might cause a short circuit in the system. Even if lightning does not strike the lines directly, just the presence of a thunderstorm is often enough to cause a surge. To

Lightning arrestors

Grounding cable

protect the cables from electricity in the atmosphere, ground lines are strung from tower to tower at the topmost point. At each tower a heavy cable leads from the ground line to conductors buried in the ground at the base of the tower.

Lightning arresters are usually connected between each transmission line and are grounded close to the transformers. The photo on page 370 shows a stack of lightning arresters. These will drain off the extraordinarily high electric current due to lightning. Lightning arresters do not allow line currents to pass at ordinary operating voltages. When lightning surges occur (some have been around 100 000 000 volts) the lightning arresters become conductors, discharging electricity harmlessly to the ground. Lightning arresters do not allow the line current to follow the surge.

Relays. A relay is nothing more than a switch which is operated by an electric current. The diagram shows how a relay works. When current flows through the electromagnet, the magnetic force closes the switch. When the current is turned off, a spring pulls the switch back to the off position.

Note two important points. First, the circuit of the electromagnet and the circuit of the switch are not connected electrically. The magnet requires only enough current and voltage to move the switch. The switch can be built to open or close circuits which carry very high voltages. Second, the switch that supplies power to the electro-magnet can be at some distance from the relay.

In every central power station, there are many generators that must be switched on and off the power lines as current demands change. The huge switches required to handle the enormous amount of current are too large to operate by hand. Relays are used to do the switching. One relay after another can be used until a relay is small enough to fit on a central control panel.

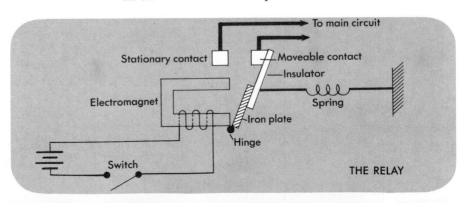

THE RELAY

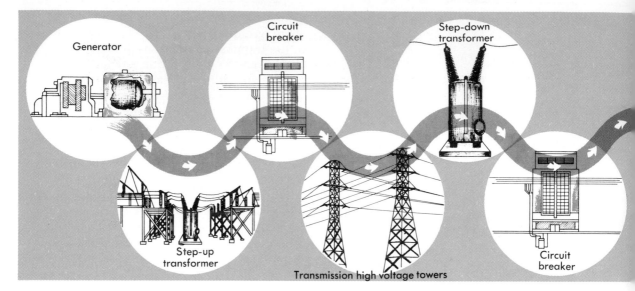

Generator

Circuit breaker

Step-down transformer

Step-up transformer

Transmission high voltage towers

Circuit breaker

Generating Stations. If electricity could go directly from the generator to the transmission line, generating stations would not have to be so complicated. Various precautions must be taken to avoid short circuits. Short circuits are dangerous to the people in the station and cause expensive damage that would take time to repair. In addition, to keep the cost of transmitting electricity down, voltage must be stepped up through a transformer. Most generators produce from 10 000 to 18 000 volts, but transformers step up electricity to 345 000 volts or more.

Substations. Electrical power is drawn off the high voltage lines to substations where the voltage is stepped down for transmission into the primary distribution system. Again, circuit breakers are provided for safety before electricity leaves the substation carried in insulated cables strung on poles or buried in underground pipes. Overhead wires are supported on porcelain insulators attached to the crossarms of the poles. Circuit breakers or fuses or both may be installed along the line to protect against overloads or short circuits.

The Consumer. When electricity is needed for a house or apartment, the primary distribution system is connected to a step-down transformer. The transformer further reduces the voltage to 110 or 220 volts. These transformers are often mounted on poles in an overhead system or they may be in a box above ground. From these transformers, the secondary distribution system carries electricity

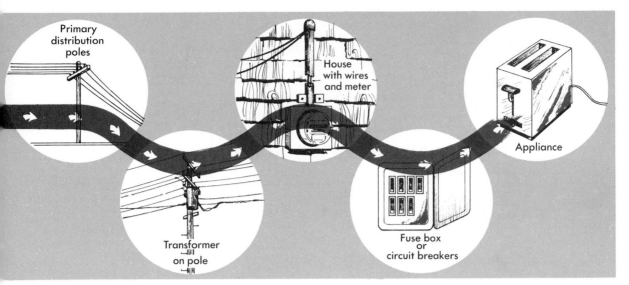

Primary distribution poles

House with wires and meter

Appliance

Transformer on pole

Fuse box or circuit breakers

directly to the customer. The electricity passes through a meter and the fuse box inside the customer's dwelling.

The Electric Meter. Consumers buy electricity in kilowatt-hours. It is a unit of measurement like kilograms of meat or meters of cloth. Each customer of the electric company has an electric meter to measure how many kilowatt-hours are used.

Within the meter there is a metal disk. The amount of electricity you use determines the speed at which the disk turns. Each revolution measures a precise amount of electric energy. A 100-watt light bulb lighted for one hour uses 100 watt-hours of electricity. Since the watt-hour is so small a unit, another unit is used equal to 1000 watt-hours. This unit is called a kilowatt-hour.

Reading the Meter. When reading an electric meter, always read the dials from left to right. Some dials run clockwise while others run counterclockwise. Write down the last number passed by the pointer on each of the dials. For example, the meter here reads 66 482. In order to figure out how much electricity you have been

Disk

Appliance	Number of Turns in 3 Minutes
Toaster	
Washing machine	
Vacuum cleaner	
Television	
Iron	

using, subtract the last meter reading from the current reading.

You can determine the amount of electricity used by the various appliances you use at home. Ask someone to plug in the toaster (or other appliance you want to test) while you count the number of turns made by the meter wheel in 3 minutes. Make up a chart such as the one here.

REVIEW QUESTIONS

1. Why did electrical engineers decide to generate alternating current rather than direct current for use in homes and factories?
2. How is it possible to make electrons flow along a conductor?
3. In what ways is a nuclear power plant different from a fossil fuel power plant?
4. What are three examples of fossil fuels?
5. How is water used in a fossil fuel power plant?
6. How are nuclear reactions controlled in the core of a nuclear power plant?
7. What factors affect the conductivity of a metal wire?
8. What are two kinds of turbines and how are they different?
9. How many coils must there be in the secondary wire of a transformer compared to the primary wire in order to reduce 10 000 volts to 2500 volts?
10. What is a brownout?
11. How are transmission lines protected against lightning?
12. Why do transmission lines carry very high voltages?

THOUGHT QUESTIONS

1. How can a generator produce higher voltage?
2. Cities use lots of electricity. Where are the transmission lines?
3. Why is the term "telephone pole" an incorrect term?
4. What invention would reduce the need for more power stations?
5. Why are relays very useful in electrical distribution systems?
6. Propose why internal combustion engines are less efficient than turbines.

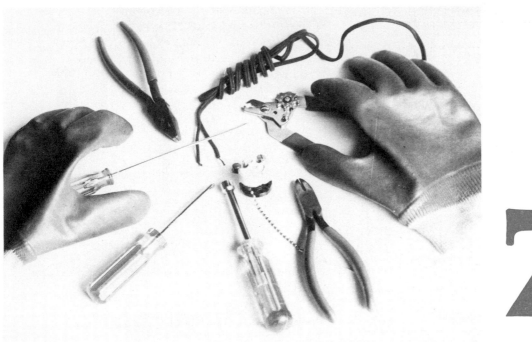

Electrical Safety

Electricity is one of the wonders of the modern world. Your great grandparents would stand in amazement at the jobs electricity performs and the pleasures it provides by running television sets, radios, stereos, and various games. Unfortunately, these benefits may also provide some danger. Electricity must be treated with care and thought.

This chapter introduces safe practices and safety devices that have been designed to offer protection from electrical accidents. But all the safety devices in the world are useless if you don't use electricity intelligently. Having studied this unit, you should understand the reasons and importance for using electricity safely.

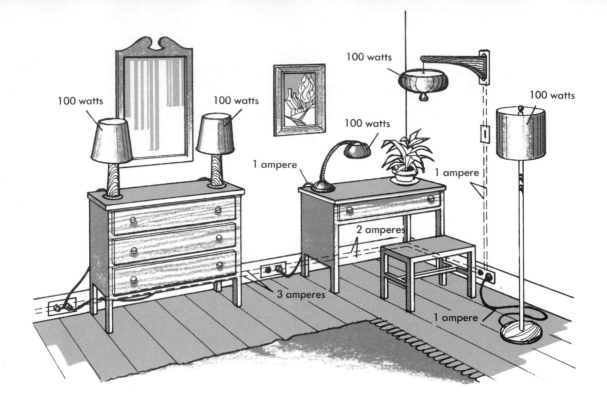

100 watts

100 watts

100 watts

100 watts

100 watts

100 watts

1 ampere

1 ampere

1 ampere

2 amperes

3 amperes

1 ampere

PROTECTING HOMES FROM ELECTRICITY

Although electricity can be very useful, it can be dangerous too. Therefore, ways for protecting homes against the dangers of electricity have been invented.

Several kinds of protective devices are in use today. All are successful when they are used properly. However, when people are careless and when they deliberately misuse electrical devices, accidents happen.

DETERMINING CURRENT. To determine the current through any device such as a toaster, divide the rating of the device by the voltage of the electricity used, usually 110 volts.

$$\text{Current} = \frac{500 \text{ watts}}{110 \text{ volts}} = 4\frac{1}{2} \text{ amperes}$$

Overloading Electric Circuits. Several 100-watt lamps are connected in parallel in the picture above. If one lamp is lighted, the current in the wires is small. When a second lamp is turned on, the current is doubled. A third lamp triples the current. If many lamps are lighted, a large current flows and the wires may become hot.

The flow of electricity is measured in *amperes*. The current through one of the 100-watt lamps shown above is about 1 ampere (if the voltage is 110 volts).

Study the picture. When all the lamps are lighted, what is the total current in the wires? Which part of the wiring has the greatest current?

Safe Currents. Most wires that carry 110-volt electricity in a home stay safely cool if the current is less than 15 amperes. This means that only one 1000-watt device should be used in a circuit at one time.

If more electric power is needed, such as for an electric stove, electricians install cables containing larger wires. How does this help?

Separate Circuits. People usually want to use several electrical devices at the same time. Therefore, homes are provided with many separate circuits, each of which can safely carry a current of 15 amperes. The diagram below shows a home that has three separate circuits.

Circuit A provides current for the lamps and outlets of the bedroom and living room. Devices that require a large current are not often used in these rooms.

Circuit B provides electricity only for the electric stove. If a clothes dryer were used, it would need a separate circuit also.

Circuit C provides electricity for the outlets in the kitchen. Devices such as toasters and waffle irons usually require a large current.

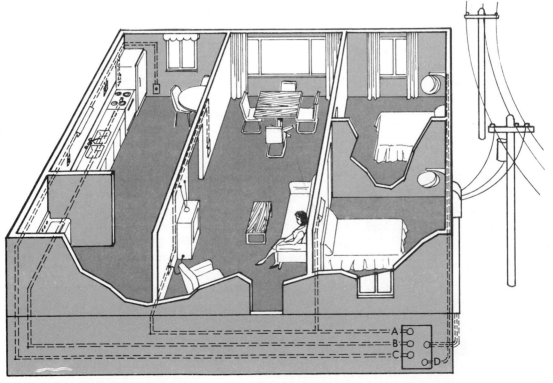

Normal

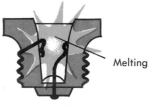

Melting

Overloaded

Fuse wire melting

Short Circuits. Circuits are sometimes overloaded because the covering on electrical wiring becomes worn or damaged. If the bare wires come in contact with each other, a new circuit is set up through the point of contact. This new circuit has a very low electrical resistance, and an enormous current is produced.

The new circuit is called a *short circuit*. It results because most of the electricity in the wires flows through the point of contact instead of through the electrical devices connected to the wires.

Electric Fuses. Homes are protected against overloaded circuits by automatic switches that open whenever the current in the wires becomes dangerously large. One of the most common types of automatic switches is called an *electric fuse*.

An electric fuse contains a strip of metal that melts at a relatively low temperature. All the current in a circuit passes through this strip of metal. If a dangerous amount of electricity flows through the circuit, the metal strip heats up and melts.

The diagrams above show what happens when a short circuit is produced in a worn electric cord. A large current flows and the metal strip in the fuse melts, breaking the circuit before any harm is done.

The photographs show the point of contact between two wires in a worn electric cord and the fuses in the same circuit. Note how hot the wires become. Note also the light given off by the melting fuse.

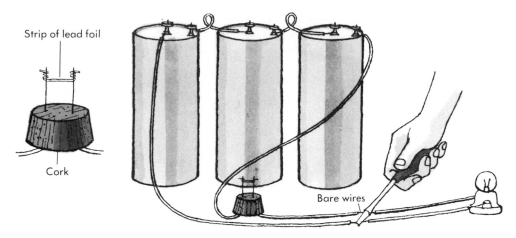

Strip of lead foil

Cork

Bare wires

A Model Fuse. Make a model fuse like that shown above at the left. Lead foil serves best for the fuse because it melts at a low temperature. Aluminum foil may be used but the strip must be very narrow or it will not become hot enough to melt.

Connect the fuse in a circuit. Produce a short circuit by laying a screw driver across bare places in the wires.

HOW MANY APPLIANCES IN A CIRCUIT? Multiply the rating of the fuse in the circuit by the voltage of the electricity. This gives the number of watts that the circuit can take care of safely.

For example, suppose that a 15-ampere fuse is used in a 110-volt circuit. Energy may be used safely at the rate of 1650 watts. Therefore, a 1000-watt toaster and a 500-watt coffeemaker can be used at the same time. How many 100-watt bulbs could be lighted by this circuit?

15 amperes $\times$ 110 volts $=$ 1650 watts

toaster	1000 watts
coffeemaker	500 watts
total load	1500 watts

Build the apparatus at the right.
How many watts can the fuse accommodate?
Plug in one lamp, then the other, then the iron.

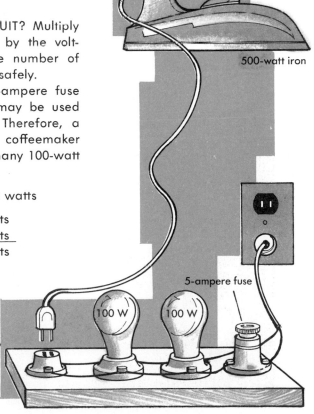

500-watt iron

5-ampere fuse

100 W

100 W

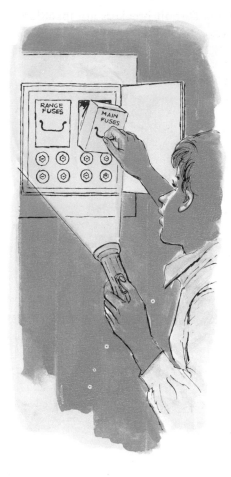

Replacing a Fuse. Everyone should know how to replace a fuse because no appliance can be used in a circuit until a melted fuse has been replaced. Below are listed the steps to be followed.

1. First, find the cause of the overload. If the overload is not removed, the new fuse will melt immediately.

If the overload was caused by too many appliances, remove some of them from the circuit. If the overload was caused by a short circuit in an electric cord, discard the cord until it is repaired.

2. Next, open the master switch that connects the house circuits to the transmission lines outdoors. It is dangerous to change fuses in a circuit that is connected to a source of electrical energy.

Opening modern fuse boxes, such as the one shown at *A*, opens the master switch at the same time. The box remains locked while the master switch is closed.

There is no master switch in the fuse box shown above. Instead, the main line fuses are pulled out to open the circuit. This is the same as opening a switch.

The fuse box shown at *B* has a master switch installed above it. This switch should be opened before changing a fuse.

3. Find the fuse that has melted. The metal strip can be seen through the glass top of the fuse.

4. Unscrew the melted fuse and replace it with a new fuse of the same rating.

5. Close the master switch.

Look for the fuse box in your own home. Find the master switch. Find out where the new fuses are kept. Ask someone to show you how to replace the fuses.

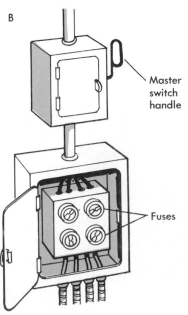

B

Master switch handle

Fuses

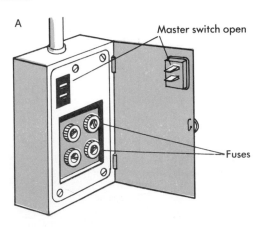

A

Master switch open

Fuses

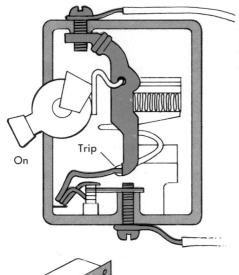

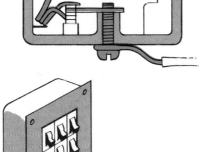

On   Trip

Circuit Breakers. Circuit breakers are installed in many new homes in place of fuse boxes. A circuit breaker is more convenient than a fuse because the automatic switch can be closed again by pushing a knob.

There are two general types of circuit breakers. One type depends upon an electromagnet to open a switch in case of an overload. The other type depends upon a metal strip which bends when overheated, thus opening the switch.

The second type of circuit breaker is shown here. The switch is held shut by a strip of metal called the trip. The path of the electricity is shown by a colored line. Note that the current passes through the trip.

A dangerously large current heats the trip which bends and releases the switch. The switch can be closed again by pushing the knob up and then down again. The trip then holds the switch closed until the next overload.

The model circuit breaker shown below operates with an electromagnet. Adjust the trip so that it springs back when the metal strip is pulled down by the coils. Produce a short circuit in the wiring and watch the circuit breaker. What happens?

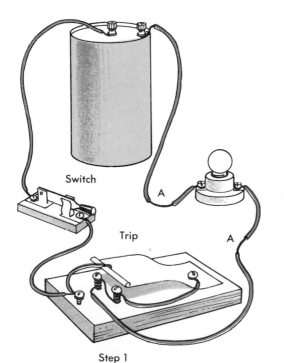

Switch

Trip

A

A

Step 1

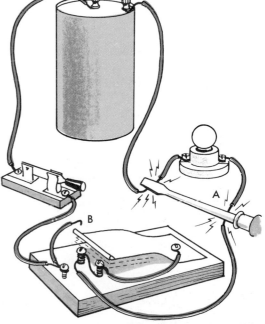

B

A

Step 2

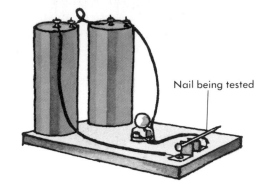

Nail being tested

This device shows whether a substance is a good insulator for 3-volt electric circuits. If the lamp lights, the substance is a good conductor and a poor insulator. If the lamp does not light, the substance is a poor conductor.

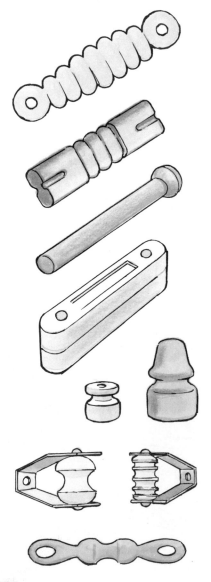

Voltage and Safety. The model house described at the beginning of Chapter 1 is supplied with 3-volt electricity. The usual home, on the other hand, is supplied with 110-volt electricity. The first voltage is harmless. The second can cause injury or death.

High-voltage electricity has more energy than low-voltage electricity. The higher the voltage, the more difficult electricity is to control and the more dangerous it becomes.

Insulators. Electricity can be controlled by surrounding wires with materials that are poor conductors of electricity. These materials are called *insulators*.

Insulation may be of different materials and different thicknesses. For low voltages, a thin layer of cloth, enamel, or plastic might be quite sufficient.

Examine the insulation on wires used for dry-cell circuits. What materials are used and how thick are the coatings?

Thicker and stronger materials are needed to insulate 110-volt electricity. Study the insulation on wires used for 110-volt circuits. What substances are used? How thick is the insulation?

Insulation for high-voltage circuits must be kept dry. Moisture in an insulating material may allow electricity to flow where it is not wanted. Why is rubber a safer insulator than cloth? How is moisture kept from the insulation inside electric cables?

At the top of the page is shown a device for discovering materials that can serve as insulators at low voltages. Test a number of different materials with it, including those shown at the left.

SAFETY RULES FOR USING ELECTRICITY

DON'T touch the metal parts of appliances or electrical fixtures when standing on a wet floor, damp concrete, or damp earth.

DON'T use frayed or broken electric cords.

DON'T use electrical appliances that cause sparks or blow fuses. Either have them repaired or throw them away.

DON'T pull out plugs by the cords. This may break the wires and cause short circuits.

DON'T handle electrical appliances while in the bathtub. DON'T put them where they may fall into the water.

Replace melted fuses with others of exactly the same rating.

Before replacing fuses or repairing electrical fixtures, open the master switch in the main line.

DON'T overload circuits.

DON'T climb the poles of transmission lines. DON'T fly kites near the wires.

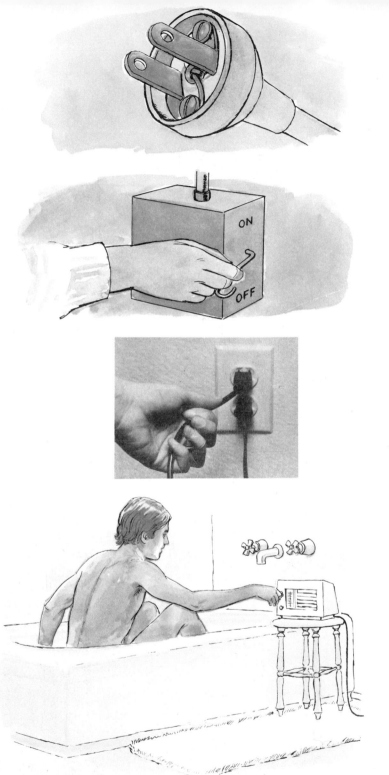

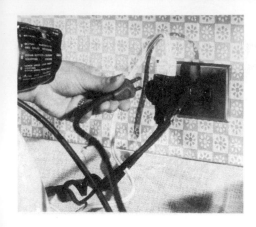

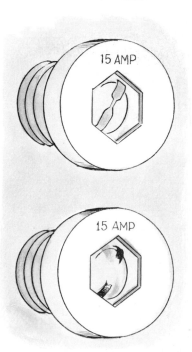

Explain the safety rules on the previous page as they relate to these illustrations.

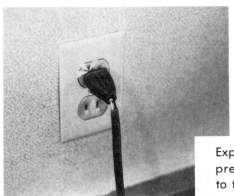

ENERGY

Work and Power

During 1973–1974 the OPEC (Organization of Petroleum Exporting Countries) nations of the Middle East cut off our oil supplies. Fuel became scarce, causing schools to close for lack of heat and American motorists to wait in line for hours at gasoline stations. When the OPEC nations resumed the sale of oil to us the price was quadrupled. In a very abrupt way these events made us aware of our large consumption of energy, something we had been taking for granted. Since we still consume so much energy and depend on foreign countries for it, the energy crisis of 1973–1974 has never been resolved. By conserving energy, exploring for more fossil fuels, and developing alternative sources of energy, the

United States is attempting to become more self-sufficient in energy.

The concept of energy is easy to describe but it is not easy to understand. We often associate energy with a fuel which is used to make something happen. Whatever happens usually makes our lives more comfortable. Gasoline makes a car go, nuclear or fossil fuels are used to produce electricity, and the food you eat supplies your body with energy to ride a bike or throw a ball. By examining how fuels are used and their limitations, we can get a better understanding of energy.

WHEN FORCE CAUSES THINGS TO MOVE

Work has different meanings to different people. A student uses the term work in reference to studying such as working on a science project or doing homework. To keep his or her muscles in shape, an athlete constantly works out. A basketball team requires team work. A radio that falls and breaks will not work. What other ways is the word work used?

$W = F \times d$
$W = $ Work
$F = $ Force exerted on the object
$d = $ Distance the object moves

Work In Science. Work has a specific meaning to the physical scientist. He or she defines work quantitatively through the interaction of mass and the forces exerted on the mass. A force is a push or a pull exerted on an object. Forces can be used to start an object moving, change the direction of a moving object, or slow down and stop an object that is moving. Work has been done only when a force affects the motion of an object.

Lifting a book, pedaling a bicycle, catching a ball or pushing a wagon are all examples of work. To push against a building is not classified as work by physical scientists, because the force applied did not move the building. Will you get tired from trying to lift a car or moving a house? What is the difference between pushing a house and pushing a wagon? The amount of work done is calculated by multiplying the force applied times the distance the force moves.

388

During the 18th century, the physicist Isaac Newton formulated a relationship between force, mass, and acceleration. He found that accelerating or decelerating an object depends on the mass of the object and the amount of force exerted on it. If two objects of unequal mass are to be accelerated or decelerated the same amount, the object with the greater mass will require the greater force. If a kilogram object moves at 10 meters per second and you want to slow it down in one second to 9 meters per second, you must exert one newton of force. One more newton of force exerted for one second will slow the kilogram mass to 8 meters per second. This measurement of force, called *newtons*, will cause a one kilogram mass to accelerate or decelerate one meter per second every second. The force of gravity pulling on a stick of butter in the palm of your hand is approximately equal to the force of one newton.

Accelerate—Speed up

Decelerate—Slow down

Comparing weight and mass. The pull of the earth's gravitational force on an object is a measurement of weight. An object's weight will vary on earth depending where the measurement is taken. A person standing on a scale in Death Valley may get a reading of 70 kg. If the same person stood on that scale high in the Rocky Mountains, the reading would only be 68 kg because of the variation in distance from the earth's center of gravity. The mass of an object remains constant not only on earth but throughout the universe. So a person with a mass of 70 kg in Death Valley will have the same mass on a mountain or out in space. Even though the scale reading varies, the person's mass will remain the same.

In this unit all problems and situations occur on the earth's surface. Therefore, for the sake of simplicity, weight and mass will be considered as similar forces. Unless stated otherwise, the weight of one kilogram may be considered equal to the force of approximately ten newtons.

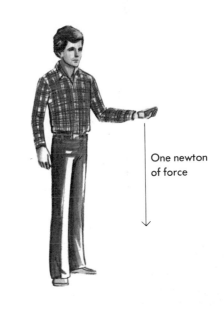

One newton of force

0 kg gravitational force
70 kg mass

Space

12 kg gravitational force
70 kg mass

Moon

70 kg gravitational force
70 kg mass

Earth

Measuring Work. Work is calculated in units called joules. One *joule* is equal to the work done by the force of one newton applied through the distance of one meter. How many joules of work are done when a bat exerts a force of 300 newtons on a ball through a distance of one meter? How many joules are required to draw back a bow 0.5 meter with a force of 200 newtons?

The unit of work is named for James Prescott Joule, a British physicist. His studies of how heat and mechanical energy are related was of great importance to the development of theories about energy.

Place a pile of books on the floor by your desk. Use a spring scale and meterstick to determine the force needed to lift each book and the distance from the floor to your desk top. Calculate the work needed to pick up each book. What is the total amount of work needed to pick up each book? What is the total amount of work needed to pick up all the books one at a time? Put the books back on the floor. How much work is done lifting all the books together? Does the amount of work differ if you pick your books up one at a time or all at once?

Measure the height of a stairway near your classroom. Calculate the work done lifting your body from one floor to the next. Tie a rope to your books and hoist them up the stairway. How much work is required to raise the books? Calculate the amount of work required for you and your books to go up the stairs. How does the amount of work necessary for you to go up the stairs with your books compare to the amount of work needed to bring yourself and your books up the stairs separately?

	Force	Distance	Work
Book 1			
Book 2			
Book 3			
Book 4			
All 4 books			

	Force	Distance	Work
Self			
Books			
Self plus books			

Power. In which of these pictures is work being done? Is the amount of work being done the same in all three pictures? Will the work be completed in the same amount of time in each picture?

Calculate the amount of work needed to run up the stairs. Was the same amount of work done running up the stairs as compared to walking up them? Did your body react as if the work had been the same? Was the interval of time required the same?

When one device does work quicker than another, scientists say it has more power. *Power* is a ratio of work done divided by the time needed to do the work. Power $= \dfrac{\text{work}}{\text{time}}$. The idea of power became necessary when James Watt wanted to sell his steam engine in the 1800's. Since most of the heavy work was done with horses, a comparison of the rate of doing work by horses and steam engines became necessary. Because of this early comparison many engines today still have horse power ratings.

A common power unit is the watt. A *watt* is equal to one joule of work per second. The kilowatt (1000 watts) and the megawatt (1 000 000) are larger power units. How many watts were used carrying your books up the stairs? Did you use the same number of watts running up the stairs? The electrical appliances in your home have a power rating in watts. Do they all use electricity at the same rate?

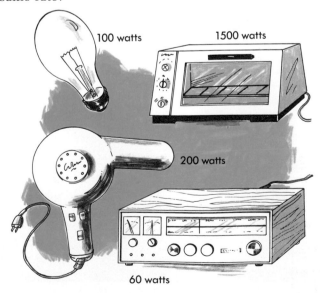

100 watts

1500 watts

200 watts

60 watts

391

0.007 kW
0.009 1 hp

SUMMARY QUESTIONS

1. In what ways can force affect mass?
2. In order to decelerate a moving object of fixed mass, what must happen to the force?
3. Define a joule.
4. How does power differ from work?

0.23 kW
0.229 hp

0.75 kW
1 hp

150 kW
200 hp

21 000 kW
27 300 hp

2.8×10^{11} kW
3.6×10^{11} hp

2.8×10^{28} kW
3.6×10^{28} hp

2 250 kW
3 000 hp

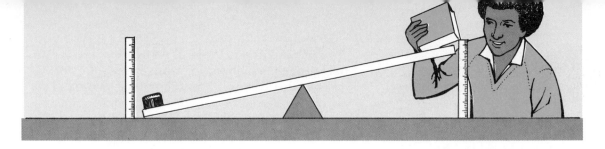

POTENTIAL AND KINETIC ENERGY

Pick up a book and hold it over your head. What force did you have to overcome in order to lift the book? What happened to the energy used in overcoming that force? How can this energy now be transformed into work?

	Mass of book	Height of book	Work	Mass of eraser	Height eraser move	Work
Trial 1						
Trial 2						

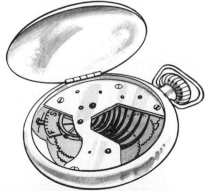

Transferring Energy. Set up a lever as shown at the left with a chalkboard eraser on one side. Raise your book 15 cm and drop it onto the other side of the lever. What happens to the eraser? Was any work done? Where did the force come from to lift the eraser?

Work is needed to raise the book against the force of gravity. The energy you used to raise the book is transferred to the book in its raised position. Stored energy or energy of position is called *potential energy*. When you let go of the book the energy of position changes to energy of motion as the book falls. This energy of motion is called *kinetic energy*. The gain in kinetic energy is equal to the loss in potential energy. When the book hits the lever, the book's kinetic energy does work on the eraser.

How would increasing the book's height affect the kinetic energy available at impact? (**Caution:** Do not drop the book more than 30 cm.) Set up an experiment to determine how increasing the mass of the book affects the kinetic energy available at impact.

Water behind a dam and the raised hammer of a pile driver both have potential energy. In each case the stored potential energy is changed to kinetic energy. The kinetic energy of motion can then be used to do work. The water may turn a water wheel or a turbine. The raised hammer may pound a post into the ground.

How does the amount of potential and kinetic energy change as a skateboarder goes down a hill? What type of energy is in a loaded mousetrap? Was work required to load the trap? Where is energy stored in a watch? How did it get there?

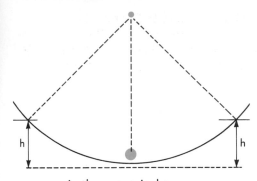

m is the mass in kg

v is the velocity in m/sec

Suspension	Length of Time Pendulum Swings
Fine thread	
Fine string	
Thick string	
Clothes line	
Rope	

Kinetic Energy. Water flowing from a dam and the falling hammer of a pile driver each do work because they have energy due to their motion. This energy is referred to as *kinetic energy*.

The amount of kinetic energy of the pile driver's hammer can be calculated by considering the work which must be done on the hammer to give it its speed. When it strikes the post and stops, the hammer gives up its kinetic energy. The formula used for computing kinetic energy is:

$$K.E. = \frac{1}{2} mv^2$$

If the hammer of the pile driver in the photograph strikes the post with a velocity of 4.4 m/sec, what is its kinetic energy? Kinetic energy is also measured in joules.

The Pendulum. Potential energy can be changed to kinetic energy and back again. Set up a pendulum by hanging a weight on a piece of thread. Hang the weight so that it just misses the surface over which it is swinging. Pull the pendulum weight to one side, and measure its height above the surface. Now release the weight and measure the height to which it goes on the other side. Would you expect the weight to swing up as high on the other side as the height from which you dropped it? Why? At what point does the swinging pendulum have maximum potential energy? Maximum kinetic energy? Where in its path is kinetic energy changed to potential energy? Where is potential energy changed to kinetic energy?

You might expect that a pendulum would swing just as high on one side as on the other and continue to swing forever. Your pendulum slowed down because not all of the original potential energy was transformed to kinetic energy. Some potential energy must have changed into another form.

Experiment with your pendulum to determine what variables might be causing it to slow down. Keep all the variables the same except the one you are studying. Use find thread, fine string, thick string, and laundry line for the pendulum. Always use the same weight, the same knot, the same length of string, and release the pendulum from the same height. Time how long each pendulum swings. Make a chart such as the one in the margin. What effect does this variable you have been testing have on a pendulum? Why? Which pendulum converts the most potential energy to kinetic energy?

394

Calculating Potential and Kinetic Energy. The potential energy of a raised hammer of the pile driver depends on three variables. These are (1) the mass of the hammer, (2) the acceleration which the force of gravity will give the hammer as it falls, and (3) the height of the hammer above the post. Potential energy (P.E.) can be calculated by multiplying together these variables. This idea is expressed by the formula: P.E. = m × g × h. The potential energy is calculated in joules if the mass (m) is measured in kg, acceleration due to gravity (g) is measured in m/s² and the height (h) is measured in meters. The acceleration due to gravity has been measured at 9.8 meters per second each second (m/s²). If the mass of the hammer is 45 000 kg and it is raised 1.5 meters above the post, then using the formula:

$$\text{Potential Energy} = m \times g \times h$$
$$= 45\ 000 \text{ kg} \times 9.8 \text{ m/s}^2 \times 1.5 \text{ m}$$
$$= 661\ 500 \text{ kg m}^2/\text{s}^2$$

The Potential Energy is 661 500 kg/m²/s² or joules.

m stands for mass in kg

g stands for acceleration due to gravity in m/s²

h stands for the height of the pile driver

The amount of kinetic energy of the pile driver's hammer can be calculated by considering the work which must be done on the hammer to give it its speed. When it strikes the post and stops, the hammer gives up its kinetic energy. The formula used for computing kinetic energy is K.E. = ½ mv². If the hammer of the pile driver in the photograph strikes the post with a velocity of 5.5 m/s, what is its kinetic energy? Kinetic energy is also measured in joules.

YOUR FOOD ENERGY CONSUMPTION

In the previous section you have seen that potential energy is stored energy. Water behind a dam, a compressed spring or a peanut butter and jelly sandwich are all similar in that they store energy. In this section you will examine the potential energy stored chemically in the food you eat. Of all the different ways you use energy, food is the only source that you see as it enters the home to be consumed.

	Amount of water	Initial water temp.	Final water temp.	Change in water temp.	Calories released
Butter					
Food sample 2					
Food sample 3					

Energy For You. In the picture above work is being done. Where does the energy come from to do this work? How does your body store energy? You need energy to move, to breathe, to keep your heart beating and to digest food. Your body requires energy for everything you do.

Counting Calories. The energy in foods is measured in food calories. A food calorie is often written with a capital C to distinguish it from the 1000 times smaller calorie used in the study of heat. A *food Calorie* is the amount of energy which can raise the temperature of one liter of water 1° Celsius.

Measure the amount of energy stored in samples of the food you eat. Measure 1 gram of butter (about one teaspoon) and place it in a small beaker. Bury a piece of string in the butter with one end sticking out to form a wick. Place one liter of water in a saucepan above the butter. Stir the water

and record the temperature. Light the wick and when all the butter has burned measure the temperature of the water again.

Repeat the experiment using other food samples and calculate the amount of Calories per gram in each food. In setting up the experiment, what variables must be held constant? What are some sources of error in this experiment? Why must you expect your Calorie determinations to be smaller than those given in the food table?

Energy in Some Common Foods

Food	Energy (kcals/serving)	Food	Energy (kcals/serving)	Food	Energy (kcals/serving)
Dairy		Vegetables		Cereals	
cheese	100	beans, green	15	bread, brown slice	100
cocoa	180	beets	35	bread, white slice	60
cottage cheese	40	broccoli	23	bread, whole wheat slice	55
egg (1)	80	cabbage	15	brownie, 5 cm × 5 cm	125
ice cream or milk	150	carrots, raw	10	cake	140
milk, whole	170	carrots, cooked	25	cereal, 15 grams	60
milk, skim	90	corn, ear	65	cookie, small	35
		corn, canned	85	cracker (1)	20
Meat		peas, frozen, fresh	55	doughnut	140
beef, lean	180	peas, canned	85	oatmeal	75
beef, medium	220	potatoes	50	pancake (1)	60
beef, fatty	290	potatoes, french fries	80	pie	300
chicken with skin	180	potato chips, per chip	10	rice	100
chicken without skin	100	spinach	25	spaghetti, noodles	100
hot dog (1)	160	tomatoes	20		
bacon (2 slices)	100			Misc	
bologna (slice)	90	Fruit		candy bar, 15 grams	75
ham	250	apple	70	gelatin	70
		applesauce	45	pickle	10
		banana	85	sherbet	120
		grapefruit	50	sugar (tspn)	18
		grapes	50	butter, margarine	20
		lemon	60	salad dressing	30
		orange	70		
		peach	35		
		pear	100		
		pineapple, slice	100		
		watermelon	65		

ENERGY FACTOR
FOR VARIOUS ACTIVITIES

Activity	Calories per hour used per kilogram of body weight
Sleeping	1.0
Awake, lying still	1.1
Sitting still, reading	1.4
Watching TV	1.4
Doing homework	1.5
Knitting, writing	1.6
Playing cards	1.6
Eating	1.8
Walking slowly	2.9
Walking fast	4.3
Bowling, playing baseball	5.0
Tennis	7.7
Running, skating	8.1
Bicycling	8.3

Balancing Calories. Make a list of the food you ate in the last 24 hours. Use the food table on the previous page to calculate the number of Calories you consumed in one day.

The table on the left gives the energy used for various activities. The energy factor listed is the number of Calories you use for each kilogram of your body weight when you perform each activity for one hour.

Calculate the number of Calories you use in one day. List your activities for 24 hours and estimate the time you spent at each activity. Multiply your body weight in kilograms by the energy used for each activity and multiply that product by the hours you spent at each activity. Compare the amount of energy your body used in these activities each day with the amount of energy your body took in as food. Assuming your eating and exercise habits remain the same, will the body build up or deplete stored energy over a period of time?

Every few months a new book is published suggesting a different way to lose weight. Most doctors agree that weight can be lost by only one general method. When the body eats fewer Calories than the body uses up in its daily activities, weight is lost. When this happens, the body consumes some of its stored fat to make up the difference in its energy needs. What happens to a person's body when more Calories are eaten than are used in the body's activities?

It has been calculated that 7700 Calories is equal to one kilogram of body weight. According to your Calorie intake and use, how long will it take before you gain or lose a kilogram?

Activity	Estimated Calories used per hour per kilogram of body weight	X	Body weight in kilograms	X	Hours spent at activity	=	Calories used
_____	_____		_____		_____		_____
_____	_____		_____		_____		_____
_____	_____		_____		_____		_____

Total _____

REVIEW QUESTIONS

1. What is the difference between force and work?

2. In each of the following statements, explain if work is or is not accomplished according to a physical scientist:
   a. sawing a piece of wood
   b. trying to move a stubborn donkey that is sitting in the road
   c. climbing a tree
   d. jacking up a car

3. How is mass different than weight?

4. How much work is done by lifting a 10 kg block 1.5 meters off the ground?

5. You are assigned the job of moving ten 1-kg bricks to the top of a wall 2 meters high. Would you do more work by lifting the bricks one at a time or by lifting them all at once? Explain.

6. What is the rating in watts of a motor that can lift a 600 kg weight 30 meters in 2 minutes?

7. How many joules of energy are used when a 100 watt light bulb is used for 1 hour?

8. What is potential energy? Give an example.

9. What is kinetic energy? Give an example.

10. What is the potential energy of a pile driver with a mass of 100 kg at a height of 6 meters?

11. Why is energy required by living things?

12. How are the Calories in a sample of food measured?

13. What would happen to your weight after 3 months if you daily consumed 2800 Calories and used only 2000 calories?

THOUGHT QUESTIONS

1. The World Trade Center is 110 stories tall. Explain how you can lose weight in that building.

2. Explain the motion of a yo-yo in terms of potential and kinetic energy.

3. Explain how a lumberjack can have a 4400 daily Calorie requirement while a bus driver has only a 2500 Calorie need per day.

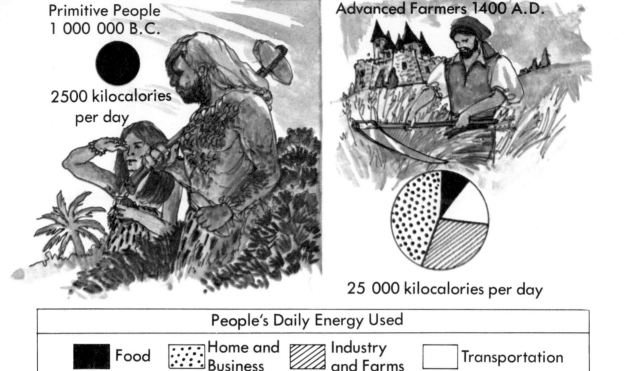

Primitive People
1 000 000 B.C.

2500 kilocalories
per day

Advanced Farmers 1400 A.D.

25 000 kilocalories per day

People's Daily Energy Used			
■ Food	⋮ Home and Business	▨ Industry and Farms	□ Transportation

Measuring the Use of Energy

The three circle graphs show the relative amounts of energy used by people at three different times throughout history. The shadings in each circle show how people used energy in each era. Compare the amount and uses of energy at each time.

Primitive people over 1 000 000 years ago consumed energy only for food. These people, smaller than we are today, probably ate about 2000 Calories (or kilocalories) per day. The energy needs of these people were met mostly from plant foods which grew quickly, replenishing human energy needs.

Technological People 1980 A.D.
250 000 kilocalories
per day

The second circle is 10 times bigger than the first. People in 1400 A.D. were consuming about 25 000 kilocalories (or Calories) of energy per day. Their daily energy needs for food increased as food was grown for livestock. Most of their other energy needs for home and industry were provided by wood. Wood is a renewable energy resource.

The third circle is ten times bigger than the second. Technological people today consume nearly 250 000 kilocalories (or Calories) of energy per day. Today, our daily food energy needs are only a small part of our daily energy consumption. Nearly all our other energy needs are provided by burning oil, gas, and coal. These three fuels are nonrenewable resources.

ENERGY CONSUMPTION IN YOUR HOME AND SCHOOL

The graph on the second page of this chapter shows that about 25% of our daily energy consumption is used in homes or businesses, including schools. Over 60 000 kilocalories of energy are consumed each day by each American for this purpose. Most of the energy is consumed for two general purposes. One purpose is for heating and cooling, usually by oil, gas or electricity. The other purpose is to operate electrical fixtures and appliances.

Energy For Heating Your School. Select a committee of students to learn about your school's heating system. What type of fuel is used? How much fuel is used each year? Compare the amount of fuel used for heating a new and an old building of about the same size. Learn what ideas the person in charge might have about methods of saving fuel.

Use the information gathered above to calculate in class the number of kilocalories of heat produced by the fuel used in your school each year. Assuming that the fuel is used equally throughout the year, calculate the fuel energy consumed each day. How many kilocalories of heat energy are used each day by every person in your school?

Energy For Heating Your Home. Use a similar procedure to determine the fuel burned each year to heat your home. Use this information to calculate the energy used in heating your home each day. How much energy is this per person per day? Compare the energy used per person per day to heat your home with others in your class. Suggest explanations for the differences observed.

Degree Days. Degree days are indicators of how much fuel your home will require to maintain a comfortable temperature. Degree days are measured by taking the average of the high and low temperature of the day and subtracting the average from 65°F. The temperature of 65°F is used as the standard for comfortable domestic temperature settings.

If the high temperature was 55 degrees and the low temperature was 35 degrees, the average temperature would be 45 degrees. Subtracting 45°F from 65°F, the result is 20°F. Therefore, that day was a 20 degree day. By keeping the records of degree days by months and years, people can anticipate their fuel needs for a typical year. How can this information be useful if a winter is more

Heat Energy of Some Common Fuels

$$coal = 12\ 000\ \frac{kcal}{kg}$$

$$natural\ gas = 7\ 500\ \frac{kcal}{m^3}$$

$$oil = 11\ 000\ \frac{kcal}{kg}$$

$$wood\ (dry) = 4\ 000\ \frac{kcal}{kg}$$

Some Common Conversion Factors

$$1\ kW \cdot h = 860\ kcal$$
$$1\ kcal = 4\ 186\ Joules$$

severe than a typical year? How would this information be useful to the person who delivers heating fuel to your home or the power companies that supply electricity for home heating?

Electric Meters. The diagram below at the left shows the dials on an electric meter. You can read the amount of electrical energy which has passed into a building by the following method. Observe carefully the position of the needle on each dial, starting at the left. Record the lower of the two numbers on each side of the dial needle. For example, the dial at the top reads 2, 4, 9, and 3, so the meter shows 2493 kilowatt-hours.

The bottom diagram shows the same meter dial one month later. What is the meter reading one month later? How many kilowatt-hours of electrical energy were used during that month?

Measuring the Electrical Energy Used in Your School. Select a committee of students to learn about the school's electrical use. Arrange interviews with the person in charge of the electrical system, the person who has worked in the building for the longest time, and the person who has taught for the longest time. From these people, learn about the history of electricity in schools. Find out how much electricity per month or year is used and when electrical improvements were made to the building.

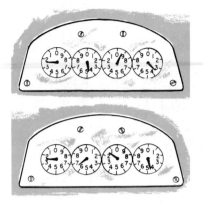

Degree Day Data
Sacramento, California

1978 Heating	1978 Cooling		30 Year Average Heating	30 Year Average Cooling
451	0	Jan	617	0
362	0	Feb	428	0
235	0	March	372	0
269	0	April	227	26
46	98	May	120	98
0	157	June	20	185
0	318	July	0	316
0	315	Aug	0	286
11	157	Sept	5	200
51	87	Oct	101	48
449	0	Nov	360	0
715	0	Dec	595	0
2589	1132	Total	2843	1159

Electronics laboratory

Find out when electric equipment such as overhead projectors, language and electrical laboratories, such as home economics, electronics, and ceramics, became available for school use. How do these devices affect electrical use in schools?

Use the information gathered above to calculate the kilowatt-hours of electricity used by your school each year. How much electrical energy is this per person each year? Per person each day?

One kilowatt-hour of electrical energy is equivalent to 860 kilocalories of heat energy. Calculate the electrical energy used per person per day in your school in kilocalories of heat energy.

The efficiency of a power generating station and electrical transmission is only 30% (.30). Divide your last answer by .30. This new value is the kilocalories of fuel used at the power generating station to deliver electricity to each person each day in your school.

The Increasing Use of Electricity in the Home. On the next page is a partial list of electrical devices which may be in your home. Number a piece of paper from 1 to 102 and survey your home to find which devices are there. If your home has more than one type of device, list it. How many total electrical devices are in your home?

Ask your parents to make a similar list indicating how many of these devices were in their homes when they were your age. Have an older person, such as a grandparent, prepare a similar list. Discuss your findings in class with the results obtained by other students.

The Electric Bill. Bring a copy of a recent electric bill to class. How many kilowatt-hours of electric energy were used? How many days does the bill cover? Calculate the energy per day for your family. Calculate the energy per day per person.

Use the equivalents given earlier to calculate electrical energy used per person each day in terms of kilocalories. Then divide the value by 0.30 to calculate the fuel energy used to deliver this electricity to you each day (at an efficiency of 30%).

Your Total Daily Energy Consumption. Add the energy used to heat your share of your school and home, and to supply you with your share of electricity in school and at home. Compare this value with that given on the first page of this chapter. Discuss reasons for any differences.

	Electrical Appliance	Number of Each		Electrical Appliance	Number of Each
1.	Air conditioner, central	_____	52.	Lawn edger & trimmer	_____
2.	Air conditioner, room units	_____	53.	Lawnmower	_____
3.	Battery charger	_____	54.	Lights, indoor night	_____
4.	Baby bottle warmer	_____	55.	Lights, indoor wall fixture	_____
5.	Beanpot	_____	56.	Lights, outdoor lawn	_____
6.	Blanket	_____	57.	Manicure set	_____
7.	Blender	_____	58.	Massager	_____
8.	Broiler (portable)	_____	59.	Mathematical calculator	_____
9.	Electric broom	_____	60.	Mirror, lighted, for makeup	_____
10.	Can opener	_____	61.	Mixer	_____
11.	Coffeemaker	_____	62.	Organ	_____
12.	Comb	_____	63.	Oven, bun or roll warmer	_____
13.	Clock	_____	64.	Oven, dutch	_____
14.	Defroster for refrigerator	_____	65.	Oven, microwave	_____
15.	Dehumidifier	_____	66.	Oven, portable	_____
16.	Dishwasher	_____	67.	Oven, toaster	_____
17.	Disposer, food waste	_____	68.	Pencil sharpener	_____
18.	Door bell	_____	69.	Popcorn popper	_____
19.	Drill	_____	70.	Projector, movie	_____
20.	Dryer, clothes (or gas)	_____	71.	Projector, slide	_____
21.	Fan	_____	72.	Radio, clock	_____
22.	Fingernail buffer	_____	73.	Radio, standard	_____
23.	Floor waxer	_____	74.	Range, kitchen	_____
24.	Food warmer tray	_____	75.	Record player	_____
25.	Freezer (independent unit)	_____	76.	Refrigerator	_____
26.	Fryer, deep fat	_____	77.	Rotisserie	_____
27.	Frypan	_____	78.	Router (tool)	_____
28.	Garage door	_____	79.	Rug shampooer	_____
29.	Griddle	_____	80.	Sander	_____
30.	Grill, outdoor	_____	81.	Saw	_____
31.	Guitar	_____	82.	Scissors	_____
32.	Hairbrush	_____	83.	Sewing machine	_____
33.	Haircurlers	_____	84.	Shaver	_____
34.	Hair curling iron	_____	85.	Shoe polisher	_____
35.	Hair dryer, standing or portable	_____	86.	Soldering kit	_____
36.	Heater, room	_____	87.	Tape recorder	_____
37.	Heating pad	_____	88.	Television	_____
38.	Hedge trimmers	_____	89.	Thermostat (oil or gas heat)	_____
39.	Hot dog cooker	_____	90.	Toaster	_____
40.	Humidifier	_____	91.	Toothbrush	_____
41.	Ice cream maker	_____	92.	Train set	_____
42.	Ice crusher	_____	93.	Typewriter	_____
43.	Intercom system	_____	94.	Vacuum cleaner	_____
44.	Iron, regular or steam	_____	95.	Vaporizer	_____
45.	Kiln, ceramic	_____	96.	Waffle iron	_____
46.	Knife	_____	97.	Washer, clothes	_____
47.	Knife sharpener	_____	98.	Water heater	_____
48.	Lamp(s), standard	_____	99.	Water pik	_____
49.	Lamp, heat	_____	100.	Woodburning set	_____
50.	Lamp, sun	_____	101.	Yogurt maker	_____
51.	Lathe	_____	102.	Other _____	_____

ENERGY USED IN INDUSTRY AND FARMS

Industries and farms are the largest users of energy each day by technological people. About 90 000 kilocalories per day of energy are used this way for each person in our technological culture. Unfortunately, this use is the most difficult one to observe directly.

This aluminum plant uses large amounts of electrical energy produced at the hydroelectric plant shown in the background.

Energy Decisions in Industry. In most regions, soft drinks can be purchased in returnable bottles, nonreturnable bottles, or throw-away cans. The energy used to produce each of these containers is about 0.5 kilowatt-hour of electricity (equivalent to 430 kilocalories). How much does the energy cost to produce one container at the rate you pay at home?

Aluminum and glass industries both require large amounts of energy, usually electricity. These industries are often located where such energy is cheap. Such industries are called *energy intensive*.

The energy cost of recycling an aluminum can is about 20 kilocalories. How many cans can be recycled with the same energy a new can used when produced from aluminum ore? The energy cost of reusing a returnable bottle is much less than recycling an aluminum can. Suggest reasons why industry would rather not recycle cans or encourage returnable bottles. Suggest ways we can encourage industries to recycle more cans and bottles.

Short-Term and Long-Term Energy Savings. The energy shortage became known to most people in the early 1970's. Many industries realized they had to conserve fuel or they would be forced to close. Industries use energy for two broad purposes. Industries provide heat, light and cooling for employees, and use energy for manufacturing processes such as ovens. One large maker of electrical appliances estimates that two thirds of its energy is used for employee environments. How might this fraction differ in a steel mill?

Industries with high-energy use for employee environments have the easiest task of conserving energy. For example, one airplane manufacturer raised the thermostat settings on all its air conditioners from 22°C to 24°C, and saved a million dollars a year in fuel. A bank removed 240 fluorescent light fixtures and found no noticeable decrease in lighting, but the amount of electricity

used for lighting *and* air conditioning dropped considerably. Another company has started the policy of hiring only new employees who live within four miles of the factory. How does this reduce energy consumption?

Industries are in business to make a profit. The owners could not afford to build well-insulated buildings to save heat when energy was cheap. Lighting was so inexpensive when the building at the left was built that the company found it cheaper to control the lights on each floor with one switch. The company saved the cost of additional wiring and switches. Today, however, switches are becoming cheaper than the additional cost of keeping all the lights going while one or two people clean at night. Some companies now have the offices cleaned during the day so that all the lights can be turned off at night.

The cost of energy is still increasing. More money is being spent on research to discover new ways to make products while using less energy. These energy savings will not be noticed as quickly as some of the others mentioned earlier.

Energy in Agriculture. Human history and culture can be thought of as the slow steady process of freeing more and more people from the task of providing our daily food. In 1900, 50% of Americans worked on farms. Today, less than 10% of Americans work on farms. In 1918, one fourth of all the land used for agriculture produced food for farm horses, donkeys, mules, and oxen. Note what happened to the number of horses and mules in the 50 years shown on the upper graph. What replaced these energy suppliers? Study the lower graph. Why have many people left farms to work in other jobs?

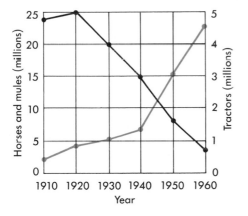

The tractor freed millions of acres of land which had been used to supply food for the energy of horses. However, the tractor uses a nonrenewable energy source which is being consumed at an alarming rate. Gasoline allows the farmer to do more work than previously, and there are many other nonrenewable energy users which help today's farmer produce more food. These include irrigation pumps, lighting in chicken coops (longer "days" cause chickens to lay more eggs), electrical milking machines, power sprayers of insecticides and herbicides, and fertilizers which require energy for manufacture and delivery.

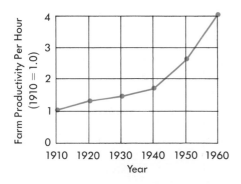

ENERGY USED IN TRANSPORTATION

When most people think of energy for transportation, they think of gasoline. Nearly all cars use gasoline as their energy source. Another important energy source in transportation is diesel oil. Buses and large trucks often use this fuel. Electricity is the source of energy for many urban train systems.

Diesel oil and gasoline are both refined from crude oil. Each fuel can be made readily from the other, but neither can be made cheaply except from crude oil.

The Energy in Gasoline. Gasoline is a high-energy fuel. Each liter produces about 9000 kilocalories of heat energy (equal to 12 000 kcal/kg). Compare the energy in a liter of fuel with the energy in your food for a day.

A worker at hard labor can do the work equal to about 2000 kilocalories of energy. How many worker-days of labor are equal to the energy in one liter of gasoline?

Crude oil is also the raw material for many other important products such as plastics, oils, and greases. Today, there are few substitutes for the jobs which grease and oil do. Resource experts state that we are wasting too much crude oil by producing gasoline to be used inefficiently in cars. Fifty years from now your children will wish your generation had conserved gasoline because cheap grease and oil may not be available for essential tasks.

Cars and Gasoline Consumption. The chart on page 409 shows the U.S. Environmental Protection Agency's estimated gas economy for 1979 automobiles. The figures were estimates of average driving if there were no wind resistance and no hills. Actual gas economy was 10–20% lower. Figures shown are for cars with a standard transmission and the small engine. Automatic transmissions, power accessories such as air conditioners, larger engines and poor driving produce even lower gas economy.

Cars are arranged in four groups according to their passenger space. Study these data. Where are the cars with the smallest engines? The largest engines?

Where are the lightest cars? The heaviest? Prepare three graphs of (1) engine economy vs. weight, (2) gas economy vs. engine size, and (3) car weight vs. engine size. Discuss what these graphs show.

EPA Ranking of 1979 Automobiles

Make and Model	Number of Engine Cylinders	Engine Size (liters)	Car Weight (kg)	Gas Economy (km/1)
sub-compacts				
Buick Opel	4	1.8	1010	11.8
Chevrolet Monza	4	2.5	1250	10.0
Chevrolet Monza	6	2.6	1310	9.1
Chevrolet Monza	8	5.0	1380	8.2
Datsun 210	4	1.4	920	11.8
Datsun 510	4	2.0	1040	10.9
Ford Fiesta	4	1.6	820	12.7
Ford Pinto	4	2.3	1110	9.5
Ford Pinto	6	2.8	1170	8.2
Honda Civic	4	1.5	800	10.5
Toyota Corolla	4	1.6	1000	10.9
VW Rabbit (Diesel)	4	1.5	1090	18.2
VW Rabbit	4	1.5	880	11.4
compacts				
AMC Concord	4	2.0	1330	9.1
AMC Concord	6	3.8	1440	7.7
Audi 5000	5	2.1	1290	8.2
Buick Skylark	6	3.8	1550	8.6
Buick Skylark	8	5.0	1620	7.3
Chevrolet Nova	6	4.1	1560	8.6
Dodge Aspen	6	3.7	1510	8.2
Dodge Aspen	8	5.2	1550	7.3
Oldsmobile Cutlass	6	3.8	1500	8.6
Oldsmobile Cutlass	8	4.3	1550	8.6
Oldsmobile Cutlass (Diesel)	8	4.3	1550	10.9
Pontiac LeMans	6	3.8	1450	8.6
Pontiac LeMans	8	4.9	1510	7.7
mid-sized				
Buick Electra	8	5.7	1850	6.8
Chevrolet Impala	6	4.1	1680	6.8
Chevrolet Impala	8	5.0	1730	7.3
Chrysler Newport	6	3.7	1700	7.7
Chrysler Newport	8	5.2	1730	7.3
Mercury Cougar	8	5.0	1850	6.8
Ford LTD	8	5.0	1670	6.3
Oldsmobile 98	8	5.7	1850	6.3
Pontiac Catalina	6	3.8	1680	8.2
Pontiac Catalina	8	4.9	1760	7.3
Ford Thunderbird	8	5.0	1880	5.9
full sized				
Cadillac De Ville	8	7.0	1970	5.4
Cadillac De Ville (Diesel)	8	5.7	2040	9.1
Lincoln Versailles	8	5.0	1740	6.3
Lincoln Continental	8	6.6	2200	5.4
Oldsmobile Toronado	8	5.7	1830	7.3

Studying Driving Habits. Set up a class project to study driving patterns near your school. Study the way cars are used in at least two different areas. One area might be a major highway carrying mostly cars traveling 10 or more kilometers. The other area might be a place where there is mostly local traffic. Plan to study the following variables: time of day, make of car, number of passengers, direction of travel. For each set of observations, also list the location, weather conditions, and date.

At some locations, too many cars may pass to observe them accurately. In such cases, devise a sampling procedure. Observe every third or tenth car, or whatever sampling rate is convenient. When presenting your data, be sure to indicate the fraction of the number of cars you sampled.

Observe one location for the same length of time at different times during the day, on different days, and under different weather conditions. Different students can do the observing. Decide in class how to handle buses, trucks, and so on.

For each set of observations, calculate the average number of passengers per car, the number of cars per hour, and the number of people per hour. Make charts and graphs of your class observations. Make graphs showing how these numbers change during the day, during the week, and under different weather conditions.

Energy for Transporting Students to School. Select a committee of students to learn about the bus system which transports students to school. How many kilometers do the buses drive each day serving your school? How many kilometers per liter do the buses get? Calculate the amount of fuel used per day. Calculate the energy used per day. Calculate the energy used per person per day.

Ask a bus driver about ideas to save fuel. How far do students have to live from school in order to ride a bus? What will happen to our energy consumption if this distance is increased? Decreased?

Energy for Transporting Teachers to School. Select a committee of students to study the transportation habits of the teachers in your school. Devise a way to determine how many teachers walk to work and how many use public transportation. How many teachers commute in car pools, drive small, relatively efficient cars alone, or drive relatively large inefficient cars alone? Prepare a

Cars Per Hour — Time of Day

Passengers Per Car — Time of Day

Name	A Car and Year	Kilometers Traveled	B Days	C Gas Used	$D = \dfrac{A}{C}$ Gas Economy	$E = \dfrac{C}{B} \times 365$ Gas Per Year

chart or graph comparing the percentages of teachers in your school who fit into each group.

Help the committee to devise a questionnaire with two goals: (1) make teachers more aware of the energy they consume getting to work, and (2) encourage teachers to use more efficient means of transportation. Interview one or more teachers from each group above. Report your findings in class.

Energy for Transportation at Home. Ask a parent to fill the gas tank of the family car the next time it is used. Also record the date and odometer reading. The next time the tank is filled, record the date, odometer reading, and amount of gasoline added. Use these figures to calculate the car's gasoline economy in kilometers per liter.

Use your data to calculate the information for a chart at school like the one shown above. Discuss the accuracy of calculating figures for the last two columns, and the factors affecting the accuracy of your gasoline economy calculations. How could the accuracy of these calculations be improved?

Divide the figure in the last column by 365 days per year to obtain the gasoline used per day by the family car. Divide this number by the number of people in your family to obtain the gasoline used per day per person in your family. Calculate the gasoline energy in kilocalories used per day by each person in your family. How will your answer be affected if your family has more than one car?

Transporting Yourself and Your Goods. The two tables at the right give estimates of the energy used transporting people and goods. Which methods are efficient and use the least amount of energy? Why are inefficient methods sometimes used? What might the government do to encourage the use of more efficient methods?

Total Daily Energy Needs for Transportation. Estimate your total daily energy needs for transportation by adding the daily energy you use being transported to school and your daily automobile energy use. Compare this value with the estimate on page 410. Propose reasons for the differences.

Gasoline energy =
11 500 kilocalories/liter

Diesel oil energy =
11 000 kilocalories/liter

Energy Efficiency for Passenger Transportation

Method	Energy Used (kcal per passenger per km) urban	intercity
Bike	30	—
Walking	45	—
Bus	580	250
Railroad		450
Auto	1 250	530
Airplane		1 300
SST		2 800

Energy Efficiency for Freight Transportation

Method	Energy Used (kcal per metric ton per km)
Pipeline	78
Railroad	115
Waterway	120
Truck	660
Airplane	7 300
SST	15 000

Degree Day Data

| | Atlanta, Georgia | | | | | Boston, Massachusetts | | | |
| | 1978 | | 30 year ave. | | | 1978 | | 30 year ave. | |
Heating	Cooling	Heating	Cooling		Heating	Cooling	Heating	Cooling
966	0	701	0	Jan.	1127	0	1100	0
714	0	560	0	Feb.	1057	0	969	0
412	2	443	12	March	885	0	834	0
137	40	144	27	April	480	0	492	0
57	144	27	154	May	209	40	218	20
0	346	0	321	Jun.	18	122	27	117
0	428	0	403	July	11	237	0	260
0	420	0	388	Aug.	11	221	8	203
0	345	8	227	Sept.	150	48	76	61
112	40	137	57	Oct.	381	0	301	0
194	7	408	0	Nov.	635	0	594	0
580	1	667	0	Dec.	916	0	992	0
3172	1773	3095	1589		5880	668	5621	661

REVIEW QUESTIONS

1. How has the amount of energy we use each day changed in the past million years?
2. How have our uses of energy changed in the past million years?
3. How have our sources of energy changed in the past 500 years?
4. What energy sources are used to heat and cool your school?
5. Using the data above, determine which city has the highest energy needs for: a) heating, b) cooling, and c) total annual requirement.
6. If a domestic oil burning furnace uses 1 gallon of oil for every 7 degree-days, how much oil will be used in an average heating year for the city of Boston? Atlanta?
7. If oil costs $1.25 a gallon, how much will it cost to heat the two homes in question?
8. How can degree-day data records be useful in planning for regional energy needs?
9. How are car weight, engine size, and gasoline economy related?

THOUGHT QUESTIONS

1. Why might a factory assembling radios have an easier time reducing its energy use by 20% than would a steel mill?
2. How would life styles in America change if farmers were unable to use large amounts of energy such as gasoline and electricity?

Transforming Energy

The paired objects in the photographs all weigh the same. The items in each pair are made of the same elements. However, one member of each pair can do something that the other cannot. What can the rear car do that the front car cannot? How do the batteries differ?

Scientists say that these pairs differ in the energy they have. A toy car with its spring wound tightly can move because it has energy in the spring. A fresh battery can run an electric motor or light a bulb because it has energy in the chemicals within it. As the car moves, the spring unwinds and loses its energy. As the battery gives up electricity, its chemicals change and lose energy.

Energy is everywhere and takes several different forms. All living things have energy and require energy to stay alive. Heat, light, sound, magnetism, gravity, and movement are some recognized forms of energy. Energy has always been produced, and is constantly being supplied to our environment in such large quantities that we cannot use it all.

ENERGY ALL AROUND YOU

Energy exists in many forms in our homes and in the natural world. Much of this energy is not found in a form that we can use. By converting the energy to other forms we are able to make it more useful. This interchangeability of various forms of energy is an important characteristic. Most scientists recognize the following forms of energy: heat, light, sound, electrical, mechanical, chemical, magnetic, gravitational, and nuclear.

Energy in Nature. The unequal heating of the earth's surface provides kinetic energy for atmospheric winds and ocean currents. Meteorologists estimate that the potential wind power at selected places throughout the world is about 2×10^{12} kW. This is about 60 times the present yearly capacity of all the electric power plants in the United States. Using present day technology 1.75×10^{10} kW could be generated from the major wind currents of the world.

The changing forms of energy can be witnessed in the natural world. An imbalance of electrons between the earth and sky creates potential electrical energy. The kinetic energy of electrons rapidly moving back to earth results in lightning. Much of the energy in lightning is converted to heat. The intense heat causes air to expand quickly, creating sound waves of thunder in the atmosphere.

The potential chemical energy of coal and oil comes from radiant energy captured by plants long ago. The changing tides of the oceans result from the gravitation pull of the moon and sun. What type of energy is stored in trees? What kinds of energy changes take place as a forest burns? What other forms of energy can you identify in nature? Make a list of other energy conversions that take place in

nature. In what way do we convert nature's energy to forms more useful to us?

Using Energy. In harnessing energy for our use, we convert the energy that exists in the natural world to forms which are useful to us. The sun's radiant energy evaporates water from lakes and oceans. The kinetic energy of the rising water vapor is converted to potential energy as the vapor cools to liquid droplets of a cloud. The kinetic energy of the falling rain is evident in streams which fill with water from the storm. Building a dam converts the kinetic energy of the stream to potential energy of a lake. The controlled release of this water turns generators which produce electricity for our use.

The pictures at the right show ways in which energy is converted around the home. Electrical energy is changed to light energy by the lamp. Is there any other type of energy produced? How many energy transformations result from turning on a television set? Describe the energy conversion which takes place in a toaster. What forms of energy result from playing a guitar? What other devices in your daily life convert energy? What is the most common energy conversion in your home?

ENERGY TRANSFORMATIONS

Although energy is around us in enormous quantities, we usually need the energy in a certain form before we can use it. What has happened to the person in the photograph below? Why won't the car go?

Automobile engines need chemical energy. The automobile engine is designed to transform the chemical energy in gasoline to mechanical energy. The engine will not operate on other sources of energy. There is research being done on automobiles which can use other sources of energy.

Converting to Heat. Energy can be converted from one form to another. Unfortunately, many energy conversions are expensive and inefficient. If we could convert energy cheaply and efficiently into other forms, the energy problem would be resolved.

What are some of the techniques for converting the other forms of energy into heat? How is chemical energy converted to heat? Nuclear energy is converted to heat energy in a nuclear reactor. Kinetic energy is converted to heat energy by friction, usually without the heat being wanted. By what technique is electrical energy converted to heat?

Converting Light to Heat. When you are sitting in the sun's rays you can feel them warming you. Therefore, you should be able to convert the light energy to heat. Half fill a black plastic bag with a known amount of water. Take the temperature of the water in the bag. Then place the bag of water on a black background and put it where it will receive the direct rays of the sun. Take the temperature of the water every 5 minutes for an hour. How many calories did the water absorb? Graph your temperature observations compared to time. Estimate on your graph what the temperature of the water will be in 1½ hours and in 2 hours. Is it possible to boil water using this device? Why?

Time	Temperature (°C)

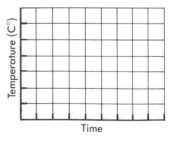

Converting Light to Chemical Energy. All living things require energy to live, but any form of energy will not do. Living things need the energy found in foods. What form of energy is stored in foods?

Animals are unable to make their own food. For their food energy, they depend on eating other animals or plants. Green plants are the only living things which are able to make their own food. Plants are able to use the energy of sunlight to chemically combine atoms of hydrogen, carbon, and oxygen into a complex sugar molecule containing energy. This process is called *photosynthesis*. A simplified formula for the process is given below.

$$6 \ CO_2 + 6 \ H_2O \xrightarrow[\text{sunlight}]{\text{chlorophyll}} C_6H_{12}O_6 + 6 \ O_2 \uparrow$$

Notice that oxygen as a gas is given off during the process. In the experiment which follows, you will estimate the rate of reaction by the rate at which oxygen is being given off.

Light and Photosynthesis. Do the following experiment. Test the effect light brightness has on producing chemical energy during photosynthesis. Set up the equipment as shown. Place *Elodea* and sodium bicarbonate solution in the test tube with the solution filling exactly half the test tube. With the clamp removed, carefully insert the stopper as tightly as possible. Attach the long rubber tubing to the "U" shaped tube called a *manometer*. Place the test tube in a flask with water.

Place a lamp at a distance of 40 cm from the *Elodea*. Measure the distance between the test tube and the light bulb. Allow the set-up to stand for 5 minutes before placing the clamp on the short rubber tube. This is done to allow the solution to become saturated with oxygen. After another 5 minutes, fasten the clamp.

As oxygen is produced, the liquid in the right side of the manometer will rise. Why? Every 2 minutes record the level of the water on the right side of the manometer. Continue recording until the rate the water rises for two or three 2-minute intervals is the same. Once this rate of the production of oxygen is constant, record the rate.

Move the light source to a distance of 20 cm from the *Elodea* and determine the rate of oxygen production. Move the light to within 10 cm and determine the rate. How is the rate of photosynthesis affected by the brightness of the light? If you continued to bring the light closer and closer, would the rate increase indefinitely? Why?

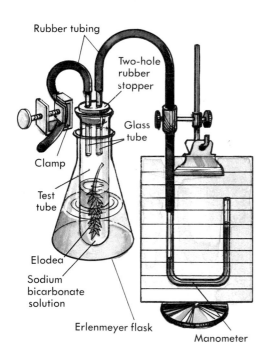

Rubber tubing

Two-hole rubber stopper

Glass tube

Clamp

Test tube

Elodea

Sodium bicarbonate solution

Erlenmeyer flask

Manometer

Chemical Energy. The bonds which hold atoms and molecules together have energy. Fill an insulated cup with 100 grams of water and take the temperature of the water. Carefully weigh out 1 g of ammonium chloride and find its temperature by placing the bulb of the thermometer into the salt. (Make sure that the bulb of the thermometer is absolutely dry when you place it in the salt.) Add the 1 g of ammonium chloride to the water and note the temperature of the mixture. Continue adding 1 g of ammonium chloride at a time until there is no change in temperature. Record your results in a table such as the one in the margin. What effect does adding ammonium chloride have on the temperature of the solution? How many calories of chemical energy were transformed?

Grams Added	Temperature of Water	Time in Minutes

In place of ammonium chloride use sodium hydroxide (lye). (**CAUTION:** Be careful when handling sodium hydroxide. Do not touch it with your bare hands, and when it is dissolved in water do not splash the solution.) Put 100 g of water in an insulated cup and measure the temperature. Measure the temperature of the sodium hydroxide you are using. Weigh out 0.5 g of sodium hydroxide and add it to the water. Note the temperature of the solution and add another 0.5 g. What effect does adding sodium hydroxide have on the temperature of the solution? How many calories of heat energy were transformed?

When these two chemicals dissolve in water, some of the bonds holding the molecule together break apart. The energy for breaking the bonds apart came from the heat in the water. In one case the bonds gave up more energy than they used. This is called an *exothermic reaction*. In the other case the bonds gave up less energy than they used in breaking apart. This is called an *endothermic reaction*. Which of the reactions is exothermic and which is endothermic?

Place 50 g of Glauber's salt in a pyrex beaker. Place the beaker into a larger beaker that contains water. Put a thermometer into the salt and begin heating. Note the temperature every minute and

exo—means outside

endo—means inside

thermic—means heat

Glauber's salt is
$Na_2SO_4 \cdot 10H_2O$

enter the temperatures in a chart like the one in the margin. Heat until the temperature reaches 33°C and continue heating for 4 minutes longer. Then remove the heat and allow the salt to cool to 26°C. Continue recording the temperature each minute. Plot the temperature against the time for heating and cooling. What happens to the salt as the temperature reaches 33°C? How does the temperature change when the salt is being heated? Why doesn't the temperature continue to change the same each minute? What could be happening to the heat energy as the salt approaches 33°C? How does the temperature change as the salt cools? What happens to the salt as it cools below 33°C?

Glauber's salt has been used to store heat in some experimental homes using solar energy for heating. How does Glauber's salt store heat? How could the experimental homes have used it?

Batteries. The chemical battery was the first device that converted chemical energy directly to electrical energy. Two hundred years ago it was the only continuous source of electricity. The battery gets its energy from using unlike materials for its electrodes. A conducting fluid or paste must be present to allow a current to flow between the electrodes. The voltage difference arises because of the different strengths of the chemical bonds. There are many materials which can be used for electrodes and as a conducting material. A few battery reactions are listed in the margin. Note that these reactions can be reversed. Reading from left to right, the reaction is discharging electricity. Reading from right to left, the reaction uses energy to become charged. The reaction producing the current will end as the chemicals are all changed.

Fuel Cells. Research stimulated by space exploration has led to the development of the fuel cell. Unlike the battery, the fuel cell has a continuous supply of fuel. The diagram at the right illustrates a hydrogen-oxygen fuel cell. It is typical of all fuel cells. In the hydrogen-oxygen fuel cell, hydrogen is supplied to one electrode (the anode) while oxygen is supplied to the other electrode (the cathode). These two gases combine to form water. The two electrodes develop opposite charges, causing a current to flow. During the Apollo missions to the moon, the water was drained off the fuel cell and used by the astronauts for drinking.

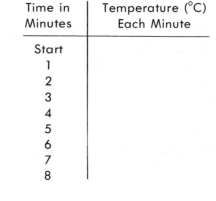

Time in Minutes	Temperature (°C) Each Minute
Start	
1	
2	
3	
4	
5	
6	
7	
8	

Time (horizontal axis), Temperature (C°) (vertical axis)

Battery Reactions

$$Pb + PbO_2 + 2H_2SO_4 \rightleftharpoons 2PbSO_4 + 2H_2O$$

$$Fe + NiO_2 \rightleftharpoons FeO + NiO$$

$$Zn + AgO + H_2O \rightleftharpoons Ag + Zn(OH)_2$$

$$Pb + Ag_2O \rightleftharpoons PbO + 2Ag$$

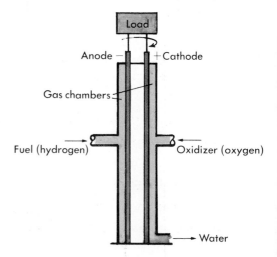

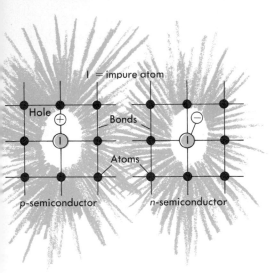

I = impure atom

Hole

Bonds

Atoms

p-semiconductor   n-semiconductor

Diagram of a Solar Cell

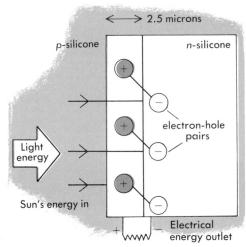

← 2.5 microns →

p-silicone   n-silicone

electron-hole pairs

Light energy

Sun's energy in

Electrical energy outlet

Solar Cells. All fossil fuels owe their existence to the stream of energy that has been striking the earth for billions of years. The power in this stream amounts to about 1400 watts per square meter at the earth's surface. Eventually, scientists hope to convert solar energy into electrical energy at the place where the energy is going to be used. In this way, consumers of energy would become independent of large electrical distribution systems.

A solar cell can convert the sun's energy into electrical energy. The cell is a thin sandwich of semiconductors made of silicon crystals. In producing the silicon crystals, small amounts of other elements are added as impurities to give the crystals their semiconductor properties. A positive or p-type semiconductor contains "impure" atoms such as boron, aluminum, or gallium that do not have enough outer electrons for surrounding atoms. The material behaves as if it had holes for electrons. These holes act as if they are positively charged. When the element arsenic is added to the silicon crystal, the wafer becomes a semiconductor with extra negative electrons. This is called an n-type semiconductor.

When two semiconductors are sandwiched together, a voltage difference is created across the junction. As shown in the diagram, the top semiconductor layer exposed to the sun is extremely thin. The sun's energy can penetrate this layer and reach the junction of the semiconductors. This energy causes the holes and electrons to join, and the pairs are forced to move because of the electric field across the junction.

Advantages and Disadvantages of Solar Cells. Solar cells convert light directly into electricity without the resulting byproducts of air or thermal pollution. Solar cells have no moving parts to wear out, so their lifespan is very long. Solar cells are thin, light wafers made of silicon. Silicon is the second most abundant element on earth.

Unfortunately, solar cells have drawbacks which prevent them from competing with fossil-fueled or nuclear-powered generating plants. Solar cells are only about 12% efficient. Some of the radiant energy that hits their surface is reflected

This array of solar cells is being prepared to power navigational lights on an offshore oil platform.

and some radiant energy that penetrates the surface is not changed into electricity. It is costly to grow, cut, polish, and mount the silicon crystals. Research is now being done to reduce the costs.

Using Solar Cells as a Photometer. Make a solar panel by mounting three or four cells in series to a piece of wood. In order to measure DC current produced by the panel, connect the wire leads to a milliammeter. This small photometer can now be used to measure solar radiation.

Solar radiation reaching the earth's surface is measured in langleys. One *langley* is equal to 1 cal/cm². In order to measure solar radiation instead of current flow, calibrate your meter to read in langleys. Assume one langley of solar radiation reaches the earth's surface each minute around noon anywhere in the U.S. Thus your maximum meter reading on a clear, sunny day around noon will equal one langley. If two hours later the meter reads ¾ of what it did at noon, the solar radiation will equal 0.75 langley.

Early in the day bring your photometer outside into an open area. Record two solar radiation readings hourly all day. One reading should be made with your panel lying flat on the ground. The other reading should be taken while you aim the face of the panel toward the sun. (**CAUTION:** Only aim your solar panel toward the sun. **DO NOT** look at the sun because sunlight can damage your eyes.) Record cloud cover and visibility along with your radiation data. Plot the readings on a graph such as the one here.

What time of day is best for collecting solar energy? In which of the two positions does the panel collect more energy? How much difference is there in the total amount of energy collected as a result of changing the panel's position? If you could not move the panel throughout the day, in what position would you fix it to attain maximum energy collection?

Time	Lying Flat		Directed Toward Sun	
	Meter reading	Langleys per minute	Meter reading	Langleys per minute
7				
8				
9				
10				
11				
12				
1				
2				
3				
4				
5				

SUMMARY QUESTIONS

1. How does lightning cause thunder?
2. Why must animals depend on plants for food?
3. What energy change takes place in a battery?
4. What prevents the wide use of solar cells?
5. What do langleys measure?

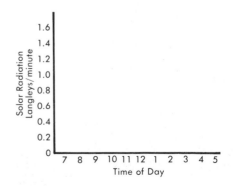

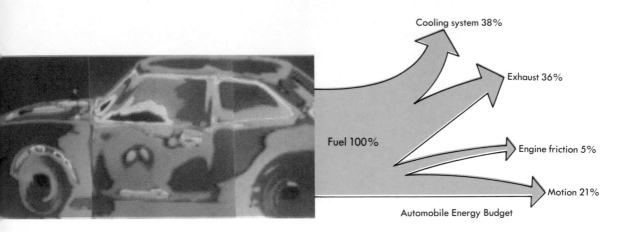

Cooling system 38%

Exhaust 36%

Fuel 100%

Engine friction 5%

Motion 21%

Automobile Energy Budget

CONSERVATION OF ENERGY

In the 19th century, physicists conducted many experiments and concluded that energy cannot be created or destroyed. Energy may be transformed from one form to another, but the total amount of it never changes. Scientists also discovered that whenever energy is transformed from one form to another, some useable energy is always lost.

You Can't Break Even. A heat engine changes heat energy into mechanical energy. The internal combustion engine is a heat engine used in automobiles. What other machines use internal combustion engines? What fuel is commonly used in the internal combustion engine? Trace the energy transformation in a moving car.

The photograph above was taken with infrared film. Infrared film is sensitive to heat energy. What parts of the car are radiating heat? Is all the heat energy in the car being converted to motion? Why do cars have a cooling system? Is the heat energy of the cooling system ever used? Where does this heat energy come from?

Place your hand on top of a television set or the cover of an electric motor which has been on for a while. What do you feel? Turn on an incandescent light bulb and place your hand next to it. Is the electrical energy converted only to light? Saw a piece of wood in half and place your hand on the blade. Was all your energy used only to cut the wood? What other forms of energy is present?

The Law of Conservation of Energy and Mass.
The 19th century law of conservation of energy states that energy can neither be created nor destroyed, but can change from one form to another. This law does not seem to be true in the case of one energy source of the 20th century, nuclear energy. Great amounts of nuclear energy come from tiny masses of matter. Albert Einstein, one of history's great physicists, showed that matter can be transformed into energy. This discovery forced physicists to change the conservation law to include Einstein's ideas, and the following general statement can be made. In any transformation, the total amount of matter and energy does not change. The law applies to everything we do from pounding a nail to launching space vehicles.

EINSTEIN'S EQUATION. During some nuclear reactions, matter disappears and an amount of energy is released in its place. The relation between the quantity of matter and the amount of energy it stores is stated by Einstein's equation, $E = mc^2$. Energy cannot appear without the disappearance of mass. When gasoline is burned in the engine of a car, one gram of gasoline will release 48 000 joules of energy. Einstein's equation states that in this case, mass disappeared in the amount of:

$$m = \frac{E}{c^2} = \frac{4.8 \times 10^4}{9 \times 10^{16}} = 5.3 \times 10^{-13} \text{ kg (½ a billionth of a gram)}$$

When an H-bomb explodes, several grams of mass are converted into energy. Scientists and engineers are constantly experimenting with safe and productive ways to convert mass into energy.

Speed of light $= 3 \times 10^8$ m/s
E is energy
m is the mass that disappears
c^2 is the speed of light squared
$\quad 3 \times 10^8 \cdot 3 \times 10^8 = 9 \times 10^{16}$

MEASURING HEAT ENERGY

Heat is a special form of energy. Whenever energy is converted from one form to another, some energy is always wasted. Much of this wasted energy is in the form of heat. Heat works to our advantage in toasters, ovens, and electric blankets, but in automobiles and power plants the thermal loss is unwanted and costly.

Heat energy is measured in calories. The calorie used by chemists and physicists is 1000 times smaller than the food Calorie you used in the first chapter of this unit. Calories are calculated from measurements of temperature and mass of the substances being studied.

A small calorie is the amount of heat needed to raise the temperature of one gram of water one degree Celsius.

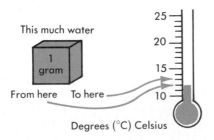

This much water

1 gram

From here   To here

25
20
15
10

Degrees (°C) Celsius

Defining the Calorie. A small calorie is the amount of heat needed to raise the temperature of one gram of water (about 20 drops) one degree Celsius. For example, one calorie of heat energy can raise the temperature of one gram of water from 19°C to 20°C.

Similarly, five calories of heat energy can raise the temperature of one gram of water five degrees C. Five calories can also raise the temperature of five grams of water one degree C.

How many calories are needed to raise the temperature of five grams of water 10 degrees Celsius? 100 degrees Celsius?

Heat Energy from Friction. Operate a bicycle pump vigorously. After one minute feel the base of the pump. When Indians used a fire bow, enough heat was generated by friction to start a fire. The nichrome wire used in toasters and electric heaters becomes warm because of the friction created by moving electrons.

Add five grams of water at room temperature to a test tube. Record the Celsius temperature of the water.

Using a steel file, smooth off the tip of a large nail. When the nail becomes blunt, quickly put it in the test tube. Record the highest temperature the water reaches. Calculate the increase in temperature and the number of calories which entered the water.

Heat from Electric Energy. Put 125 grams of water (one-half cupful) into a can. Record the temperature in degrees Celsius. Hold the can so that the bottom of it is against a glowing 50-watt light bulb. While heating the water, stir it gently with a thermometer.

How much time is needed to raise the temperature of the water 4 degrees Celsius? How many calories of heat energy entered the water during this time?

Repeat the experiment with 25-watt and 100-watt bulbs, each time heating the water 4 degrees Celsius. Compare the heating rates of the lamps.

Heat from Chemical Energy. A fuel contains chemical energy which is changed to heat energy during burning. The energy of a sample of fuel is measured by burning the fuel and using the heat to raise the temperature of water.

Make a small cup of aluminum foil and set it in a pan of sand. Place over the cup a small metal can containing 125 grams of water at 20°C.

Put one gram of alcohol into the aluminum cup. Burn the alcohol and determine the change in the temperature of the water. Calculate the calories of heat energy gained by the water.

Test one-gram samples of other fuels. Compare your results with the figures given in the table. Try to explain why your results are different.

Sand

Heat of Combustion

substance	calories per gram
Natural gas	12 500
Gasoline	11 500
Furnace oil	10 500
Lignite	9 000
Coal	8 000
Charcoal	7 200
Peat	7 000
Alcohol	6 500
Wood	3 500

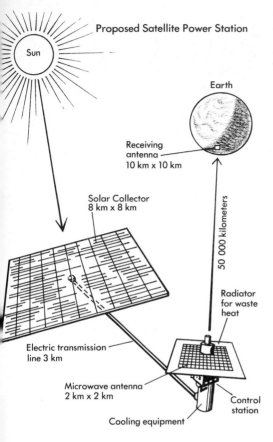

Proposed Satellite Power Station

Sun

Earth

Receiving antenna
10 km x 10 km

Solar Collector
8 km x 8 km

50 000 kilometers

Radiator for waste heat

Electric transmission line 3 km

Microwave antenna
2 km x 2 km

Control station

Cooling equipment

REVIEW QUESTIONS

1. Use another piece of paper and fill in the blank boxes in the chart. Name a device or process by which the form of energy at the top of the column is transformed into the form of energy at the left of a row. For example, heat energy is transformed into light energy in the filament of a light bulb.

2. How many forms of energy can you identify that are associated with your refrigerator?

3. Explain why some scientists believe that the best place for a large photoelectric power station is out in space.

4. What is the Law of Conservation of Energy?

5. What is Einstein's equation relating matter and energy?

6. How did the Law of Conservation of Energy have to be changed when nuclear energy was discovered?

7. How does the value of a calorie differ depending on whether a person is a nutritionist or a physicist?

8. How many calories of heat are necessary to bring 100 grams of water at 25°C to boiling?

THOUGHT QUESTIONS

1. Why is there a shortage of energy when there is a great deal of energy all around us?

2. Why are the two Laws of Thermodynamics called "You can't win" and "You can't break even"?

Chart of Energy Conversions

To From→	Heat	Light	Electrical	Mechanical	Chemical	Gravity	Nuclear
Heat						unknown	
Light						unknown	
Electrical						unknown	
Mechanical							
Chemical						unknown	
Gravity	unknown	unknown	unknown		unknown		unknown
Nuclear	unknown		unknown	unknown	unknown	unknown	

Energy Economics

In the previous chapters you have learned that energy is necessary to do work and that the use of energy is governed by certain rules. These rules state that energy exists in many forms which can be converted from one form to another. The rules also state that whenever energy is converted to other forms, some energy is wasted. These rules are known as the first and second laws of thermodynamics.

In this chapter you will study how energy is transferred to make it convenient for our use. The flow of energy from origin to use, whether pertaining to food, heating or appliances, is still subject to the laws of thermodynamics.

Although we cannot defy the laws of thermodynamics, we can work to lower our energy wastes. As consumers we will have to examine the total energy cost of living in a throw-away society which is adapted to convenience. Energy-wise decisions and life styles will only result when we are able to understand the laws and systems which govern the use of energy.

ACCOUNTING FOR ENERGY LOSSES

In the previous chapter, you observed that energy can be transformed from one form to another and that work often accompanies these changes. By using a jack to lift a car, mechanical energy is converted to potential gravitational energy. The potential chemical energy of the food you eat is converted to kinetic energy when you ride a bicycle. Unfortunately, during these transformations some of the energy is always wasted.

Calculating Efficiency. Place a piece of rope over a broom handle. Tie a stack of books to one end and a spring scale to the other end. Calculate the amount of work needed to raise the books. The force exerted on the rope multiplied by the distance the force moves is equal to the amount of work put in by you. The weight of the books multiplied by the distance the books moved is equal to the work done. How much work did you put in to raise the books? How much work was done? Was the amount of work done equal to the amount of work you put in?

Place a piece of sandpaper over the broom handle and repeat the experiment. Was the amount of work done the same? Did you apply the same amount of force as before? Explain how much work you put in to raise the books. How did the sandpaper affect the efficiency?

Remove the sandpaper and coat the broom handle with soap. Again calculate the amount of work you put in and the amount done. How does using the soap affect the amount of work put in to raise the books? Was the amount of work done the same as before? What was the purpose of the soap and how did it act?

The efficiency or useful work of a machine is the ratio of the amount of work done to the work put in. Efficiency is usually expressed as a percent so the ratio is multiplied by 100. How much did the efficiency of your machine change when you altered the handle?

	F_a = Force put in	D_a = Distance moved	W_a = Work put in ($W_a = F_a \times D_a$)	F_b = Force exerted (weight lifted)	D_b = Distance moved	W_b = Work done ($W_b = F_b \times D_b$)	Efficiency = $\dfrac{W_a}{W_b}$ x 100
Broom handle							
Sandpaper							
Soap							

Table of Energy Consumption Efficiencies

System	Energy In	Energy Out	Efficiency (in %)
Photosynthesis	light	chemical	.1-1 in field (30 in lab)
Incandescent lamp	electrical	light	5
Steam locomotive	chemical	mechanical	8
Solar cell	light	electrical	10
Fluorescent lamp	electrical	light	20
Automobile engine	chemical	mechanical	25
Nuclear power plant	nuclear	electrical	30
Diesel engine	chemical	mechanical	38
Steam electric generator	chemical	electrical	40
Steam turbine	heat	mechanical	47
Fuel cell	chemical	electrical	60
Small electric motor	electrical	mechanical	63
Home oil furnace	chemical	heat	65
Storage battery	chemical	electrical	73
Home gas furnace	chemical	heat	85
Large steam boiler	chemical	heat	88
Dry cell battery	chemical	electrical	90
Large electric motor	electrical	mechanical	92
Electric generator	mechanical	electrical	99

Friction. Friction is the resisting force which opposes the sliding or rolling motion between two objects. Your ability to exert large forces between your shoes and a sidewalk enables you to speed up or slow down as you walk. What happens to your ability to exert large forces when you walk on ice? What happens to your ability to speed up or slow down? Give other examples of friction which affect the accelerating and decelerating motion of objects.

When you rub your hands together, work is done to overcome the friction between your palms. What two forms of energy result from rubbing your hands together? When the moving parts of a machine rub together, energy is lost, efficiency is reduced, and parts wear out.

The force of friction can be reduced by using wheels or ball bearings to decrease the area of contact. Friction can also be reduced by sanding or polishing to smooth the contact areas, or by using lubricants, such as oil, grease, soap or graphite to form a low-friction layer between the sliding surfaces.

Initial energy → T R A N S F O R M A T I O N → Useful energy + Waste energy

Efficiency of Energy Systems. Before electrical energy is used in your home, many transformations take place to produce and transmit the energy to you. Burning fossil fuels transforms chemical energy to heat. The heat changes water to steam under great pressure. The energy in the steam is transformed to mechanical energy as the steam spins the turbines. This mechanical energy is transformed to electrical energy by the generator. The electrical energy is then transmitted to your home. Each time energy is changed from one form to another the efficiency of the change can never be 100%.

Follow the path of energy in the diagram. Note the efficiency of each step along with the efficiency of the whole system. What happens along each step of the process? Does each transformation reduce the efficiency of the system? Do all transformations have the same efficiency? What steps are least efficient? Can you do anything to increase the efficiency of the system?

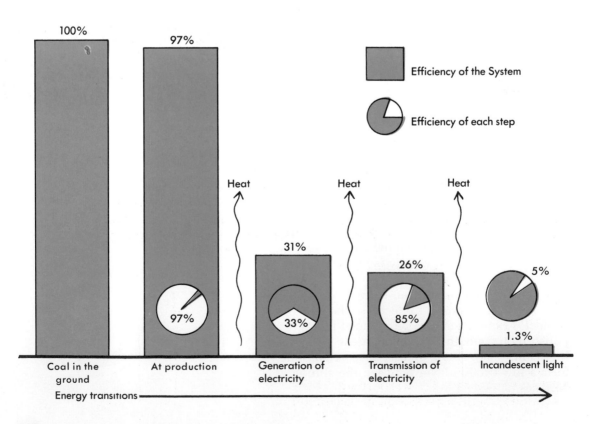

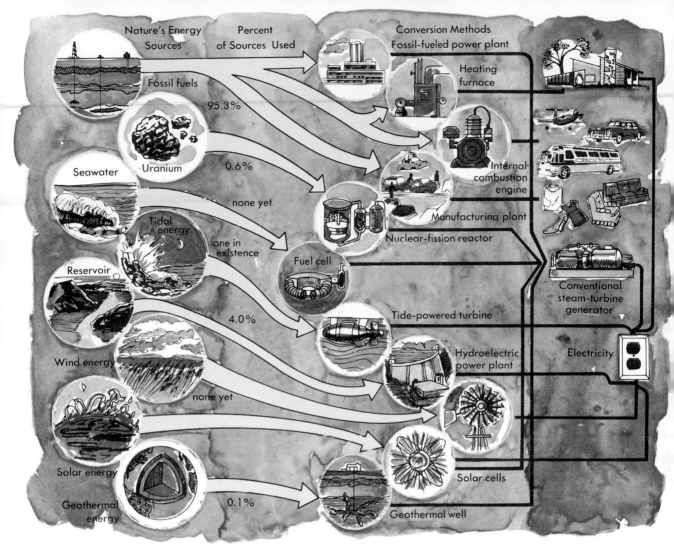

Nature's Energy Sources — Percent of Sources Used — Conversion Methods

Fossil fuels
Seawater
Uranium
Tidal energy
Reservoir
Wind energy
Solar energy
Geothermal energy

95.3%
0.6%
none yet
one in existence
4.0%
none yet
0.1%

Fossil-fueled power plant
Heating furnace
Internal combustion engine
Manufacturing plant
Nuclear-fission reactor
Fuel cell
Tide-powered turbine
Hydroelectric power plant
Solar cells
Geothermal well

Conventional steam-turbine generator

Electricity

Transforming Energy to a Useful Form. Study this chart. It will give you an idea of how much we depend on transformations of energy. On which source do we depend most for obtaining energy? Which of the sources will finally be used up? Which can be renewed indefinitely?

Modern civilizations require enormous amounts of electricity because there are so many machines and devices that are designed to use electrical energy. Our transportation needs require large amounts of chemical energy such as that stored in oil. Since most methods of transportation use engines designed to run on oil, we are very dependent on oil.

SUMMARY QUESTIONS

1. Why do baseball players, football players, and golfers wear special shoes?
2. How do lubricants increase the efficiency of machines?
3. What happens to the waste energy which results from the generation of electricity?

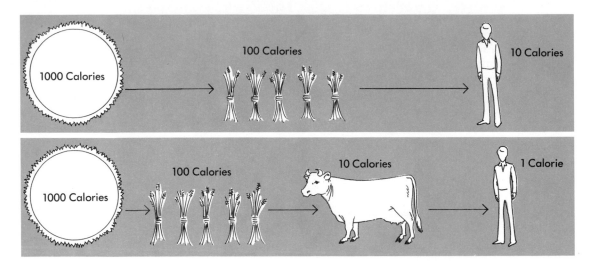

AGRICULTURAL EFFICIENCY

Agricultural efficiency is a ratio of the amount of food energy produced compared to the energy required to produce it. The energy used in manufacturing and using farm machinery, fertilizers, and insecticides must be considered in calculating this efficiency. The high degree of technology used in the United States results in a high number of food Calories per hectare, but with a very low energy efficiency.

Food Chains. Animals feed on plants either directly by grazing or indirectly by consuming other animals that have fed on plants. The simplest and most efficient food chain is to gather plants and eat them. Before fire, primitive people probably ate a diet rich in fruits, nuts, and grains. Most of the food that primitive people gathered was consumed directly by them. By keeping the food chain short more protein was made available to the people.

People in technological societies consume about 3 300 Calories a day. Approximately 10 000 Calories per person per day are required in producing meat, milk, and eggs which constitute much of our diet. Over half the land farmed in the United States is used to grow crops for feeding cattle, poultry, and other animals. We would get more energy from eating these plant crops than we receive by eating the animals and their products.

Compare the two food chains. What is always lost when energy is transformed? How much useable energy is lost at each step? What is the form of the lost energy? What happens to the percent of available energy as the chain gets longer? How can we get more food without growing more crops?

The Energy Price of Food. In primitive agricultural societies farming is done by the farmer's whole family. The human energy required to clear the land and plant the seeds is small compared to the amount of food energy harvested. This food energy surplus provides the farmer with enough energy for living and working. In high population areas a large number of farmers would be necessary to provide for everyone.

In a technological society large amounts of energy are used for growing, marketing, and preparing food. This agricultural system depends on fossil fuels for producing fertilizers, pesticides, and machinery. The farm equipment used in irrigating, planting, and harvesting crops consumes large amounts of energy as does the processing, transportation, and storage of foods. Though fewer people are farmers, large crops can be harvested and food becomes plentiful. Technological farming does not use the sun's energy any more efficiently than primitive farming. It is the supplemental use of fossil fuel energy which supports the farming efforts. Some of our high-protein crops use three to six times more energy per hectare in farming than the plants themselves yield. This type of agriculture consumes more stored solar energy than it captures from daily sunlight. Can we ever get an energy surplus with this type of agriculture? Are fossil fuels unlimited? How can technological agriculture become more efficient?

Wet rice farmer

50 Calories of food out

1 Calorie of work in

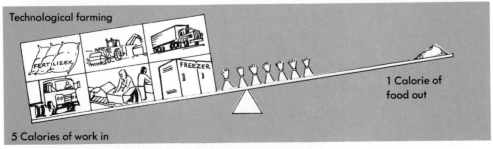

Technological farming

FERTILIZER

FREEZER

1 Calorie of food out

5 Calories of work in

ENERGY EFFICIENCY AT HOME

Our demand for electrical energy is growing rapidly. Every ten years we double the amount of electrical energy we use. The amount of energy wasted during the production of electricity will be equal to 30% of our total national energy consumption by the year 2000.

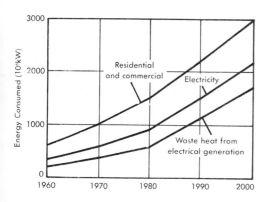

The Use of Energy in Appliances. Households are using more appliances than ever before. Many of the new appliances have special features that use more energy than older models. Frost-free refrigerators and freezers use twice as much energy as the standard models. Self-cleaning ovens raise the oven temperature to 425°C for one to two hours after use to completely break down food which has built-up in the oven. An "instant-on" television is always on and no time is necessary for the television to warm up when you use it. The convenience of self-cleaning ovens and instant-on televisions have a high energy cost. The cost of using the convenience feature may be as much as using the appliance itself.

Energy Efficiency Ratio. The *energy efficiency ratio* (EER) is a measure of how effective an appliance uses the energy it consumes. The higher the EER the more efficient the appliance. The Federal Government now requires many appliances, such as air conditioners and refrigerators, to have an efficiency rating mark on them.

Calculating the Energy Use of an Appliance. Find the power rating in watts on an electrical appliance such as a toaster. Push down the switch and time how long the toaster uses electricity before the switch pops up. The energy used by the toaster is determined by multiplying the power at which the toaster uses electricity (in watts) times the time (in minutes) the power is used. The result is electrical energy in watt-minutes.

Electrical energy is measured in units of power-time. In the above case, power was in watts and time was in minutes, so electrical energy was in watt-minutes. Divide your answer above by 60 minutes per hour to obtain the energy used in watt-hours. Divide this answer by 1000 watts per kilowatt to obtain the energy in kilowatt-hours. Most electrical energy used for building is described in terms of kilowatt-hours. Why is this unit used instead of watt-minutes or watt-seconds?

Toaster oven uses 800 watts of power.

Bread toasts in 2 minutes. Energy used is:

$$E = \text{power} \times \text{time}$$

$$E = 800 \times 2$$

$$E = 1600 \text{ watt-minutes}$$

$$1600 \text{ watt-minutes} = 1600 \text{ watt-min} \times \frac{1 \text{ kW}}{1000 \text{ W}} \times \frac{1 \text{ hr}}{60 \text{ min}} = 0.026 \text{ kW} \cdot \text{h of electrical energy.}$$

Home Heating. Heating houses with electricity has recently become popular. If the electrical energy used to heat a house is compared with the energy burned in oil or gas furnaces, electricity might seem like a good method. No heat escapes up the chimney of a house with electric heat. Electricity, however, involves many energy transformations before it enters the house. Heat energy is lost up the chimney at the power plant as well as in the wires on the way to your house.

In which energy chain shown below is the home heating unit more efficient? Which energy chain is most efficient overall? What makes one system more efficient than the other? What type of energy chain is in your home? In your school?

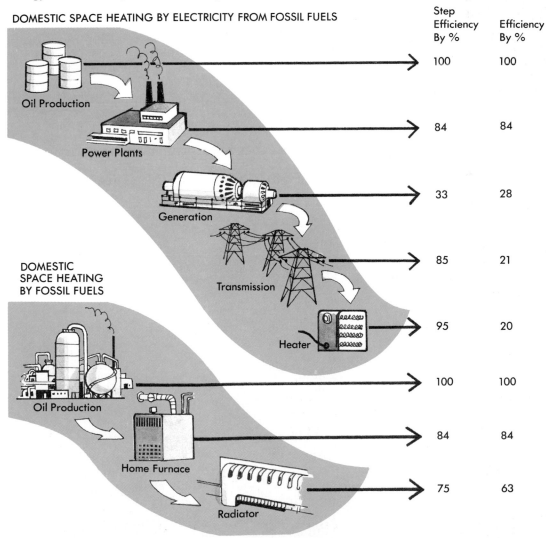

DOMESTIC SPACE HEATING BY ELECTRICITY FROM FOSSIL FUELS

DOMESTIC SPACE HEATING BY FOSSIL FUELS

	Step Efficiency By %	Efficiency By %
Oil Production	100	100
Power Plants	84	84
Generation	33	28
Transmission	85	21
Heater	95	20
Oil Production	100	100
Home Furnace	84	84
Radiator	75	63

Efficiency in Appliances. Many appliances sold today have built-in devices that save energy. Gas ranges and dryers are now equipped with electronic ignitions. By using a spark to ignite the burner instead of using a continuously burning pilot light, 20% of the energy is saved. The original cost of these appliances may be a little higher, but their operating costs are much lower.

A standard 10 000 BTU air conditioner costs $200. A similar 10 000 BTU air conditioner with a built-in device that saves energy costs $250. The energy-saving device makes the more expensive air conditioner two times as efficient. The cheaper air conditioner may cost $50 per year to operate. How much will the air conditioner that is twice as efficient cost you to operate under the same conditions? How much money will you save during the first year of use? How long will it take for the more efficient model to pay for its higher cost? If both air conditioners have a life expectancy of 10 years, how much money did the more efficient model save you? Which air conditioner really costs more?

Efficiency in Lighting. The light of an incandescent or fluorescent lamp is the most easily recognized form of energy used at home or at school. The light of an incandescent bulb results from the lamp's tungsten filament resisting the flow of electrons through it. This resistance causes the wire to become white-hot. A fluorescent bulb is filled with mercury vapor. When a current flows through the tube the mercury atoms move about and collide with each other, giving off light.

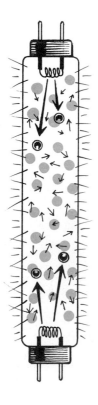

An incandescent lamp works more like an electric heater than a lighting fixture. Only 5% of the electrical energy entering the lamp is converted to light, the rest is given off as heat energy. A fluorescent lamp produces almost four times the amount of light per watt and at an increased efficiency of 20%. What type of lighting fixtures do you find in your school or local department store? Explain why.

Efficiency of Your Home. Improvements that save energy in your home will also save money. Heat losses in a home can be reduced by insulating ceilings and walls, weather stripping seams, and installing storm windows and doors. The cost of these home improvements eventually will be offset with the money saved by using less energy.

Storm windows are a good example. Heat escapes readily from house and apartment windows. Suppose your heating bill was $900 last year. Storm windows can be installed for $600. By installing the storm windows you save 15% on your heating bill. How much money will be saved on this year's heating bill? In how many years will the storm windows pay for themselves?

As the demand for energy increases, the cost of energy also increases. Suppose over the next five years the cost of heating your home increases to $1100. Do the storm windows still save 15% of your home heating cost? How much money will the storm windows save for you then? Is 15% of your annual heating cost at $900 the same as when the annual heating cost is $1100? As the cost of heating your home increases, what happens to the time it takes to recover the cost of your storm windows?

Insulation. A house will lose less heat if the ceiling and walls are insulated. The principle of insulation is based on the fact that air is a very poor conductor of heat. By using a material which traps air, such as fiberglass, wool, vermiculite or styrofoam, a dead air space is created. This dead air space prevents hot air from entering in the summer or leaving in the winter. The quality of an insulating material is measured in R values. The R stands for resistance to air flow. A material with a rating of R-19 has more insulating capability than a material with a R-11 value.

SUMMARY QUESTIONS

1. What happens to the amount of available energy as a food chain gets longer?
2. Why is it important to check the energy efficiency ratio when shopping for new appliances?
3. Why do most large office buildings and schools use fluorescent instead of incandescent lamps?
4. Compare the houses shown here and give an explanation for the different amount of snow cover on each.

AMERICAN LIFE STYLES

The conservation of energy takes many forms, from the personal level to the cultural level. Americans make up only six percent of the world's population but we consume a third of the world's animal protein, half the world's natural resources, and a third of the world's total energy.

Are We Energy Wise? Five metric tons of coal are required to produce one metric ton of nitrogen fertilizer. Some of this fertilizer is spread on our lawns along with pesticides and herbicides in order to produce a nice green carpet. In order to control this green carpet, the average home owner uses 20 liters of gasoline each year to cut grass.

You have seen how the production and distribution of food require large amounts of energy. Storage and preparation also consume vast quantities of energy. How many appliances are used in preparing a hamburger, French fries, and a drink at a fast-food restaurant? How are these items wrapped when they are served to you? What happens to all the packaging material?

Heat is lost more readily where there is a greater surface-to-volume ratio. Approximately 2.5 times as much energy is required for small houses in a row than for one apartment building with an equivalent amount of floor space. Tall apartment buildings with common walls, ceilings, and floors utilize escaping heat. The heat radiating from one apartment is beneficial to the other apartments.

Compare the pictures on these two pages and examine the way energy is used in each pair. What are the energy benefits and drawbacks shown in each pair? Which picture in each set is more closely related to your life style? What other energy-conserving options exist if you were to change your life style?

REVIEW QUESTIONS

1. A system of pulleys requires that 500 joules of work be put in to it in order to accomplish 350 joules of work. What is the efficiency of the pulley system?

2. Give five examples where the force of friction is useful in your daily living.

3. Why do some people refer to food chains as energy chains?

4. How much power is used by the electrical device shown at the left?

5. Calculate the electrical energy used if the device here is turned on for 24 hours.

6. Calculate the cost of the energy used in item five if electricity costs five cents per kilowatt hour.

7. How does electrical power (measured in watts) differ from electrical energy (measured in watt-hours)?

8. Using your home's electric bill, estimate how much energy your home uses in one year. If the electrical energy used comes from a fossil fueled power plant, how much fuel is required for your electrical needs? Calculate the amount of fuel used in your city or town, assuming your requirements are average.

9. Explain why it is said that people who believe that technology will solve the energy crisis have forgotten basic laws of physics.

10. Can Americans save energy without changing their standard of living? Explain your answer.

THOUGHT QUESTIONS

1. Why do the world's largest whales and elephants exist on a diet comprised mainly of green plants?

2. Some people are very concerned about the high birth rates in the developing countries. It is said that these large families will consume the world's resources. Which presents a greater danger to these resources, a typical American family having two children or a Central American peasant family having ten children?

3. The cost of fuel has doubled in the past several years and is expected to increase more. How will this affect the design of new buildings and size of the average car?

Investigating on Your Own

Experts in career education say that half of today's students will spend most of their lives working at jobs which don't exist today. Energy is one field which will have many workers doing jobs which are unknown today. The energy field also provides opportunities for many different kinds of projects. By carrying out different projects on your own, you can explore the possibilities of a career in the energy field.

Many of the investigations in this section require more than one class period, or special conditions not found in school. For these reasons, you should identify several activities from these pages to carry out on your own outside of class. Ask your teacher for class time to plan your work. Plan to share your findings with others during later class periods. Remember, the ideas on these pages are only suggestions; the best ideas are usually the ones you create.

The students in the photograph are trying to melt a pipe by using a Fresnel lens to focus the sun's rays onto the pipe.

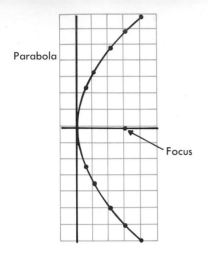

Parabola

Focus

SOLAR ENERGY CONSTRUCTION PROJECTS

Few new fields of science capture people's imaginations as does solar energy. If you can use simple woodworking tools, the projects below will allow you to use some solar energy for practical purposes. In the process, you will discover some of the joys and frustrations which are a part of working with solar energy.

Material	Melting Point, (°C)
Paraffin	55
Sugar	186
Solder	200
Lead	327
Zinc	419
Aluminum	660
Copper	1080
Steel	1450
Chromium	1890

1. Construct a solar hot dog cooker as shown in the photograph above. The curved pieces labeled *A* should be cut in the shape of a parabola. Ask a math teacher or librarian for help in constructing the parabola. The hot dogs are held at the focus of the parabola (where the sun's rays meet after reflecting off the curved aluminum). Cooking time can be decreased by wrapping hot dogs in aluminum foil, then blackening the foil. (Note: The larger the reflector, the faster the meat cooks.)

2. Use a Fresnel lens (the type used in overhead projectors) to construct a solar double boiler. The blackened can should be placed so that all the sun's rays which pass through the lens hit the can. Fill the blackened can one-fourth full of water. Place a second, smaller can containing soup or water inside the blackened can. Keep records of the time required to heat the contents of the inner can. (**CAUTION:** Never allow skin, clothing or flammable material near the place where the sun's rays focus.)

3. Construct an apparatus like the one shown on page 441 for use as a solar furnace. Find out what metals can be melted where the sun's rays meet. Use the chart at the left to estimate the highest temperature reached by the furnace.

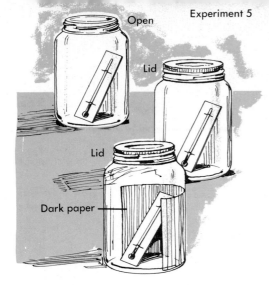

Experiment 5

Open

Lid

Lid

Dark paper

Experiment 6

Same size black paper

4. Devise ways of increasing the temperature attained in number 3. Try adding more lenses or concave mirrors such as those used for shaving and applying cosmetics.

5. The remainder of the solar projects make use of a principle called the greenhouse effect. Projects 5 and 6 quickly demonstrate this principle. Set up the experiment shown at the left. Place the three jars in a sunny place. Record the temperature changes for 30 minutes. Explain your findings. Try other conditions for the jar. Which ones cause the air in the jar to heat up fastest?

6. Set up the experiment shown below at the left. Put the jars in a sunny place. Record the temperatures every 2–3 minutes for at least 30 minutes. Explain your observations.

7. Construct the solar ovens shown below and on page 441. The oven parts should be large enough to bake a loaf of bread or brownie mix. This part consists of two halves of cardboard packing boxes with insulation between them. The inside of the smaller box is painted black. The slanting edge of the oven should be straight and smooth so that the double-thickness glass cover and frame make a tight seal with the oven. The reflectors around the oven are made of cardboard with aluminum foil glued to them. The reflectors should fold down when not in use. Use spring-type clothespins to hold the reflectors together. Put an oven thermometer in the oven to determine the temperature. Temperatures of 160–175°C can be reached in the spring, summer and fall, even in northern latitudes. If used in winter, shield the solar oven from the wind and tilt it so that the sun's rays hit the oven more directly. Display in the class some foods baked in the oven.

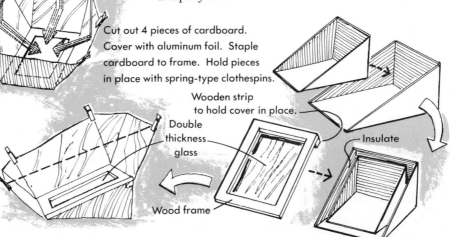

Sunlight

Cut out 4 pieces of cardboard. Cover with aluminum foil. Staple cardboard to frame. Hold pieces in place with spring-type clothespins.

Wooden strip to hold cover in place.

Double thickness glass

Insulate

Wood frame

8. The photographs above show two examples of solar heated homes. The panels collect solar energy to provide hot water and heating for the homes. The solar collectors consist of different types of clear covers and a dark metal absorber in an insulated frame. Study the factors which make good solar collectors. Use two similar cigar boxes. Cover the bottom of one box with a piece of sheet metal painted black. Put a thermometer in the box and then cover the front of the box with a piece of glass. Use masking tape to seal the glass to the box. This solar collector is a control box. Prepare a second or experimental box with one variable: a different transparent cover, collector surface, or type of insulation on the other five sides of the box. Also put a thermometer in this box. Place the boxes facing the sun. Record the temperatures in the boxes for 30–45 minutes. Explain your findings. Try other variables.

If located in Philadelphia, the sun's energy could provide this solar-heated mobile home with 80% of its heating and air conditioning and 95% of its hot water energy needs.

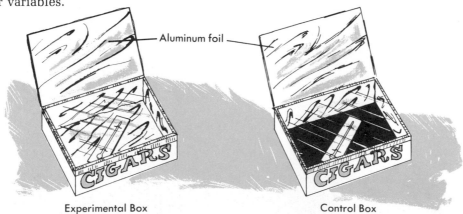

Aluminum foil

Experimental Box   Control Box

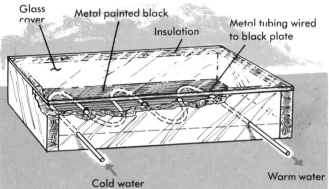

Glass cover — Metal painted black — Insulation — Metal tubing wired to black plate

Cold water — Warm water

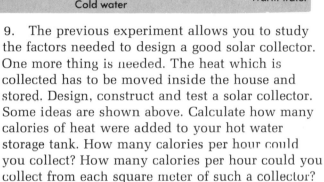

9.   The previous experiment allows you to study the factors needed to design a good solar collector. One more thing is needed. The heat which is collected has to be moved inside the house and stored. Design, construct and test a solar collector. Some ideas are shown above. Calculate how many calories of heat were added to your hot water storage tank. How many calories per hour could you collect? How many calories per hour could you collect from each square meter of such a collector?

10.   Modify the solar collector in number 9 to deliver hot air instead of hot water. Drill several large holes in the bottom and top of the collector frame. Use thermometers to measure the temperature of air going into or coming out of each set of holes as shown at the right. Explain the circulation of air.

11.   Prepare a model of a solar house from a description in a magazine or other source. Describe the house to your class. Show how the heat is collected, stored and distributed, as well as the added cost of this type of heating system and the expected savings per year in fuel.

12.   Using a magazine description or other source, prepare a model of a solar farm to collect solar energy. Describe its operation in class.

13.   Do library research to learn how space stations might be used to supply the earth with solar energy. Prepare a model of such a plan. Describe its operation in class.

EXPERIMENTAL RESEARCH

Many problems related to energy conservation cannot be understood unless you first learn some important ideas about heat and other forms of energy. In this section you can carry out experiments to learn many of these important ideas.

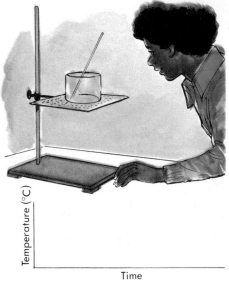

1. Study the way a hot object cools. Heat a container of water to 60°C above room temperature. Remove the heat and record the temperature of the water each minute for at least thirty minutes. Make two graphs, one showing time vs. temperature, and the other showing time vs. difference between room temperature and the water temperature. Use the second graph to determine how long the water took to lose half its heat (to drop to 30°C above the surrounding temperature). How long does the water take before its temperature drops from 30°C to 15°C above room temperature? From 15°C to 7.5°C? Isaac Newton first studied this problem. He said that under ideal conditions, the time interval should be the same for all your answers above. Propose explanations for any differences you observed.

2. Many people say that cooling your house at night does not save fuel. Study the graphs below and plan your own experiment to test this idea. Find out whether the graph labeled *A* (house cooled at night) has its furnace on longer or if graph *B* (house kept warm at night) has its furnace on longer. To accurately compare your data, do experiment *A* first, then carry out experiment *B* for the same length of time as experiment *A*.

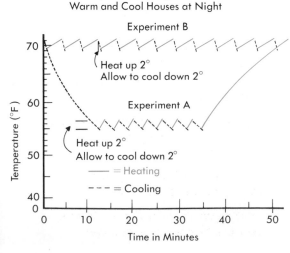

Comparing Heating Times in Warm and Cool Houses at Night

Data for Experiment B

Time (minutes and seconds after the hour)	Elapsed Time (minutes and seconds)	Start of Heating or Cooling	Heat Time	Total Heat Time
10:00	0	C	—	—
10:00–12:30	2:30	H	1:30	1:30
10:00–14:00	4:00	C	—	—
10:00–16:00	6:00	H	1:10	2:40
10:00–17:10	7:10	C	—	—
10:00–19:05	9:05	H	:50	3:30
10:00–19:55	9:55	C	—	—

(continues for 50 minutes)

3. Set up ice cube contests of two types. First, place identical ice cubes inside small plastic bags. Wrap the covered ice cubes with different types of insulation to just fill small identical boxes. Find out which types of materials make good insulators. In the second experiment, wrap covered ice cubes in different amounts of insulating materials. What effect does increasing the thickness have on the insulating properties?

4. Set up cooling races. Use identical flasks of boiling water with one-hole stoppers and thermometers. Wrap the flasks with insulating material and set them in identical cardboard boxes. Which materials make good insulators?

5. Compare the amount of hot water used when taking a shower or bath. Plug the drain of a shower and take a shower for your normal length of time. Measure and record the amount of water used. Next time you take a bath, measure and record the amount of water used. Which requires less water?

6. Plug the drain and take a Navy shower. Turn the water on just long enough to get wet. Turn off the water and soap yourself. Turn on the water only long enough to rinse off. Compare the amount of water used this way with other ways of bathing.

7. Investigate your furnace! Record the length of time it stays on and off, the outside temperature, time of day, sky cover and wind conditions. Repeat under different conditions and note any differences.

8. With your parents' permission, find out how long the furnace stays on and off when it cycles off and on (not when it is getting to temperature) at both 18°C and 24°C. Discuss your findings in class.

9. Compare the temperature of a pan of water on your stove when the water is boiling vigorously and when it is just barely boiling. Which requires less energy? Which cooks faster?

10. Compare the times to boil a cup of water in a large frying pan and a small saucepan when each is placed on a large burner.

11. Compare the times to boil one cup and two cups of water under identical conditions.

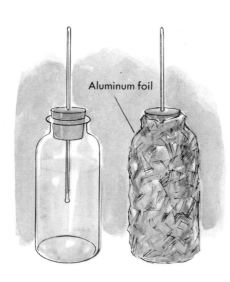

Aluminum foil

12. Set up two bottles with hot water as shown. How does foil affect the temperature change?

HELPING YOURSELF AND OTHERS SAVE ENERGY

1.   Note the Energy Saver's Checklist on the next page. How many energy savers in the list do you currently practice? How many more could you begin practicing without difficulty?

2.   Use the Energy Saver's Checklist for a family discussion of ways to save energy (these ways also save money).

3.   Find another use for the Energy Saver's Checklist. Report on your results to the class.

4.   Add a switch or special switch cord as shown below to your instant-on television. This saves about $30 a year on a large vacuum tube color set.

5.   Interview your neighbors using the checklist on the next page. Find out how many energy-saving ideas your neighbors use. Which ones might they consider trying?

6.   Survey several adults in your neighborhood. How far does each worker travel to work? What methods of travel are used? What is each worker's experience with and thoughts about car pools?

7.   Interview an adult bicycle rider. Find out how many kilometers per week the cyclist rides. What does the cyclist feel are some of the advantages of this method of transportation?

8.   Interview someone who heats part or all of their house with wood. What are some advantages and disadvantages of this method of heating?

9.   Keep a record of ways you have saved energy during a week. For example, you may have suggested that potatoes be cooked in a toaster oven instead of in a regular oven.

10.   Make a list of three ways you might save energy in your house that will also save time. Two ways are presented here: (1) Bake two items in the oven at the same time, then freeze one for later use; (2) Radar ranges use much less energy to cook foods than electric or gas stoves.

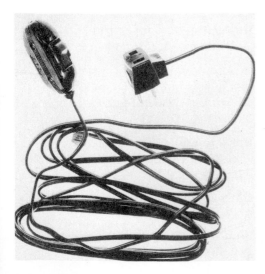

ENERGY SAVER'S CHECKLIST

_____ 1. Set hot water heater at 43°C.

_____ 2. Set refrigerator on a not-so-cold setting.

_____ 3. Set up cartons to recycle paper, glass and metals.

_____ 4. Use lower-wattage bulbs.

_____ 5. Remove extra bulbs in ceiling fixtures.

_____ 6. Replace two 60W bulbs with one 100W; this gives more light and uses less electricity.

_____ 7. Do not use long-life bulbs; they give less light per watt.

_____ 8. Replace incandescent lights with fluorescent lights, where possible.

_____ 9. Use light-colored walls, curtains, rugs.

_____ 10. Check efficiency ratings on appliances such as refrigerators and air conditioners before purchasing.

_____ 11. Use returnable bottles.

_____ 12. Caulk and weatherstrip around doors and windows.

_____ 13. Turn off lights when leaving room.

_____ 14. Set furnace at 20°C or lower during the day, 16°C at night; set air conditioners at 25°C.

_____ 15. Clean furnace filter and get the burner serviced yearly.

_____ 16. Wash clothes in cold water.

_____ 17. Add storm windows and doors or plastic sheets over windows.

_____ 18. Insulate the house—10 cm in walls, 15 cm in ceilings.

_____ 19. Keep garage door closed in winter if attached to house.

_____ 20. On nice days dry clothes out-of-doors instead of in a dryer.

_____ 21. Dry full loads in dryer.

_____ 22. Check gasoline economy of car; tune regularly.

_____ 23. Drive under 90 kilometers per hour.

_____ 24. Buy products made of recycled materials such as recycled paper.

_____ 25. Walk or ride a bike for short trips.

_____ 26. Buy products of natural materials instead of manufactured materials such as plastic (which require more energy to produce).

_____ 27. Shop at businesses which make an effort to conserve energy and resources.

_____ 28. Encourage parents and others to use a car pool whenever possible.

_____ 29. Turn down the thermostat if everyone is leaving the house for several hours.

_____ 30. Shut off the furnace pilot light during the summer.

ADDITIONAL INVESTIGATIONS AND PROJECTS

1. Arrange for a class speaker from one of the energy industries. What is the speaker's industry doing to provide for future energy needs?

2. Prepare a library report about perpetual motion machines designed to produce energy.

3. Write the Federal Energy Office, Washington, D.C. for information about research projects for new energy sources which the government is financing. Prepare a chart comparing the amount of money being spent to develop each energy source. Propose possible reasons for the differences.

4. Prepare a report describing how aluminum is produced. Show why large amounts of energy are needed by this industry.

5. Study the cost of a dripping hot water faucet. Measure the water collected in 15 minutes from a dripping faucet. Calculate the water which drips in an hour, a day and a year. Calculate the cost per year if hot water costs 1.25 cents for 4 liters. (Note: A faucet washer to fix the drip costs only a few pennies!)

6. Visit the office of a home heating oil company. Find out how to calculate heating degree-days. How is this idea used by the company to estimate when it is time to deliver oil to a customer?

7. Prepare a bulletin board display of newspaper and magazine articles dealing with alternate sources of energy.

8. Prepare a display of different insulating materials for houses. Visit a building supply store and find out which materials are the best, least expensive, easiest to install, and so on.

9. Study the heating system in your house. Make a diagram showing the various water or steam pipes or hot air ducts and cold air returns. Use arrows on your diagram to show the direction that heat flows through the system. Also label any pumps or fans in the system.

10. Take apart and study an old thermostat. Prepare a display of the parts.

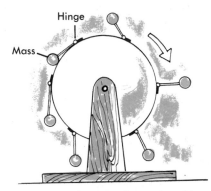

One type of perpetual motion machine

11. Prepare a report on thermos bottles. Include a diagram showing how a thermos bottle cuts down on heat losses by reducing heat conduction, convection, and radiation.

12. Ask a parent to assist you in removing the cover from a home thermostat. Learn how it works and report your findings to the class.

13. Prepare a debate with a friend for your class. Each person should be allowed to speak once for several minutes, in turn. Then each person should be given one minute to point out weaknesses or arguments against the other speaker's talk. Some possible topics are: A dam, raising the water 6 m, should be built on the stream or river nearest your school to supply your region with electricity; or debate the pros and cons of building a nuclear power plant, oil refinery or airport near your school.

14. With your mother's permission, study the insulating characteristics of your refrigerator. Put a thermometer inside. When the refrigerator is not running, pull out the plug. Open the door briefly three times, once each 15 minutes, to determine the temperature inside. Then plug in the refrigerator. Make a graph of your findings.

15. Repeat the previous experiment. This time open the door for 15 seconds once each minute. Graph the drop in temperature during 15 minutes. Discuss the differences between the graphs in numbers 14 and 15 in terms of wasted energy.

16. Visit the local office of your gas company. Find out the calorie content of the gas as well as the percentage composition of the gases in the mixture of gases they use.

17. Prepare a design for an urban area using the "Total Energy Concept." For example, trash could be sent to power companies, waste hot water from power companies to apartments, and so on.

18. Energy scientists say that about two calories of sunlight enter each cm^2 of the earth's atmosphere each minute. About half this amount reaches sea level. Calculate the surface area needed to collect sunlight to heat 200 liters of water each day from 20°C to 60°C. Assume that sunlight falls on the collector for six hours. List some assumptions in your calculations and how they affect your estimate.

REDUCING HEAT ENERGY LOSSES AT HOME

There are two important reasons for reducing the energy losses which occur while heating your house. One reason is to reduce the fuel which is wasted; the second reason is to reduce the cost of heating. Often, the cost of improving your house so as to reduce heat losses is more than paid for by the lower fuel cost.

1. Count your windows and doors which are not insulated. Visit or call a local building supply company. Ask for a free estimate of the cost of storm windows and doors for your home. Also ask for an estimate of the fuel savings per year if storm windows and doors are installed. Calculate the number of years needed to pay the cost.

2. Make a survey with one of your parents to determine the thickness of insulation in the walls and ceilings of your home. Visit or call a building supply company. Discuss with them ways of increasing the insulation in your home. Also determine the cost and savings per year.

3. On a cold and/or windy day, check for drafts around windows and doors. Install weather stripping or caulking, and observe the difference in the amount of cold air leaking into the home.

4. Find out when your furnace was last cleaned. (Most furnaces should be cleaned each year, or they waste fuel.) Arrange to observe the next time the furnace is cleaned. Discuss with the furnace cleaner ways in which fuel is wasted in a dirty furnace.

5. Find out if your furnace has an air filter. If it does, ask your parents to assist you in removing, cleaning or replacing it.

6. Survey your home to see if room heat sources are obstructed with radiator covers, or if furniture is placed in front of radiators or hot air ducts. Ask your parents to remove these obstructions.

7. Encourage your family to wear sweaters and lower the thermostat. Also lower the thermostat at night. (Lowering the thermostat at night saves fuel.)

8. Check for drafts around unused fireplaces. Discuss ways of reducing heat losses here.

6

Energy as a Resource

Our way of living has changed since colonial times. We eat a larger variety of foods which are often transported from distant places. We travel farther, faster, and more often. We also have much more leisure time to spend on entertainment, hobbies, sports, and other interests.

Each person today is using 30 times as much energy as early colonists used. Furthermore, our use of energy continues to increase faster and faster. Study the photographs of energy sources on this page. What is the source of energy in each photograph? In which photographs is the source of energy limited or limitless?

Presently, we use naturally occurring materials to provide us with energy. Since we will use up these resources, energy must be considered as a limited resource. There are other ways of obtaining energy. Some of these other ways do not use up natural resources at such a rapid rate. By using our resourcefulness, we can make energy become an unlimited resource.

HISTORY OF THE USES OF ENERGY

The ability of people to transform energy to meet their needs has been an important factor in helping us to live comfortably. When primitive people lit the first fire and learned to control it, they took the first big step in the use of an energy resource. The use of energy depends on two factors: available sources of energy, and the know-how to transform it into a useful form. There have always been energy sources, but devices to put energy to work have only recently been invented.

Energy in Colonial America. The energy needs of colonial America were fairly simple. People required heat, transportation, preparation of building materials, and food. Heat was obtained from wood fires. Transportation was provided by animal power. The animals received their energy from grasses and other plants. The principal building material was wood. Saw mills used waterwheels to cut logs into boards. Waterwheels and windmills were also used to grind wheat and other grains into flour. Farming also depended on animal power for the jobs requiring more power than one person could supply.

A colonial family was constantly busy. Lacking the labor-saving devices of today, all the necessary tasks to make living comfortable required much more time than they do today. There was very little leisure time available.

Preparing wood for burning at home meant cutting with an axe and a hand saw. The wood was loaded on a horse-drawn cart and transported to the home. The logs then had to be cut smaller and split. Keeping warm required much time and effort. Each task required more time than the same task requires today. Think of several jobs you must do at home and compare them with what a colonist had to do.

Development of Prime Movers. Any machine which converts a source of energy into work is called a *prime mover*. Water power takes advantage of the kinetic energy of moving water. Water power has been used to run waterwheels since the first century B.C. What are some prime movers today?

The waterwheel was probably the first prime mover ever invented. By the ninth century A.D., there were about 8000 waterwheels in England. The waterwheels served about a million people.

Following the waterwheel, the windmill was invented. Although windmills could be built in more places, they had the disadvantage of not producing power constantly because of variations in the wind.

The development of the steam engine marked the first prime mover that could produce more or less power as needed. The steam engine was also the first prime mover that could provide energy for transportation because it could be mounted on a platform. A new fuel had been discovered at the time the steam engine was invented.

The history of the steam engine is closely related to the use of coal. The original steam engine was designed to pump water out of coal mines. Coal was needed for the production of iron, and iron was being used for making steam engines.

As the steam engine was improved, it was used for locomotives on the country's railroads. Now with a prime mover that could transport itself and was adaptable to many different types of jobs, the number of steam engines in use multiplied quickly. During the same time, new devices were being developed (steam shovels, steam tractors and steam automobiles) for turning the energy of fossil fuels into mechanical energy. In 1908, the gasoline-powered Model "T" Ford was mass-produced.

Study the graph which shows the growth of power output by some machines. Machines designed to deliver more power usually require more fuel to do so. By how much has the steam engine been improved? What is one of the most recently invented machines? Which machines are powered by fossil fuels?

Growth of Electrical Energy Requirements. As recently as 100 years ago homes were lighted by flames from candles, coal oil, gas, and other fuels.

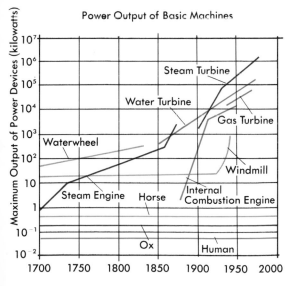

Power Output of Basic Machines

Kitchen at the turn of the century

Kitchen of the 1940's

Kitchen of today

When Thomas Edison invented the light bulb, the demand for electricity took a big jump. In 1882, Edison established the first electrical generating station in New York City. The electric light and the electric power distribution system changed our lives as no other technological development has.

Prior to the development of electricity, energy in the home consisted of small wood or coal stoves and small lamps. After the development of electricity, huge amounts of heat energy were converted to enormous amounts of electricity in distant plants and distributed to thousands of homes. The power of huge machines was available at the flick of a switch. Many appliances were invented to use this power. Name some electrical devices which provide basic necessities at home?

Electricity was produced by large diesel engines which used oil as fuel. Some electricity was also being generated by steam turbines. The steam for the turbines came from water heated by coal fires. Industry was also calling for enormous amounts of energy. The energy was needed to separate metals from their ores. The energy was also used for manufacturing large numbers of machines, appliances and other devices. Many of these devices were being used for the jobs previously done by human and animal power.

Note the graph tracing the growth of energy use. Between 1875 and 1925, how much had the use of energy increased? What new discoveries and inventions could account for this increase? Compare this graph to the graph on page 427. What factors can account for the rapid increase in energy use between 1950 and 1975? What do you think will happen to the rate of use in the future?

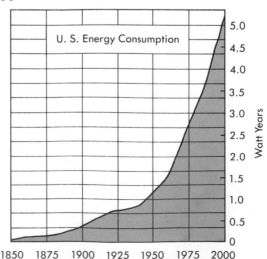

U. S. Energy Consumption

Watt Years

THE CONSUMPTION OF ENERGY

The previous section showed how the United States is gulping energy at faster and faster rates. As you study this section, think of ways that you could cut back on the amount of energy you use.

How You Use Energy. List machines which do work for you, estimate how long they work for you, and calculate the amount of energy they use in that time. Fill out a chart of your own similar to the one in the margin. Add up the amount of energy you calculated for one week. Compare your total use with the amount of energy stored in a cord of wood, barrel of oil, cubic meter of gas, and liter of gasoline. How much wood would you have to cut to satisfy your energy needs? How much time do you think it would take you to prepare that much wood?

Substance	Definition	Amount of Energy
Cord of wood	Pile of wood 4' X 4' X 8' 128 cubic feet	5.7×10^3 kilowatt-hours
Barrel of oil	159 liters	1.6×10^3 kilowatt-hours
	46 kg	10.5 kilowatt-hours per m³
	3.8 liters	10.5 kilowatt-hours

Energy Use Chart

Appliance	Total time used per week	×	Energy used per hour (from table)	=	Total energy used per week

Electric Appliances	Energy Used (kilowatt-hours)
Air conditioner	1.5
Bulbs: Read wattage and divide wattage by 1000. For Example: 40-watt bulb ÷ 1000 = 0.04	
Clothes dryer (electric)	4.8
Coffee maker	0.9
Dishwasher	1.2
Fan	0.2
Food freezer	0.4
Food mixer	0.2
Food waste disposer	0.4
Frying pan	1.2
Hair dryer	0.4
Iron	1.1
Oil burner	0.3
Oven	1.2
Radio	0.1
Range (per burner)	0.8
Refrigerator	0.6
Sewing machine	0.07
Television, black & white	0.2
Television, color	0.3
Toaster	1.1
Vacuum cleaner	0.6
Washing machine	0.5
Water heater	2.5

Gas Appliances	
Clothes dryer	1.0
Furnace (for 7 rooms)	38.8
Range (per burner)	0.7
Water heater	2.0

Oil and Gasoline	
Oil furnace (for 7 rooms)	38.8
Automobile (for every 8 km traveled)	10.0
Lawn mower	2.6

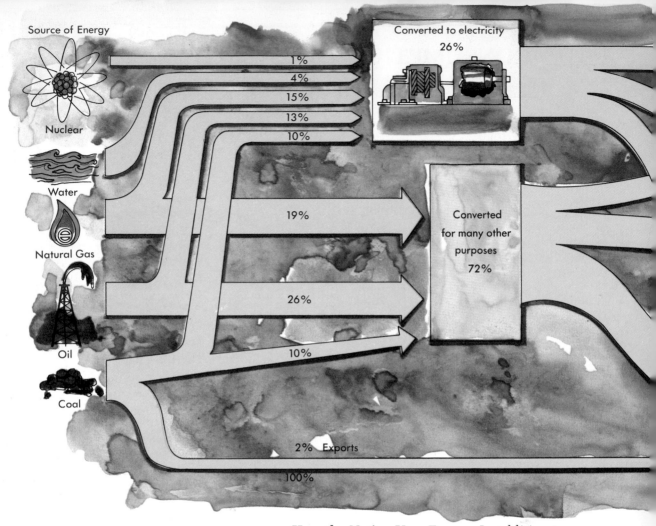

Source of Energy

Nuclear

Water

Natural Gas

Oil

Coal

1%
4%
15%
13%
10%

Converted to electricity
26%

19%

26%

10%

Converted
for many other
purposes
72%

2% Exports

100%

How the Nation Uses Energy. In addition to calculating the energy you used, there is energy being used by the nation for transportation, manufacturing, entertainment, and services. Think of the energy required to bring gasoline to your local gasoline station. Energy is used to pump oil from the well, to transport the oil to the refinery, and to operate the refinery. Energy is also used to transport the refined gasoline to the gasoline station, and finally to pump the gasoline into the tank of an automobile.

The chart across the top of these pages traces the pathways of energy from their sources through our society to their end uses. The arrows indicate the direction the energy is flowing. The thicker the arrow, the more energy that is flowing along the path. Which energy source is used in the largest quantity for producing electricity? Of the total energy obtained from all sources, how much is

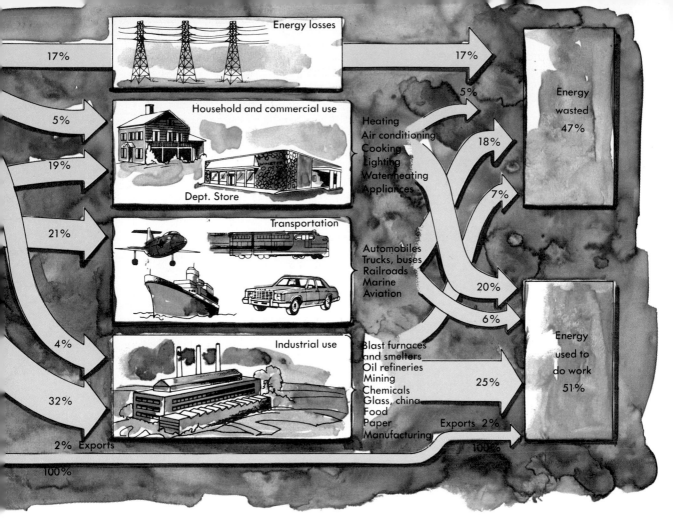

Energy losses

17%   17%

5%   Energy wasted 47%

Household and commercial use

5%

Dept. Store

19%   Heating
Air conditioning
Cooking
Lighting
Water heating
Appliances

18%

7%

Transportation

21%   Automobiles
Trucks, buses
Railroads
Marine
Aviation

20%

6%

4%   Industrial use

Energy used to do work 51%

Blast furnaces
and smelters
Oil refineries
Mining
Chemicals
Glass, china
Food
Paper
Manufacturing

25%

32%

2% Exports   Exports 2%

100%   100%

converted for purposes other than producing electricity?

Transportation uses about 21% of the total energy. Trucks, buses, trains, taxis, automobiles, and planes transport food, mail, and people. Industry, including manufacturing, uses about 36% of all the energy produced. Your clothes, the appliances you use, and the cars you travel in all used energy in their manufacture. In the process of converting energy from one form to another and transmitting it from one place to another, approximately 17% is lost. The remaining 24% is used for residential and commercial purposes. In every case there is some energy that is lost by the machines that convert energy from one form to another. Of the original energy produced, only 51% is used to do work while 47% is wasted. The remaining 2% is exported. What is the efficiency of our system's use of energy?

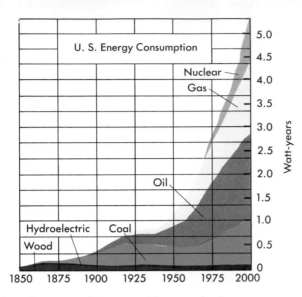

U. S. Energy Consumption

Watt-years

Nuclear
Gas
Oil
Hydroelectric   Coal
Wood

5.0
4.5
4.0
3.5
3.0
2.5
2.0
1.5
1.0
0.5
0

1850   1875   1900   1925   1950   1975   2000

Changing Sources of Energy. The graph shows how our energy sources have changed as energy demands have increased. There are various reasons why our sources of energy have changed. What was the type of fuel most used in 1850? For what purpose was the fuel being used?

After 1850, the growing demand for steam engines called for more fuel. Not only was fuel needed to make steam engines, but fuel was also needed to run them. Wood was becoming scarce. Coal was plentiful, and engineers were finding that coal delivers more heat per kilogram than wood. By 1875, coal and wood were supplying our energy needs in equal quantities. Fifty years later, how did the amount of energy being supplied by coal compare to that supplied by wood?

The first oil well was drilled in September, 1859 in Titusville, Pa. However, oil would not become a major source of energy until later. Refineries had to be built to process the oil, and engines had to be designed to run on oil.

In the late 1800's, the internal combustion engine was being developed. Diesel and other internal combustion engines require fuels which are refined from oil. Gasoline and diesel engines became very successful for providing transportation as well as generating electricity. Their success caused a rapid growth in the demand for oil. During what years were oil and coal being used in equal amounts?

From 1902 to 1907, the demand for electricity increased by two and a half times. By 1917, the rate of consumption was 17 times the rate it had been in 1900. This large increase was due to the inventing, manufacturing, and selling of more and more products that used electrical energy. To meet the demands for electrical energy today, enormous amounts of oil are required to fuel the generators. With oil becoming scarce once more, we will have to be prepared to change our source of energy.

Predicting. There is no doubt that our energy needs are increasing. What will our needs be in the future? What plans can we make now to provide for those needs?

In order to predict conditions in the future, researchers look for trends which they think will continue. Researchers collect information about present and past conditions, and predict the future based on the trends they discover. Predicting the future is not an easy task. Take, for example, the decisions and predictions facing the industry in the example below.

Between 1880 and 1900, ranchers and homesteaders had to depend on windmills for pumping underground water to the surface. Study the graph. How did the windmill business change between 1880 and 1900? If you had been president of a windmill company, what would you predict your business to be during the next ten years?

Assumptions. Before you made the prediction, you would want to consider all the factors which might affect your sales. Some of these factors might be the numbers of new ranchers, the numbers of windmills needing to be replaced, and the development of some other device capable of doing what the windmill had been doing. Other factors might include the continued improvement on the design of windmills, and the ability of windmill manufacturers to keep up with the demand.

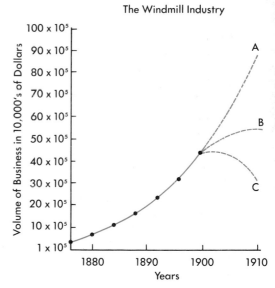

The Windmill Industry

Your interpretations on how these factors will affect the business are called your *assumptions*. What assumptions would you, as the president of the windmill company, have made if you predicted the business would follow the trend of line A, line B, or line C?

The graph of our energy consumption on the facing page predicts our consumption of energy up to the year 2000. What assumptions have to be made to arrive at this prediction? What does this prediction mean to oil suppliers? To electric companies? To gas companies?

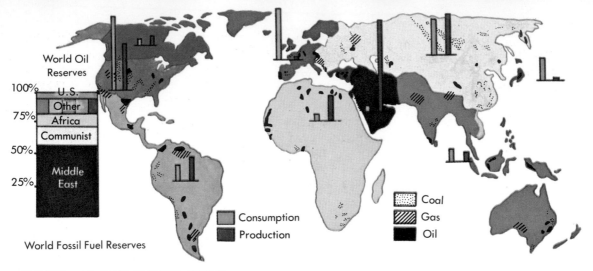

World Oil Reserves

100%
75%
50%
25%

U.S.
Other
Africa
Communist
Middle East

World Fossil Fuel Reserves

Consumption
Production

Coal
Gas
Oil

OUR RESOURCES

Study the charts on these pages. They are based on estimates made by geologists of the known reserves of coal and oil over the earth.

In addition to the resources of coal and oil, we have our own resourcefulness to consider. Not all of our energy needs are being met by coal and oil. Nuclear energy is being used increasingly while other sources of energy are being investigated.

Our Reserves of Coal. Coal, which used to account for 75% of the energy used in 1910, is not the leading supplier of energy now. Coal is a dirty fuel, and leaves ashes and soot after it burns. As coal burns, it releases poisonous gases which pollute the atmosphere. It is bulky to transport, and mining it ruins many square kilometers of land. However, coal is the most abundant fuel in the U. S. Until cheaper ways of controlling pollution are developed, coal will continue to supply only a small portion of our energy needs.

On the average, an electric power plant must burn 0.5 kilograms of coal to produce one kilowatt-hour of electricity. How much electrical energy did you calculate that you used in a week? How much coal was needed to supply you with that much electricity? If coal was being used at the rate of 20×10^9 metric tons in 1975 and continued to be used at that rate, how long would our coal last?

Oil. Oil is less expensive to take out of the ground than coal, and oil can be transported at less cost. Oil burns cleaner and is less damaging to the environment than coal. In order to meet the demands for oil in the U. S., a large amount must be imported from other countries. Study the chart

World Coal Resources
7637×10^9 Metric Tons Total

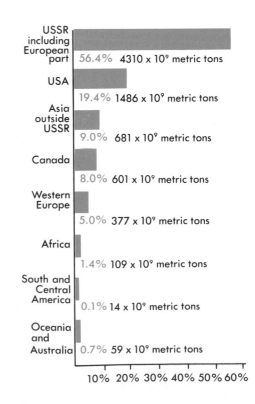

USSR including European part	56.4%	4310×10^9 metric tons
USA	19.4%	1486×10^9 metric tons
Asia outside USSR	9.0%	681×10^9 metric tons
Canada	8.0%	601×10^9 metric tons
Western Europe	5.0%	377×10^9 metric tons
Africa	1.4%	109×10^9 metric tons
South and Central America	0.1%	14×10^9 metric tons
Oceania and Australia	0.7%	59×10^9 metric tons

10% 20% 30% 40% 50% 60%

World Supply of Recoverable Fossil Fuels		
Coal and lignite	7.6×10^{12} metric tons	55.9×10^{15} kWh
Petroleum	2.0×10^{12} barrels	3.25×10^{15} kWh
Tar sand oil	0.3×10^{12} barrels	0.51×10^{15} kWh
Shale oil	0.2×10^{12} barrels	0.32×10^{15} kWh
Natural gas	1480×10^{12} cubic meters	2.94×10^{15} kWh

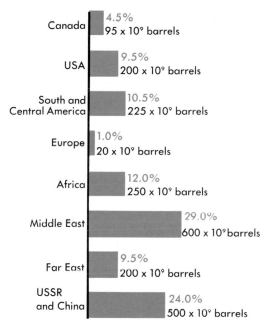

World Petroleum Resources
2090×10^9 Barrels Total

Canada — 4.5% — 95×10^9 barrels

USA — 9.5% — 200×10^9 barrels

South and Central America — 10.5% — 225×10^9 barrels

Europe — 1.0% — 20×10^9 barrels

Africa — 12.0% — 250×10^9 barrels

Middle East — 29.0% — 600×10^9 barrels

Far East — 9.5% — 200×10^9 barrels

USSR and China — 24.0% — 500×10^9 barrels

of oil resources. Where in the world is the most oil to be found? How much is found in the U. S.? If oil is used at the rate it was used in 1975, 18×10^6 barrels per day, how long would our oil resources last? What will we do when oil becomes scarce?

Estimating Quantities. The estimates of reserves on these pages are the best guesses of geologists. There is always hope that other reserves are yet to be discovered, or that the estimates are less than what is actually available. Estimates are calculated to provide a basis for making plans. Why should we investigate alternative sources of energy?

Our Resourcefulness. We must depend on our ability to design new ways of obtaining energy for at least two reasons. One is obvious, the old resources will run out. The second is that we are becoming more and more dependent on other countries for our energy needs. Why is this unwise? As we improve our ability to generate energy through newly discovered techniques, we may be able to export our ideas to other countries.

When considering alternatives for supplying energy, there are several factors to be taken into account. Time is one factor. When oil was discovered, scientists at that time knew it could be a source of energy. However, many years passed before oil was supplying a significant amount of energy. These years were spent finding the necessary funds, training engineers and technicians, and building the manufacturing plants to supply equipment for oil-fueled power plants. Time was also spent identifying sites and constructing power plants. A similar time scale is being followed in the establishment of nuclear power plants.

Another consideration is the amount of fuel available in a resource. The chart above lists the abundance of the world's supply of fossil fuels. Which fuel has the most energy to offer? Which has the least energy to offer? Engineers are developing methods for obtaining oil from tar sand and shale, but this source will be short-lived.

Another important factor in choosing a source of energy is the affect the new energy source will have on the environment. If the source is a fuel, it will have to be mined, processed, and transported. These activities require land which might be put to better use. Also, there might be a problem in disposing the wastes after the fuel has been used.

The use of the sun's energy is a promising alternative. At this time, techniques for using solar energy are being developed. Today, the cost of gathering and converting the sun's energy is too high to compete with other sources. However, engineers are developing new and cheaper ways. In the future, we should expect to get useful energy from the sun.

ENERGY AND THE ENVIRONMENT

Historically, the coal produced in the U.S. came from underground mines. During the past two decades, however, surface strip mining of coal has become popular and now accounts for half our coal production.

Production and Transportation of Coal and Oil. During the strip mining process, huge shovels carve into the earth's surface, removing topsoil and subsoil to expose seams of coal. The coal is then mined by smaller shovels and transported to railroad loading stations. This process results in large barren tracks of land. The nutrient-ladened topsoil ends up buried beneath the sterile subsoil and new vegetation cannot grow. Without vegetation to stabilize the soil, erosion takes place. Rain runs off and combines with sulfur from the unearthened areas to produce sulfuric acid. This acid water prevents streams, ponds, and the land from supporting life, rendering the land useless and hazardous to us.

So that huge tracts of land do not become lifeless pits, government regulations require strip miners to repair the land. This process is called *reclamation*. Reclamation means that the land will be returned to its original contour with topsoil being replaced and vegetation replanted. After a few years, the reclaimed land stabilizes and becomes habitable.

Much of our present and future domestic oil is to be found offshore along the coasts and in the arctic region. The delicate balance of these environments and their geographic isolation make them vulnerable to catastrophic accidents during the production or transportation of oil. Although the oil companies have good safety records, oil spills from blowouts at offshore platforms, tanker accidents or pipeline leaks are always a threat to the environment. Large disasters testify to the reality of major accidents taking place. Examples include the tanker Amoco Cadiz (1978) which lost 220 000 metric tons of crude oil to the sea off the coast of Brittany, France, and the Mexican oil rig mishap which for four months released 20 000 barrels of oil daily into the Gulf of Mexico.

Air Pollution From Fossil Fuels. Coal, oil, and natural gas are made from hydrogen and carbon in various percentages. The hydrocarbons and the impurities in them combine chemically with oxygen from the atmosphere and release heat. Air pollution results when chemical by-products of this reaction are released into the atmosphere.

Carbon monoxide is a toxic gas resulting from the incomplete combustion of carbon atoms. The exhaust of internal combustion engines, such as automobiles, contains large amounts of carbon monoxide. Once in the atmosphere the carbon monoxide will combine with oxygen and form carbon dioxide. Carbon dioxide, which is harmless to people and essential to plants, makes up 0.03% of our atmospheric gases. Carbon dioxide also forms when complete combustion of fossil fuels takes place. As we burn more fuels the amount of carbon dioxide in the atmosphere will increase. Scientists are in debate whether the increase of carbon dioxide in the atmosphere is acting as a shield to entering sunlight, thus causing the earth to cool off or whether the carbon dioxide is acting as a one-way mirror allowing sunlight to enter the earth's atmosphere but preventing the earth's heat from escaping.

Nitrogen oxides are formed when combustion takes place at a high temperature, combining oxygen and nitrogen. This toxic reddish-brown gas forms in electric power plants, large industrial boilers, and internal combustion engines. Much of the smog seen in urban areas, such as Los Angeles, is attributed to the nitric oxide emitted from automobiles.

Sulfur dioxide is produced when oil and coal containing sulfur is burned. Oil which contains

0.5% to 2.0% sulfur is more desirable to burn than coal which may contain up to 7.0% sulfur. Unfortunately, high sulfur coal (7.0%) is one of our most abundant energy resources. Electric power and industrial plants are primarily responsible for the formation of sulfur dioxide. Sulfur dioxide in the atmosphere reacts with oxygen to produce sulfur trioxide. Sulfur trioxide reacts with water vapor and forms sulfuric acid. Inhaling this acidic atmospheric vapor can be damaging to the human respiratory system. Rain showers cleanse the atmosphere and transfer the acid to earth. In areas of high sulfuric acid concentration, plant growth is affected while lakes and rivers become acidic.

Strip mining for coal

REVIEW QUESTIONS

1. Name some activities which you take part in today that colonial people could not enjoy.
2. Why did the colonists have much less leisure time than you do today?
3. What was the first prime mover invented? How did it work?
4. Why is coal a better fuel than wood?
5. Before electricity, what was used to provide light and heat in the home?
6. What three factors are responsible for the tremendous increase in the demand for electrical energy between 1900 and 1917?
7. What is an assumption? How are assumptions
8. What are three factors to be considered when choosing among possible energy sources?

THOUGHT QUESTIONS

1. Make a list of five modern prime movers and indicate what they use for fuel.
2. How does the development of prime movers affect the use of energy?
3. What historical developments might have been responsible for decreasing energy consumption between 1925 and 1945?
4. How is the history of the demand for coal like the history of the demand for electricity?
5. Study the photograph of strip mining for coal. How does this affect the environment?
6. Some power plants which generate electricity burn fossil fuels. How does the power plant shown here affect the environment?

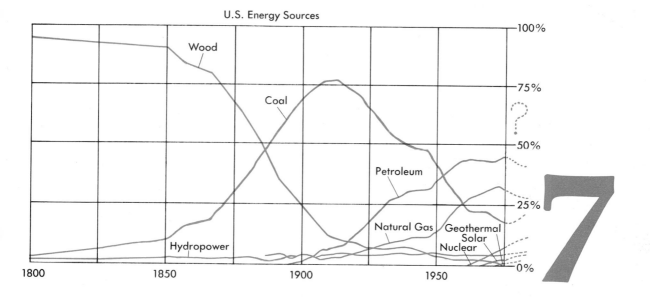

U.S. Energy Sources

Wood

Coal

Petroleum

Natural Gas Geothermal
              Solar
Hydropower   Nuclear

100%

75%

50%

25%

0%

1800   1850   1900   1950

Future Energy Sources

The chart above shows how our energy sources have changed in the past 180 years. About 100 years ago, much of the land in the eastern United States was bare because many of the trees had been cut. We were already running out of firewood for heating homes and schools.

By 1900, coal had replaced wood as a major source of energy. Many of your parents and teachers remember when coal was an important part of their daily lives. Coal trucks were common sights on the streets.

The graph above shows that wood once met 95% of our energy needs and coal once met about 77% of our energy needs. The graph also shows that today, oil fills about 45% of our energy needs. The percent of each fuel's use has dropped because our society became more complex and we had to use other energy sources. Fifty years from now, petroleum and natural gas probably will not be our most important energy sources. A good guess might be that we will use still more sources of energy than in the past. Probably no single energy source will supply more than about 25% of our total energy needs.

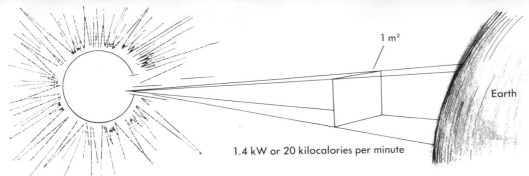

1 m²

Earth

1.4 kW or 20 kilocalories per minute

ENERGY FROM THE SUN

Measurements from satellites and rockets have determined that solar energy passes through each square meter of space in the earth's orbit at a rate of 1.4 kilowatts. Stated another way, 20 kilocalories of solar energy pass through each square meter of space every minute.

There are several problems associated with using this energy. Not all the energy passes through the atmosphere, and clouds block the sunshine on overcast days. Part of the energy collected during the day must be stored for use at night. The energy-giving sun constantly moves across the sky. Thus, collectors must constantly change their position to remain pointed at the sun.

Solar Energy for Cooking. A small amount of our daily energy needs is consumed in cooking. In some countries, however, the fuel for cooking is supplied from the manure of farm animals. The manure could be better used as fertilizer to improve the crops and diets of the people. What is needed is another source of energy for cooking.

You probably already know one way to get very high temperatures from sunlight. Perhaps you discovered it when a friend focused the sun's rays on your arm! The same idea is used to make solar cookers like the one shown below. Curved mirrors are used instead of curved lenses. These mirrors reflect nearly all the sunlight which strikes them so that the rays converge at one point. Note the two types of solar cookers shown on this page. The large mirror cooker is used to cook midday meals in

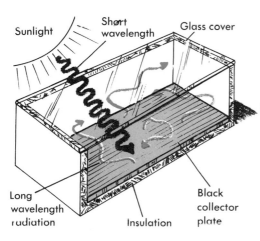

The sun heats water which is piped from the roof to rocks in the cellar. The circulating hot water heats the rocks. At night, a fan blows air through the rocks. The heated air is piped through air ducts and circulates throughout the house.

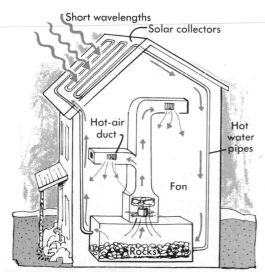

underdeveloped countries. Why doesn't this device completely replace other ways of cooking?

Solar Energy for Heating. The cheapest way to gather the sun's energy is to use solar collectors such as those shown in the photograph above. These collectors store heat by a process called the *greenhouse effect*. To understand how this process works, study the drawing at the left. Sunlight with radiation of very short wavelength passes through the glass cover of the collector. The sunlight is absorbed by the black collector plate in the box, thus heating it. The heated collector plate loses heat by radiating very long-wave infrared (or heat) radiation. The glass cover is not transparent to this long-wave radiation, so these waves reflect back inside the collector. As a result, the temperature inside increases. Heat can be removed by pumping cold water through the collector plate. The water gets heated as the temperature inside the collector is lowered.

Study the lower diagram. How is the heat from the solar collector used to heat the house in the daytime? At nighttime? Why are solar collectors usually insulated?

The greenhouse effect is also used to heat greenhouses. Only on very cold or cloudy days is heat necessary in a greenhouse. Usually, greenhouse operators are more concerned with ways of removing too much heat. Why do greenhouse operators sometimes paint their glass roofs white in the summer?

You have probably noticed a dramatic example of the greenhouse effect. Do you recall opening a car which had been left in the sun with the windows closed? The temperature inside a car under these conditions can approach the boiling point of water. Such high temperatures can kill pets left in the car.

Electricity Directly from Sunlight. The photograph at the left shows a 19th century solar steam engine. The sun's rays heated water to boiling and the steam was used to run small machines. Later, when turbogenerators were invented, the same idea was used to move a steam turbine and generate electricity.

A more efficient way of producing electricity is to use solar cells. Study the diagram of a solar cell. Each cell consists of a disc of silicon about 2-3 cm in diameter and about 2 mm thick. A second substance such as boron is allowed to diffuse into the surface of the silicon disc. This surface is called a p-layer and gives the disc the characteristics of a semiconductor. Now electrons can travel only in the direction from the p-layer to the high-purity silicon region, or n-layer. Electrons cannot flow in the other direction. When this disc is placed in sunlight, photons of sunlight knock electrons out of the p-layer making it positively charged. The electrons are forced into the n-layer making it negatively charged. The electrons can return to the p-layer only by flowing through a circuit as shown in the diagram.

Turbogenerator = turbine + generator

A turbine is a wheel with vanes. The shaft of the wheel turns as steam or water exert force against the vanes. The turbine's spinning shaft also turns the shaft on a generator, producing electricity.

The lower photograph shows thousands of solar cells mounted on two panels, each about 8 square meters in size. This system produces 1500 watts of electricity. Calculate the efficiency of these solar cells if 1.4 kilowatts of sunlight falls on each square meter.

The original cost of equipping satellites with solar cells was about $300 000 per kilowatt. By 1980, improved manufacturing methods had reduced this cost to about $15 000 per kilowatt. Experts agree that solar cells costing $500 per kilowatt would allow homeowners to meet their electrical needs for less money than they now pay electric companies.

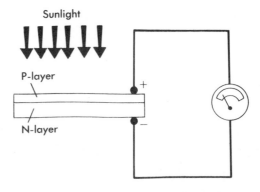

Sunlight

P-layer

N-layer

EARTH, SEA, AND AIR: FUTURE ENERGY SOURCES?

You can probably think of many potential energy sources, such as wind, garbage, trash, swamp gas, peat moss, manure, wood and so on. Although many different sources exist, about 95% of our energy today comes from the fossil sources: coal, oil, and gas. Our technology developed these sources because they were plentiful and cheap. The price of these fuels is increasing and their abundance is decreasing. Tomorrow's energy needs may be met by using many different sources of energy. Some of these sources are described here.

Geothermal Energy. The shaded area on the map shows the regions of the U. S. where hot springs and geysers are found. These features usually occur in regions of recent earthquakes and volcanic activity. These areas often have rock formations at unusually high temperatures close to the earth's surface. Such rocks are a potential source of energy.

The blue square on the map shows the location of the photograph above. This location is called Geysers, California. In this area, steam from wells is piped to turbine generators and is producing 1000 megawatts of electricity.

The process seems simple but it is not. Hot steam often contains many dissolved materials which corrode and ruin pipes and turbines. Fortunately, the steam at Geysers, California is much cleaner than the steam in most of the other areas where geothermal power is possible. Also, removing large amounts of steam and water often causes the surface of the ground to collapse and shift. In spite of these problems, some experts estimate that 10–20% of our energy needs might be met by this method early in the next century.

mega = million

471

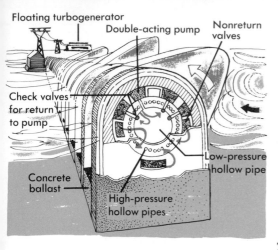

Floating turbogenerator
Double-acting pump
Nonreturn valves
Check valves for return to pump
Low-pressure hollow pipe
Concrete ballast
High-pressure hollow pipes

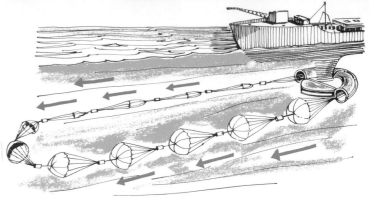

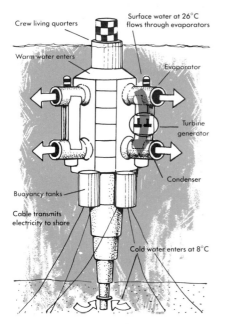

Crew living quarters
Surface water at 26°C flows through evaporators
Warm water enters
Evaporator
Turbine generator
Condenser
Buoyancy tanks
Cable transmits electricity to shore
Cold water enters at 8°C

Energy from the Oceans. The oceans cover over 70% of the earth's surface, but except for off-shore oil wells, almost no energy today is supplied by the seas. However, much energy is there for an inventor to use. This page describes proposals for four different methods to obtain energy from the seas.

Using Energy in Waves. The upper left diagram shows a huge floating device. Its vanes rock up and down as waves break against them. This motion pumps water at high pressure past the turbine vanes in the floating turbogenerator. Low-pressure water from the turbogenerator returns through the large center tube to be repressurized by wave action.

Using Energy in Ocean Currents. The upper right diagram shows a turbogenerator ship anchored in an ocean current. A device operates like a waterwheel beneath the ship. The parachutes of this 'waterwheel' are pulled out from the ship by the strong ocean current. The parachutes collapse on their return loop, greatly reducing their resistance to the flow of the ocean current. A powerful, slow turning of the ship's generator would result. Gears could speed up the turning of the generator.

Using Ocean Temperature Differences. The device at the left uses ocean surface water at 27°C and water 500 meters deep at 7°C. Warm sea water in the evaporator vaporizes a low-boiling liquid to a gas, under pressure. This gas moves past the turbogenerator, producing electricity. Then the gas is cooled by cold seawater back to a liquid, and is returned to the evaporator where the cycle repeats.

Using Tidal Power. People who have spent a half day or more at the seashore are amazed at the movement of the tides. The photograph at the left shows the world's first tidal power plant. Discuss in class how it works. What are some problems with such an electrical generator?

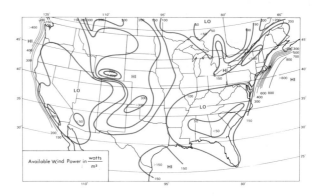

Available Wind Power in $\frac{watts}{m^2}$

Energy from the Wind. The map shows the average annual wind power in watts per square meter available in various regions of the U. S. Where is the wind power greatest? Are these areas where the population is greatest?

The photograph shows a windmill used for pumping water. These were common in the U. S. until the 1930's. At that time, the Federal government provided money to help build power lines to supply nearly every home with electricity. When wires arrived with low-cost electricity, farmers lost interest in windmills. Almost none are seen today.

The lower diagram shows one engineer's idea of a way to capture wind energy. Thousands of units like this one would be built. They would be anchored off the northeast coast of the U. S. Why would this region be chosen?

The movement of the propellers would generate electricity. The electricity would be used to break down water into hydrogen and oxygen. These two gases would be sent by pipeline or barge to power generating plants ashore. Here the hydrogen would be burned, combining with the oxygen and releasing much heat. The heat energy would run turbogenerators and produce electricity.

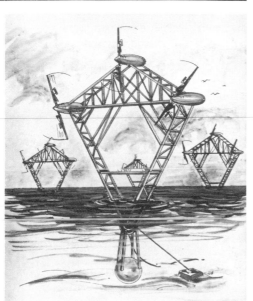

The burning of hydrogen in an oxygen environment to produce electricity has several advantages. When hydrogen burns, no chemical pollutants form. Also, electricity is a favorite form of energy for consumers. In addition, burning hydrogen produced from wind generators solves some of the problems associated with using wind or solar energy. For example, what happens when the wind doesn't blow or the sun doesn't shine? Suggest some possible problems of developing a wind-generating system such as this.

473

Energy Equivalents

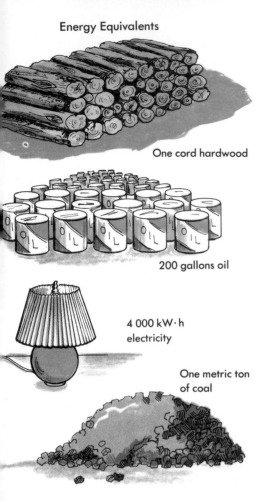

One cord hardwood

200 gallons oil

4 000 kW·h
electricity

One metric ton
of coal

Energy from the Forest. The technological problems associated with alternative energy sources are centered around converting energy to a more useable form and storing it until needed. Nature solved these problems by creating the tree. A tree converts solar energy to chemical energy by photosynthesis and stores the energy in the form of wood.

Wood as a fuel is sold by volume in units called cords. A cord of wood is a pile four feet wide, four feet high, and eight feet long. The amount of energy in a cord of wood depends upon the weight of the cord. This weight will vary depending on the density of the wood which makes up the cord.

When trees are used as fuel, they are grouped as hardwoods or softwoods according to their density. The denser the wood, the better the fuel. Hardwoods (oak, beech, hickory) are characterized by having broad leaves and are good fuel woods. Because of their dense cellular structure, there is more burnable fuel material per cord than in the less dense softwoods. Softwoods (pine, spruce, fir) are characterized by having needles, are light, and burn quickly. They are high in resins which can build up in a chimney and create a fire hazard.

Regional Energy Alternatives. The energy problems that exist in this country and the world will not be solved quickly or easily. There is no new, abundant, totally safe, and environmentally sound fuel waiting to be discovered. Some scientists believe that hydrogen fuel or nuclear fusion will someday fulfill the major portion of our energy needs. Unfortunately, the development of them is far off in the future.

The alternative forms of energy that are presently available to us will not by themselves solve the energy problem. But by developing these alternatives we could supplement our present fuel sources until other forms are available.

The composite map on the next page shows the location of renewable energy resources in North America. Examine the map and select one or two supplementary forms of energy which would be used for different regions of the country. By using alternative energy forms where their resources are available, we would be able to conserve our limited fossil fuels.

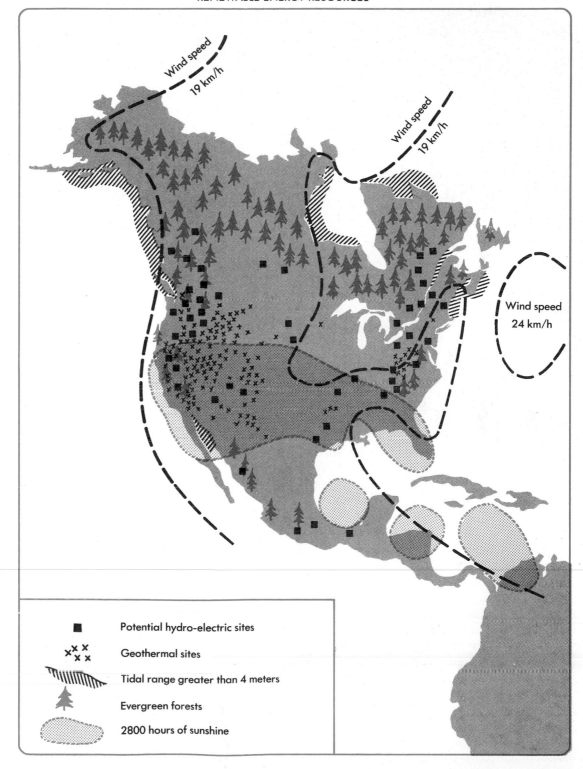

Potential hydro-electric sites

Geothermal sites

Tidal range greater than 4 meters

Evergreen forests

2800 hours of sunshine

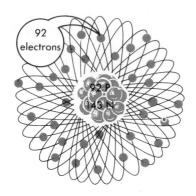

NUCLEAR ENERGY:
THE NEXT ENERGY SOURCE?

Many of the alternative energy sources described in the last section have not been built. Even if they are built, it will be many years before they can be built cheaply enough or in large enough numbers to solve our energy problems.

Only one energy source, nuclear power, has been built to solve our increasing needs for energy. There are many problems associated with this energy source. Many experts say nuclear energy is the only answer; other experts say we would be better off without it. You can make better decisions about nuclear energy if you understand how it works and the problems associated with its use.

The Fission Reaction. Atomic scientists describe atoms as light electrons in orbit around a dense, central nucleus. The nucleus consists mostly of particles called protons and neutrons. For example, uranium-235 atoms have 92 protons and 143 neutrons in each nucleus. Protons and neutrons have about the same mass, so the nuclear mass of U-235 is 92 + 143 or 235.

If a source of neutrons is placed near an atom of uranium-235, a uranium nucleus may absorb a neutron. This nucleus is unstable. It splits or *fissions*. Two smaller atoms are produced. Together, these two atoms account for between 233 and 234 of the 236 nuclear particles before fission (235 particles plus one for the neutron). During fission, two or three fast-moving neutrons are produced. The high energy of these neutrons is *nuclear energy*. These neutrons can now be absorbed by other uranium atoms and the fission process repeats, releasing more energy.

Moderating Neutrons. The reaction described above is oversimplified. One problem not mentioned above is the energy of the neutrons.

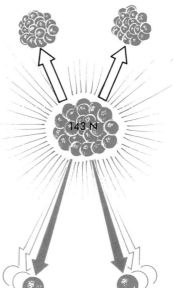

Two smaller nuclei

2-3 high-energy neutrons

U-235 nuclei fission only by capturing neutrons of very low energy. Nuclear fission produces neutrons of high energy. A second material must be present whose atoms absorb most of the energy as neutrons collide with it. Graphite, a form of carbon, is used to slow down or moderate neutrons.

Critical Mass. When an atom of U-235 splits, neutrons must be slowed down before they can react with the nuclei of other U-235 atoms. If there is not much U-235, the chances of it continuing to split are very small. The amount of U-235 necessary for this nuclear reaction to continue is called the *critical mass*.

Chain Reactions: Uncontrolled and Controlled. U-235 captures a neutron, fissions and releases more neutrons, and then slows down the neutrons to repeat the process. This process takes place very, very quickly. One estimate is 5 billionths of a second. At this rate, the process could repeat itself nearly 200 times in 1 millionth of a second. That is, one neutron causes one fission reaction, releases 2-3 neutrons which cause 2-3 more fission reactions. This releases 4-9 neutrons which cause 4-9 more fission reactions and so on 198 more times in 1 millionth of a second. This process is called a *chain reaction*. The example described above is called an *uncontrolled chain reaction*. Its common name is an atomic bomb!

In the process described above, 1 kilogram of U-235 could be completely fissioned in 72 repeats of the capture-fission-moderation process (less than 0.5 millionth of a second). The fissioning of 1 kilogram of U-235 releases heat energy equal to that of 17 000 metric tons of TNT. Three kilograms of U-235 have as much heat energy as 40 000 barrels of oil or an entire trainload of coal.

The diagram at the left shows a nuclear furnace or *reactor* which uses U-235 as its fuel in a *controlled chain reaction*. Note the graphite in the reactor. What purpose does it serve? Note that this reactor also has boron control rods. Boron is an excellent absorber of neutrons. The control rods can be raised or lowered to absorb just enough neutrons so that fission occurs at a controlled rate. If engineers want the reactor to produce more heat, they raise the control rods to absorb fewer neutrons. If less heat is needed, the control rods are lowered to absorb more neutrons. Thus the rate of fissioning drops.

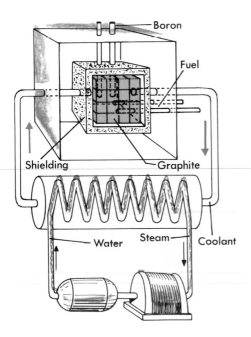

The Source of Atomic Energy. Large amounts of energy are trapped in the nuclei of large atoms. This energy keeps the atomic particles together. Smaller atoms require much less energy than even their smaller size might suggest. This excess energy is released to the extra neutrons during fission. The movement of neutrons or other atomic particles is considered as part of the heat energy.

The original mass of the U-235 atom cannot be accounted for by adding up the parts which remain. A small part of the original mass is missing. It was proven that this mass was converted to energy according to Einstein's famous equation: $E = mc^2$. How does this equation suggest that a small amount of mass produces a large amount of energy?

Atomic Energy and Safety. An energy source which has demonstrated that it can destroy whole cities is not considered to be a good neighbor by many people. Suppose that a reactor's control rods accidentally move up and then the motor fails, or the pumps which remove heat from the reactor fail. In either case, the reactor and shielding could melt, allowing harmful radioactive material to escape.

Because of such possibilities, engineers have designed and built many safety devices for nuclear power stations. However, no manufactured device is 100.00% perfect. Even a very slight risk of a nuclear accident seems much greater to the people living nearby. As a result, power companies have a difficult time getting permission to build and operate nuclear-powered electrical generating stations.

A second problem of nuclear reactors is disposing of the elements which form when U-235 fissions. The elements are radioactive and give off radiations that are harmful to living things. Thus, these elements must not get into the air we breathe, the water we drink or the soil in which we grow our plants. Neither must these substances get into rivers and oceans because the fish would be contaminated. These elements remain dangerously radioactive for hundreds of years. As a result, any accidents will be felt by many future generations of people.

Breeder Reactors. Less than 1% of uranium is U-235. Over 99% is U-238 (92 protons and 146 neutrons). U-238 does not fission when it is bombarded with neutrons. However, if U-238 is placed in a reactor, its atoms absorb neutrons and

Nuclear power station

When construction is complete, these storage tanks for radioactive wastes will be covered with more than 3 meters of earth.

change to uranium-239 (92 protons and 147 neutrons). U-239 is radioactive. A negatively charged beta particle shoots out from each nucleus, changing each atom to neptunium-93 (93 protons and 146 neutrons). Np-93 is radioactive. Another negatively charged beta particle shoots off from the nucleus, and the atom is changed to plutonium-239 (94 protons and 146 neutrons). Pu-239 is fissionable. In this way, a reactor containing U-235 and U-238 can actually produce more fissionable fuel than it consumes. Why is this type of reactor called a *breeder* reactor?

Fusion, the Sun's Way of Releasing Energy. In fission, large atoms split, producing smaller atoms. The total mass of the smaller atoms is less than the total mass of the larger atoms from which they formed. The missing mass is converted to energy in the process.

In fusion, small atoms are combined or *fused* into larger atoms. The new, larger atoms have less mass than the total mass of all the smaller atoms from which the larger atoms formed. The missing mass changes to energy during this process, called *fusion*. The sun's energy comes from fusion.

Enormous energy is required to fuse atomic nuclei. For example, if two hydrogen nuclei are to fuse, they must be moving fast enough to overcome the repelling force of their two positively charged nuclei. Another problem is that each hydrogen nucleus is very small compared to a uranium nucleus. At present, scientists use an atomic bomb as a trigger to reach the high temperatures necessary to fuse hydrogen into helium atoms, producing a hydrogen bomb. These scientists are now trying to attain a *controlled fusion reaction*. The world's supply of fusionable hydrogen atoms is very large and the energy produced is also very large. When, and if, researchers successfully control fusion, the world's search for energy sources will be solved for at least a century. This is about as long as oil solved most of our energy needs.

Albert Einstein

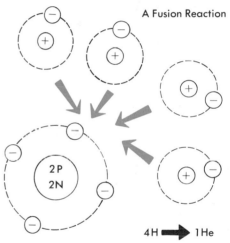

A Fusion Reaction

2P
2N

4H → 1He

Major controlled thermonuclear research is done here at the Princeton University Plasma Physics Laboratory.

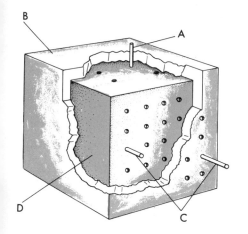

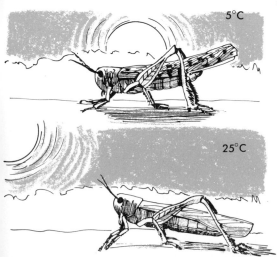

REVIEW QUESTIONS

1. Use the diagram above to explain the greenhouse effect.
2. Why is the greenhouse effect important in solar heating?
3. How do solar-heated homes get heat energy at night?
4. Why are solar batteries used on satellites?
5. What are some ways of using energy from the oceans?
6. Where is the best place in North America for using the wind as a source of energy?
7. Define chain reaction, critical mass, fission, fusion, and reactor.
8. Why are atomic-powered generators being built more slowly than electric companies thought they would?
9. What are some problems in developing controlled fusion reactors?
10. Identify the parts labeled A, B, C, and D in the drawing at the left.

THOUGHT QUESTIONS

1. Solar batteries and fuel cells convert hydrogen and oxygen directly into electricity. Why would they probably be more desirable than using the sun's heat to make steam and turn a turbogenerator, producing electricity?
2. Biologists have noted that grasshoppers often sit in the positions shown at the left. Propose an explanation for this behavior.
3. Suggest reasons why it is difficult to predict what our major sources of energy will be 50 years from now.
4. Name five inventions in the energy field which would make you a millionaire if you were the first to invent them?
5. Three drawings and one photograph of ocean devices for obtaining energy are shown on page 472. Why weren't more photographs used?

WORKING WITH GLASS TUBING

Cutting Glass Tubing. Lay the tube flat on a table top. Make a scratch by *pushing* the edge of a triangular file across the tube. Do not pull the file toward you; this breaks the teeth of the file.

Hold the tubing in both hands with your thumbs opposite the scratch. Give the tube a quick bend and it will break apart at the scratch. If you fear being cut, wear gloves until you find out how to break tubing.

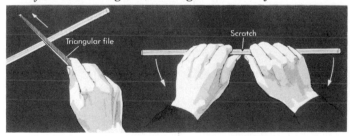
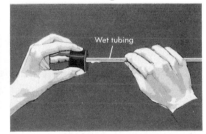

Smoothing Sharp Edges on Tubes. Sharp edges remain on a glass tube after it has been cut, as shown above. These edges can cut your hands.

To smooth the edges, hold the tube in a hot flame, such as that of a gas stove. Turn the tube slowly so that all sides become hot. In a short time the thin edges will melt and flow. Then let the glass cool. (**CAUTION:** Glass stays hot for a long time. Do not touch it for five minutes.)

Putting Tubing in Rubber Stoppers. Dry glass tubing does not slip easily into a rubber stopper. It may break and cut the palm of your hand.

Always wet glass tubing before trying to push it into the hole of a rubber stopper. Then push it gently with a twisting motion.

Do not leave glass tubing in rubber stoppers. After a few days, the glass cements to the rubber; then it is dangerous to separate them.

Making a Nozzle Tube. Heat the middle of a 15-cm piece of glass tubing. When it softens, remove it from the flame and quickly pull the two ends apart. Break the tubing in half and use either end for the nozzle tube.

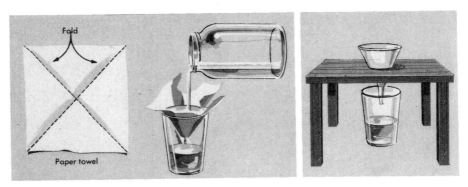

WORKING WITH CHEMICALS

Filtering Liquids. Solid particles are removed from a liquid by the process of filtering. Filter paper can be bought to remove all but the smallest particles. However, ordinary paper towels will serve for many purposes.

Cut the paper towel in squares and fold each square as shown here. Set a piece of the folded paper in the top of a drinking glass and open it somewhat to give it the shape of a cone. Pour the liquid to be filtered in the top of the cone.

Filtering is somewhat easier if a funnel is used. A stand to hold it can be made of wood or heavy cardboard.

Distilled Water. It is usually wise to use distilled water for experiments with chemicals. Tap water may have a number of minerals dissolved in it. These minerals represent uncontrolled factors in an experiment and no one can be certain what effect they may have on the results.

Distilled water is prepared by boiling tap water, collecting the steam, and condensing it. The apparatus shown below will provide small quantities of distilled water. Steam from the teakettle is led through a glass tube into a clean glass jar. The glass jar is kept cold by setting it in a pan of cold water.

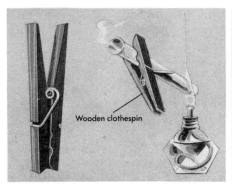 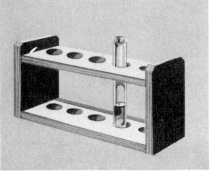

Wooden clothespin

Heating a Test Tube. Some kind of a holder is needed when heating a test tube. Wire holders can be bought or made from stiff coat hanger wire. A wooden clothespin is also good.

Hold the test tube slanted in the flame as shown above. Always point the open end away from yourself and other people; sometimes a bubble of steam throws out some of the hot water forcefully.

Heat the tube slowly, moving it in the flame to warm all parts of the liquid. When the liquid begins to boil, raise the test tube higher in the flame.

A Test Tube Rack. Test tubes need some type of support so that they do not tip over and spill their contents. Drinking glasses or a jar will help. It is well to put a pad of thick paper or cotton in the bottom of the container to help prevent breakage.

An excellent rack can be made from a few pieces of wood as shown above. Note that there are holes bored part way through the base to hold the bottom of the test tubes. Pads of paper or cotton can be put in the bottoms of these holes.

Evaporating Liquids. Sometimes it is desired to evaporate a liquid from a substance without causing a chemical change through overheating. At the left below is shown a steam bath evaporator. The temperature of the saucer cannot rise above 100° C. The metal can has a series of holes punched around the edge by a beverage can opener. Water boils in the can and heats the liquid in the saucer on top.

A heat lamp (infrared lamp) can also be used to evaporate liquids as shown below at the right. The temperature can be controlled by adjusting the distance between the lamp and the liquid.

Saucer
Holes
Tin can of boiling water
Heat lamp

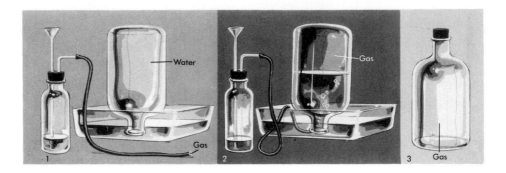

WORKING WITH GASES

Collecting a Gas. The diagram above shows how to collect a gas that has been produced from chemicals. Fill a bottle with water and stand it upside down in a large pan of water as shown above in No. 1. Let some of the gas escape from the tube for a minute or two to remove any air that was in the gas generator.

Push the end of the tube into the mouth of the bottle. The gas will bubble up into the bottle and drive out the water as shown in No. 2. When the bottle is full of gas, put on a stopper and remove the bottle from the pan (No. 3).

When a large bottle of gas is collected, be sure that the pan is large enough to hold the water that comes from the bottle.

Measuring the Volume of a Gas. To measure the amount of gas given off by a certain quantity of chemicals, collect the gas as shown below. The bottle should be large enough to hold all the gas.

When the gas has stopped flowing, cap the bottle and turn it right side up. There will probably be some water in the bottom. Pour in enough water to fill the bottle again. The amount of water put in will equal the volume of gas collected.

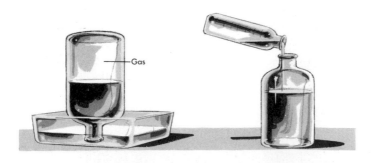

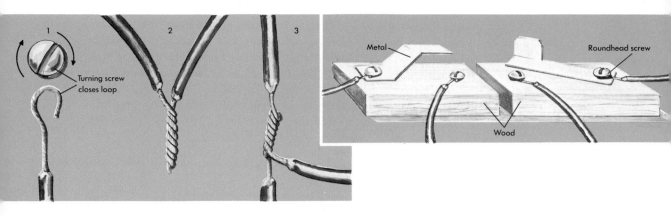

MAKING ELECTRICAL EQUIPMENT

Making Electrical Connections. First remove the fiber or plastic covering from the end of the wires to be used. This can be done with a knife blade.

Figure 1 shows how to fasten a wire to a screw. Note that the wire is bent in a clockwise direction when making the loop. If the loop is made this way, the loop will be closed even more as the screw is tightened.

Figure 2 shows one way to connect two wires together. A drop of solder on the wires makes the connection stronger and a better conductor.

Figure 3 shows how to connect one wire to the center of another wire. Insulation is taken from 2 or 3 centimeters of the second wire without cutting the metal. Again, solder is helpful.

Making Electric Switches. Most electric circuits need switches. There are two types; those that can be left either open or closed like those used in lamps, and those that remain open except when held closed like the pushbuttons in doorbell circuits.

Above are shown switches of both types. The metal strips are cut from tin cans or from aluminum food trays. The holes may be punched with a nail and the rough edges smoothed off with a file, or they may be punched with a nail set which leaves a smooth hole.

Note that roundheaded screws are used.

An Electroscope. Fit a rubber stopper to a wide-mouthed bottle. Insert a wire of large diameter through the stopper. Bend a hook in the lower end of the wire and a loop in the upper end. Wrap the loop with foil to form a knob.

Cut two strips of aluminum foil, the thinner the better. Hang these from the hook by means of a ring of fine wire passed through the strips. The strips, called leaves, should swing freely on the ring.

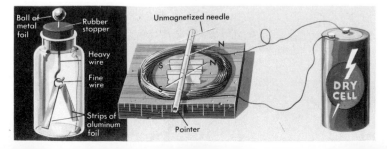

To charge the electroscope, bring a charged object toward the knob until the leaves spread apart the desired distance. Then touch the knob briefly with a finger and take away the charged object. The electroscope will have a positive charge if the object has a negative charge, and vice versa.

To identify a charge on an object, bring the object near the knob. If the leaves spread farther apart, the object and the electroscope have the same charge. If the leaves collapse, the two charges are unlike.

A Galvanometer. Push two razor blades 2 or 3 centimeters apart into a block of wood. Cut a soda straw slightly longer than the block. Magnetize two small needles. Push the needles through the straw as shown on page 485, balancing the first needle added until the straw rolls freely on the blades. Now balance the second needle. The N-poles of the two needles should point in the same direction.

Push an unmagnetized needle through one end of the straw in line with the others and balance it. Push a second unmagnetized needle through the other end; this needle should be slightly off balance.

Encircle the razor blades with a 50-turn coil of fine wire as shown. Connect the coil to a dry cell and note the direction in which the needles turn. Mark the direction of the current as indicated by this movement.

A Reversing Switch. Connect diagonally opposite terminals of a double-throw, double-pole switch with short lengths of insulated wire as shown here. Attach leads from the energy supply to the center terminals. Leads to a device may be attached to either of the other two pairs of terminals.

An Electrophorus. Fasten a wooden handle to the center of a metal pie tin by means of a flat-headed screw driven flush with the bottom of the pan. Lay a discarded phonograph record on a flat surface and charge

Insulated wire

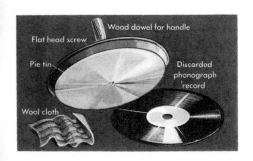

Wood dowel for handle
Flat head screw
Pie tin
Discarded phonograph record
Wool cloth

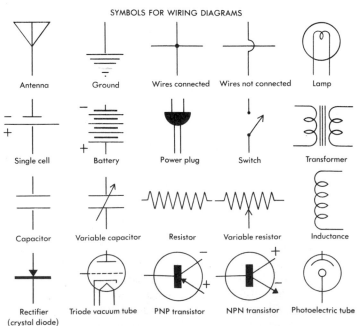

SYMBOLS FOR WIRING DIAGRAMS

Antenna   Ground   Wires connected   Wires not connected   Lamp

Single cell   Battery   Power plug   Switch   Transformer

Capacitor   Variable capacitor   Resistor   Variable resistor   Inductance

Rectifier (crystal diode)   Triode vacuum tube   PNP transistor   NPN transistor   Photoelectric tube

it by rubbing a woolen cloth over it. Set the pan on the charged record, touch it briefly with a finger, and lift it with the insulated handle. Since the record has a negative charge, the pan has a positive charge.

The charge may be transferred from the electrophorus pan to other objects by contact. The pan may then be recharged by placing it on the record as before. The record need not be recharged until it has lost its charge by leakage to the surroundings.

PROJECTS

If you want to understand the real adventure of scientific research, undertake a science project. Such a project can be almost anything you do by yourself, model-making or library investigations, but the most exciting projects are those of discovery.

There are two types of discovery projects. One type includes descriptions and comparisons of conditions, such as a study of plants in burned-over and unburned forests. Another type includes experimentation to find out how a change in one condition affects another condition, such as a test of the effect of light on the opening and closing of stomata (pores) in green leaves.

Questions that begin with "what" and "how" are generally most suitable for projects. Examples include, "How do ants find their way back to their nest?" and "What effect does temperature have on the rise of liquid in celery stalks?" Questions that begin with "why," on the other hand, may be impossible to answer. Many "why" questions, such as "Why do some birds fly south in autumn?" have been answered only with theories and hypotheses which may or may not be correct.

To find questions, make contact with the things of science. Take a walk through the woods or a field. Go fishing in a plant-filled stream with a kitchen strainer. Dig up and sort a pail of soil. Lie on your stomach and look closely at the plants and animals in a lawn.

Try out some of the experiments described in books. Then ask yourself what might happen if you changed one of the conditions of the experiment.

Once started, you will find that an attempt to answer one question leads to new questions, some of which may never have been investigated before. That is the challenge of science. There are always new areas to explore and new discoveries to be made.

Laboratory Studies

1. Study the effect of dissolved substances, such as soap and sugar, on the surface tension of water as determined by a floating needle.
2. Test the effects of soaps and detergents on the strength of cloth.
3. Make a study of the effects of temperature on the height to which a golf ball bounces.
4. Compare the bending strength of wood when green and when dry as measured by the force needed to break sticks which are supported at their ends.
5. Wind a coil of stiff wire on a pencil or round stick. Then suspend the coil and measure the change in length for increasing loads until the coil no longer springs back to its original length when unloaded. Plot the results for coils of different metals.
6. Study the viscosity of motor oil at different temperatures by measuring the amount that flows per minute through a small hole in the bottom of a can. Test other types of oil.
7. Measure the effect of changing the emitter voltage on the emitter current of a transistor.
8. Measure the rate at which a marble sinks through cooking oil at different temperatures.
9. Analyze the light reflected from various types of surfaces using polaroid lenses to determine the extent of polarization.
10. Study the effect of board thickness (or length) on the maximum load which can be supported without breaking.
11. Study the effect of temperature, dissolved materials, or type of liquid on surface tension. Measure surface tension by counting the drops of liquid which drain from a 30-centimeter length of glass tubing.
12. Find out what effect the amount of baking soda (used with vinegar) has on the amount of carbon dioxide gas produced.
13. Compare the rate at which manganese dioxide (MnO_2) catalyzes the production of oxygen and the rate catalyzed by other metal oxides (copper oxide, iron oxide.)
14. Find out what effect the amount of MnO_2 has on the rate of oxygen production.
15. Find out what effect the distance from a light to a convex lens has on the distance from the lens to the light's image.
16. Find out how the focal length of convex lenses affects the images they produce. See if the diameter affects the images produced.
17. Find out how the specific gravity of a cylindrical jar affects the fraction of the floating jar which is below the surface of the water.
18. Find out how the thickness of the wire affects the strength of an electromagnet.
19. Compare the volume of different rocks with the weight the rock loses when in water.
20. Use salt-ice mixtures to measure the freezing point of different alcohol-water mixtures.

ANALYZING EXPERIMENTAL RESULTS

The physical sciences have probably produced more changes in our lives than any other branch of science. One hypothesis for this success is that experiments are used extensively. The independent and dependent variables can usually be measured quite accurately. Other variables are usually easier to control than in the biological and earth sciences.

Experimental results are usually graphed; then the graph is analyzed to develop an understanding of how the independent variable affects the dependent variable. In many cases, the relationship can be expressed mathematically. Mathematical relationships (called *equations*) permit accurate predictions in designing new instruments and inventions. For this reason, engineers often make use of the ideas discovered by physical scientists.

As you carry out experiments in class or on your own, use the techniques described below to analyze your graphs. From your analyses, you will develop a better understanding of the experimental results.

I. **Straight line graphs which pass through the origin** (point 0,0)
   A. Example: Effect of slingshot's stretch on distance paper wad travels.

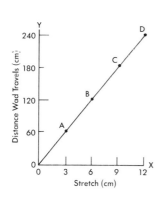

   B. Graph
   C. Analysis: Select several points (A to D) at random along the line. At point A (3,60) the ratio $Y/X = 20$. The ratio, Y/X, at points B, C, and D (120/6, 180/9, 240/12) is also 20. The ratio at all other points along this line is also 20.

   Therefore, for this experiment:
   $$\frac{Y}{X} = 20$$
   $$= \frac{\text{Distance paper wad travels}}{\text{Stretch of slingshot}}$$
   or, Dist. wad travels = 20 × stretch.

   D. Predicting from the equation (for stretch = 9 cm)

   Distance = 20 × stretch
   Distance = 20 × 9 cm
   Distance = 180 cm

   E. Other experiments whose results usually produce similar shapes:

   1. Effect of weight on friction
   2. Effect of baking soda on the production of CO_2
   3. Effect of weight on the stretch of a spring
   4. Effect of weight on the bend of a beam

II. Curved line graphs which pass through the origin (point 0,0)

A. Example: The effect of time on the distance a can rolls down a slope.

B. Data   C. Graph

Time (sec)	Distance (cm)
0	0
1	18
2	72
3	162

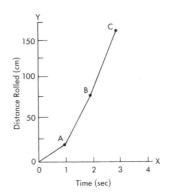

D. Analysis: The ratio, Y/X, is not always the same. At point A, the ratio is 18, at B the ratio is 36, at C the ratio is 54. Y/X is not a constant. Therefore, you can see that X does not increase as rapidly as Y. Graph Y vs X^2. X^2 is calculated and graphed below.

Time	(Time)^2	Distance
0	0	0
1	1	18
2	4	72
3	9	162

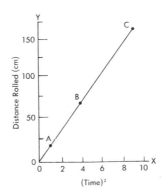

If the graph is analyzed as before, the following equation is obtained:

$$\frac{\text{Distance rolled}}{(\text{time}^2)} = 18 \text{ or, Distance} = 18 \times (\text{time})^2$$

E. Other experiments whose results usually produce similar graphs:

1. Effect of circle's radius on its area
2. Effect of twig diameter on strength

III. Curved line graphs which do not pass through origin (point 0,0)

A. Example: The effect of pendulum length on its rate of swing.

B. Data   C. Graph

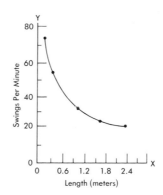

Length (meters)	Rate of Swing
0.15	78
0.30	54
0.90	32
1.50	25
2.40	19

D. Analysis: In the graph above, Y decreases as X increases. To make Y increase as X increases, graph Rate vs 1/Length, as shown at the left below.

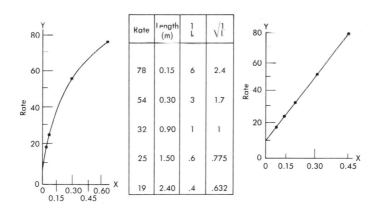

Rate	Length (m)	$\frac{1}{L}$	$\sqrt{\frac{1}{L}}$
78	0.15	6	2.4
54	0.30	3	1.7
32	0.90	1	1
25	1.50	.6	.775
19	2.40	.4	.632

The line on the graph at the left passes through the origin, but X increases faster than Y. Therefore, graph Rate vs $\sqrt{1/\text{Length}}$, as shown at the right above. From this analysis, the equation relating length and rate of pendulum is:

$$\frac{\text{Rate}}{\sqrt{\dfrac{1}{\text{Length}}}} = 32 \text{ or, Rate} = 32\sqrt{\frac{1}{\text{Length}}}$$

THE METRIC SYSTEM

Unit	Length	Volume	Mass
METRIC NAME	METER	LITER	GRAM
Approximate Amount	from your waist to the floor	4 cups	a dime has a mass of 2 grams

METRIC PREFIXES

Prefix	Meaning	Commonly Used Lengths	Commonly Used Volumes	Commonly Used Masses
micro-	$\dfrac{1}{1\,000\,000}$	micron	—	microgram
milli-	$\dfrac{1}{1\,000}$	millimeter	milliliter	milligram
centi-	$\dfrac{1}{100}$	centimeter	—	—
deci-	$\dfrac{1}{10}$	—	deciliter	—
(no prefix)	1	meter	liter	gram
deka-	10	—	—	—
hecto-	100	—	—	—
kilo-	1 000	kilometer	—	kilogram

COMPARING LENGTH TO VOLUME TO MASS

This distance |←——→| is 1 centimeter of length

(100 of these equals 1 meter)

A cube 1 centimeter on each side has a volume of 1 milliliter

(1 000 of these 1-milliliter cubes equals 1 liter)

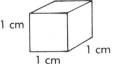

1 cm
1 cm
1 cm

One milliliter of water has a mass of one gram

CHEMISTRY

Asimov, Isaac, *Building Blocks of the Universe*. New York, N.Y.: Abelard-Schuman, 1961, 280 pp.

Asimov, Isaac, *Inside the Atom*. New York, N.Y.: Abelard-Schuman, 1966, 96 pp.

Asimov, Isaac, *The Search for the Elements*. New York, N.Y.: Basic, 1962, 158 pp.

Cobb, Vicki, *Science Experiments You Can Eat*. Philadelphia, PA.: Lippincott, 1972, 127 pp.

Coulson, E. H., A. E. J. Trinder and Aaron E. Klein, *Test Tubes and Beakers: Chemistry for Young Experimenters*. Garden City, N.Y.: Doubleday, 1971, 134 pp.

Ellis, R. Hobart Jr., *Knowing the Atomic Nucleus*. New York, N.Y.: Lothrop, Lee and Shepard, 1973, 127 pp.

Frisch, O. R., *Working With Atoms*. New York, N.Y.: Basic, 1966, 96 pp.

Gray, Charles A., *Explorations in Chemistry*. New York, N.Y.: Dutton, 1965, 224 pp.

Kelman, Peter and A. Harris Stone, *Mendeleyev: Prophet of Chemical Order*. Englewood Cliffs, N.J.: Prentice-Hall, 1969, 68 pp.

Pollack, Philip, *Careers and Opportunities in Chemistry*. New York, N.Y.: Dutton, 1960, 147 pp.

Smith, Richard Furnald, *Chemistry for the Millions*. New York, N.Y.: Scribner, 1972, 175 pp.

Sootin, Harry, *Easy Experiments with Water Pollution*. New York, N.Y.: Grosset and Dunlap, 1971, 95 pp.

BALANCED AND UNBALANCED FORCES

Ackerman, John H., *Mountain Climbing Trains*. New York, N.Y.: Ives Washburn, 1969, 110 pp.

Bronowski, Jacob, Gerald Barry, James Fisher and Julian Huxley (eds.), *The Doubleday Pictorial Library of Technology: Man Remakes the World*. Garden City, N.Y.: 1964, 367 pp.

Froman, Robert, *Baseball-istics: The Basic Physics of Baseball*. New York, N.Y.: Putnam, 1967, 128 pp.

Grey, Jerry, *The Facts of Flight*. Philadelphia, PA.: Westminster, 1973, 135 pp.

Jackson, David, *The Wonderful World of Engineering*. Garden City, N.Y.: Doubleday, 1970, 96 pp.

Kohn, Bernice, *Communications Satellites: Message Centers in Space*. New York, N.Y.: Four Winds, 1975, 58 pp.

Linberg, Peter R., *Engines (A First Book)*. New York, N.Y.: Watts, 1970, 88 pp.

Munch, Theodore W., *Man the Engineer—Nature's Copycat*. Philadelphia, PA.: Westminster, 1974, 125 pp.

Rosenfeld, Sam, *Science Experiments With Air*. Irvington-on-Hudson, N.Y.: Harvey House, 1969, 192 pp.

Valens, Evans G., *The Attractive Universe: Gravity and the Shape of Space*. Cleveland, OH.: Collins-World, 1969, 188 pp.

Von Braun, Wernher and Frederick I. Ordway III, *History of Rocketry and Space Travel*. New York, N.Y.: Crowell, 1975, 308 pp.

Walton, Harry, *The How and Why of Mechanical Movements*. New York, N.Y.: Dutton, 1968, 297 pp.

HEAT, SOUND, AND LIGHT

Adler, Irving, *The Story of Light*. Irvington-on-Hudson, N.Y.: Harvey House, 1971, 123 pp.

Adler, Irving, *The Wonders of Physics: An Introduction to the Physical World*. New York, N.Y.: Golden, 1966, 165 pp.

Bulman, A. D., *Models for Experiments in Physics*. New York, N.Y.: Crowell, 1968, 202 pp.

Chedd, Graham, *Sound: From Communications to Noise Pollution*. Garden City, N.Y.: Doubleday, 1971, 187 pp.

Froman, Robert, *Science, Art and Visual Illusions*. New York, N.Y.: Simon and Schuster, 1970, 127 pp.

Gregg, James R., *Experiments in Visual Science: For Home and School*. New York, N.Y.: Ronald, 1966, 158 pp.

Kock, Winston E., *Lasers and Holography: An Introduction to Coherent Optics*. Garden City, N.Y.: Anchor, 1969, 103 pp.

Meethan, A. R., *The Depth of Cold*. New York, N.Y.: Barnes and Noble, 1967, 173 pp.

Tannenbaum, Beulah and Myra Stillman, *Understanding Sound*. New York, N.Y.: McGraw-Hill, 1972, 176 pp.

Tolansky, S., *Revolution in Optics*. Baltimore, MD.: 1968, 222 pp.

Webster, David, *How To Do a Science Project*. New York, N.Y.: Watts, 1974, 61 pp.

ELECTRICITY

Basford, Leslie and Joan Pick, *Lightning in Harness: Foundations of Electricity*. London, England: Sampson Low, Marston, 1966, 127 pp.

Dunsheath, Percy, *Giants of Electricity*. New York, N.Y.: Crowell, 1967, 200 pp.

Gregor, Arthur, *Bell Laboratories: Inside the World's Largest Communications Center*. New York, N.Y.: Scribner, 1972, 125 pp.

Herron, Edward, *Miracle of the Air Waves: A History of Radio*. New York, N.Y.: Messner, 1969, 191 pp.

Jahns, Patricia, *Joseph Henry: Father of American Electronics*. Englewood Cliffs, N.J.: Prentice-Hall, 1970, 143 pp.

Kogan, Philip and Joan Pick, *The Silent Energy: Foundations of Electrical Technology*. London, England: Sampson Low, Marston, 1966, 127 pp.

Kyle, James, *Electronics Unraveled: A New Common Sense Approach*. Blue Ridge Summit, PA: Tab, 1974, 229 pp.

Martin, Clifford, *Basic Electricity and Beginning Electronics*. Blue Ridge Summit, PA.: Tab, 1973, 256 pp.

Noll, Edward, *Science Projects in Electronics*. Indianapolis, IN.: Sams, 1971, 144 pp.

O'Conner, Jerome, J. *The Telephone: How It Works*. New York, N.Y.: Putnam, 1972, 94 pp.

Sootin, Harry, *Experiments in Magnetism and Electricity*. New York, N.Y.: Watts, 1962, 244 pp.

Sootin, Harry, *Experiments with Static Electricity*. New York, N.Y.: Norton, 1969, 86 pp.

ENERGY

Brinkworth, B. J., *Solar Energy for Man*. New York, N.Y.: Halstead/Wiley, 1973, 251 pp.

Daniels, Farrington, *Direct Use of the Sun's Energy*. New York, N.Y.: Ballantine, 1974, 271 pp.

Ellis, R. Hobart Jr., *Knowing the Atomic Nucleus*. New York, N.Y.: Lothrop, Lee and Shepard, 1973, 127 pp.

Halacy, D. S. Jr., *The Coming Age of Solar Energy*. New York, N.Y.: Harper, 1973, 241 pp.

Halacy, D. S. Jr., *The Energy Trap*. New York, N.Y.: Four Winds, 1974, 143 pp.

Halacy, D. S. Jr., *Fuel Cells: Power for Tomorrow*. New York, N.Y.: World, 1966, 160 pp.

Jaworski, Irene D. and Alexander Joseph, *Atomic Energy: The Story of Nuclear Science*. London, England: Sampson Low, Marston, 1966, 128 pp.

Kondo, Herbert, *Albert Einstein and the Theory of Relativity*. New York, N.Y.: Watts, 1969, 182 pp.

Millard, Reed, et al., *How Will We Meet the Energy Crisis? Power for Tomorrow's World*. New York, N.Y.: Messner, 1971, 189 pp.

Sterland, E. G., *Energy into Power: The Story of Man and Machines*. New York, N.Y.: Doubleday, 1967, 252 pp.

GLOSSARY

Absorbed light: Light taken in and not reflected.

Accelerating (ak SELL er rayt ing): An object that is increasing in speed.

Acid: A substance which tastes sour and turns blue litmus red.

Action: A force exerted on any object.

Actual mechanical advantage: A comparison between the resistance and the effort of a simple machine.

Alkali (AL kah lie): Another name for a base.

Alkali metals: A family of metals which lose one electron easily.

Alpha particle: A nuclear particle which consists of 2 protons and 2 neutrons (the same as a helium nucleus).

Alternating current (AC): Electrons that flow one way and then the other way.

Alternator: A generator which produces alternating current.

Ammeter (AM eet ur): A device which measures current in amperes.

Ampere (AM pir): A current of one coulomb per second.

Amplify: To increase the strength of a sound.

Apogee (AP uh gee): The point in a satellite's orbit which is furthest from the earth.

Apparent image: A mirror image where the rays of light seem to be behind the mirror where no light passes.

Atom: The smallest particle of an element.

Atomic mass: The average number of protons and neutrons in the nucleus of an atom.

Atomic number: The number of protons in the nucleus of an atom.

Assumptions: Interpretations of how factors will affect other factors.

Base: A substance which tastes bitter and turns red litmus blue.

Battery: Two or more cells connected into a single circuit.

Bimetallic bar: Two different metals riveted together which expand and contract as the temperature varies.

Beta particle: A nuclear particle which resembles an electron; negatively charged.

Blackout: Cutting electrical power off completely to some customers.

Breeder reactor: A nuclear reactor that produces more fuel than it can use.

Brownout: Decreasing voltage on power lines to save electric power during a shortage.

Brushes: Pieces of wire used as a conductor between a circuit and a rotor.

Catalyst (KATA list): A substance that changes the speed of a chemical reaction but does not take part in the reaction itself.

Celsius temperature scale: The international temperature scale used by scientists.

Chain reaction: A series of very rapid nuclear fissions.

Charged: Refers to an electrically active object that attracts other objects.

Chemical equation: The symbols chemists use to show how chemicals react and the products produced.

Chemical properties: Properties of substances which depend on the substances' reaction with other substances.

Chemical reaction: A reaction in which the original substances become new substances with new physical and chemical properties.

Circuit (SUHR kuht): The complete path through which electricity flows.

Closed circuit: A switch that is "on," or closed.

Coefficient (koe uh FISH unt) **of friction:** A number which is constant for friction (0.32) used to multiply any change in the property of friction under certain conditions.

Commutator: A device on a generator which reverses the current from AC to DC.

Compression: Something made smaller by applied pressure.

Compressor (kuhm PRES uh): Large air pump used for raising the pressure of air.

Concave reflector: A reflector with a surface that curves away from the light.

Condensation: The process of changing a gas to a liquid.

Conduction (kuhn DUHK shuhn): The transfer of heat through a substance from molecule to molecule.

Conductor (kuhn DUCK tuhr): A material which carries electric current.

Continuous spectrum (SPEK truhm): A spectrum that has no gaps or missing colors.

Controlled chain reaction: A chain reaction which is slowed down enough to be utilized in the production of electricity.

Controlled experiment: Experiment performed where all variables are controlled but one.

Controlled fusion reaction: The goal scientists are trying to attain in which a fusion reaction can be made to take place slow enough so that it can be managed.

Convection (kuhn VEK shuhn): The transfer of heat caused by moving the material that is heated.

Covergent light beams: Beams of light that come together and cross.

Convex reflector: A reflector with a surface that curves toward the light.

Cord: A pile of wood four feet wide, four feet high, and eight feet long. A cord is an English system unit and there is no metric equivalent.

Corpuscles (CORE puhs als): The particles that make up a stream of light.

Coulomb (KOO lohm): Unit expressing a quantity of electricity, equal to 6×10^{18} electrons.

Crankshaft: An arm or rod having one or more bends for transmitting motion.

Critical mass: The amount of U-235 necessary for a nuclear reaction to continue.

Cylinder: The round wall inside a pump or automobile engine in which the piston travels.

Decelerating (dee SELL er rayt ing): An object that is slowing down.

Decompose: The breaking down of a compound into other elements or compounds.

Dependent variable: The variable which responds to other variables.

Diffraction: The ability of waves to bend around objects in their paths.

Diffusion: The spreading out of molecules of gases and liquids.

Direct current: Electrons that move in one direction only.

Discharged: A charged object which can no longer attract.

Dissolving: Solid substances that pass into solution.

Distillation (dis tuh LAY shun): The process of separating or purifying liquids.

Divergent light beams: Beams of light that spread apart.

Efficiency (ih FISH un see): The ratio of useful work to actual work.

Effort: In mechanical advantage activities, the force required to move an object.

Electric cable: Wires which are encased in different layers of protective insulation which strengthens and protects the wires.

Electric current: A flow of electrons.

Electric fuse: An automatic switch which opens whenever the electrical current in a wire becomes dangerously large.

Electric potential: The energy of the electrons at the negative terminal of a dry cell.

Electron (ih LEK trahn): A negatively charged particle of an atom.

Electrolysis (ah lek TRAHL uh suhs): The passing of an electric current through a compound to separate the compound into simpler substances.

Electrostatic (ih lek truh STAT ik) induction a)in DUCK shuhn): Charging one object from another object without the two objects touching each other.

Element: A substance which cannot be separated into simpler substances using ordinary methods.

Endothermic (en doe THERM ik) **reaction:** A chemical reaction that absorbs energy.

Energy efficiency ratio: A measure of how effective an appliance uses the energy it comsumes.

Energy intensive industries: Industries, such as the aluminum and glass industries, that require large amounts of energy, usually electricity.

Escape velocity: A speed of 40 000 km/h which a rocket is accelerated to which allows it to completely escape from the earth's gravitational pull.

Evaporation (in vap uh RAY shun): The disappearance of water as things dry out.

Exhaust valve: The valve which allows burned gases to escape from the engine.

Exothermic (ex oh THERM ik) **reaction:** A chemical reaction that gives off heat.

Family: Term used to describe elements in the periodic table which are in the same column.

Filament: The fine wire that glows inside an incandescent lamp.

Fission: The splitting of an atom's nucleus.

Flame: Burning vapors.

Fluorescence (flur ES ence): Emitting light only when exposed to radiation.

Focal length: Distance from the focus point to the center of a lens.

Focus: The point at which reflected rays come together and cross.

Follow-through: In athletics, the act of continuing a swing or stroke to its natural end after releasing or striking an object.

Food calorie: See *large calorie*.

Fossil fuels: Coal, lignite, oil, or natural gas formed from the matter of animals and plants that live hundreds of thousands of years ago.

Free fall: Term used to refer to object which is accelerating constantly and which has only the force of gravity acting on it.

Frequency: The rate at which vibration takes place; also the rate at which an electric current changes direction.

Friction (FRIK shun): A force that opposes motion.

Fulcrum (FUL krum): The pivot point of a lever.

Fusion: Combining two or more atomic nuclei into a single heavier nucleus.

Galvanometer (gal vuh NAHM uh tuhr): An instrument which measures the flow of electricity through a circuit.

Gamma ray: A high-energy wave that has no charge or mass.

Generator: A machine which converts mechanical energy into electricity.

Greenhouse effect: Process by which short-wavelength radiations are trapped and radiated again as longer wavelengths.

Half-life: The time that it takes for a radioactive element to decay to half its original amount.

Halogens: A family of elements whose atoms all have less electrons than the stable number.

Ignition coil: An induction coil which interrupts direct current and produces alternating current.

Incandescent: Something that glows because it is hot.

Independent variable: The variable that is manipulated by the experimenter.

Infrared radiation: Light with longer wavelengths than red.

Inner transition elements: Elements 57-71 and 89-103 which are grouped in two rows at the bottom of the periodic table.

Insoluble: A substance which does not dissolve in a liquid.

Ions (I ahns): Electrically charged atoms.

Isotope (I suh tohp): Atoms of the same element with different numbers of neutrons in their nuclei.

Insulator: Materials that are poor conductors of electricity.

Intake valve: The valve which allows air and fuel to pass from the carburetor into the engine.

Interference: The behavior of waves where the motion of the waves cancel each other out.

Interrupted direct current: A flow of electrons that starts and stops, but does not change its direction.

Inverted image: An image that is upside down.

Joule: A unit of work equal to the work done by the force of one newton applied through a distance of one meter.

Kilowatt-hour: A large unit which measures thousands of watts of energy converted by a device in one hour.

Kinetic energy: Energy of motion.

Langley: A unit for measuring solar radiation, equal to one small calorie per cm^2.

Large calorie: The unit for measuring the amount of energy found in foods, equal to 1000 small calories (often spelled with a capital c).

Larynx (LAHR inks): Section of the throat which contains the vocal cords.

Level: A device for determining if a surface is on an even horizontal plane.

Magnetic field: The attractive force that a magnet exerts around it.

Magneto (mag NEE tow): A machine that has a magnet which produces a magnetic field to generate an electric current.

Manometer (muh NAHM uh ter): An instrument resembling a U-shaped tube used to measure the pressure of gases or liquids.

Mass: The property of objects which cause differences in acceleration.

Mho: The unit for measuring conductance.

Miniature: Greatly reduced, something made on a small scale.

Mirror image: An image that is reversed but not turned upside down.

Mixture: Two or more chemicals mixed together.

Molecule: The smallest part of a substance, such as salt, that can be identified as that substance.

Negatively charged atom: An atom which has more negatively charged electrons than there are positive charges in the nucleus.

Network circuit: An arrangement in an electrical substation where several lines run from the substation to supply electricity to the primary system.

Neutral atom: An atom which has the same number of negative and positive charges.

Newton: The metric unit of force needed to accelerate a 1-kilogram mass one meter per second faster every second.

Nichrome (NY krohm): A mixture of nickel, iron, and chromium.

Noble gases: A family of gases which do not combine chemically in nature.

Nonpolar: Term used to describe substances whose molecules have a similar shape.

Nuclear energy: The high energy possessed by neutrons during fission.

Nucleus (NOO klee us): The core of an atom which is made up of protons and neutrons.

Observing: Looking at something carefully.

Ohm: The unit for measuring electrical resistance.

Open circuit: A switch that is "off," or open.

Orbital velocity: A speed equal to 29 000 km/h where a projectile will stay aloft.

Oscilloscope: An instrument that visually displays an electrical wave on a fluorescent screen.

Outlet box: A steel box which is screwed shut, protecting electrical connections and preventing anything from touching the wires.

Oxidize: To unite with oxygen.

Parallax: The apparent change in the position of an object resulting from the change in the position from which it is viewed.

Parallel circuit: A circuit where the two conductors are connected to provide two separate paths for the current.

Perigee (PER ih gee): The point in a satellite's orbit which is closest to the earth.

Periodic table of the elements: A chart which organizes all the known elements.

Periodicity: Producing the same repeating pattern.

Photon (FOH tahn): A packet of radiant energy (light).

Photosynthesis (foe toe SIN the sis): Process by which plants use the energy of sunlight to chemically combine atoms of hydrogen, carbon, and oxygen into a complex sugar molecule containing energy.

Physical properties: The differences which can be used to separate chemicals in mixtures.

Plumb bob: A mass hung at the end of a string which is used to determine whether a structure is vertical.

Piston: That part of a pump which slides back and forth; also the round part within the cylinder of an automobile engine which moves up and down.

Pitch: The tone of a sound determined by the frequency of the wave; the higher the frequency, the higher the pitch.

Polar: Term used to describe substances whose molecules are not similar in shape.

Positively charged atom: An atom which has fewer negatively charged electrons than there are positive charges in the nucleus.

Potential drop: The loss in energy from one level to a lower level.

Potential energy: Stored energy.

Power: A ratio of work done divided by the time needed to do the work, of $\frac{work}{time}$.

Power stroke: The movement of a piston in a gasoline engine resulting from the explosion of the fuel.

Precipitate (prih SIP uh tayt): A solid which forms and separates from a solution.

Pressure: A push exerted on something.

Primary circuit: A connection made to an electrical current and a switch.

Prime mover: Any machine which converts a source of energy into work.

Protons: A positively charged particle in the nucleus of an atom.

Pulsating direct current: Direct current which has regular variations in magnitude.

Qualitative (KWAL li tay tiv) **analysis** (uh NAL uh sis): A process by which chemists find out what chemicals are in a substance.

Quantitative (KWAN to tay tiv) **analysis** (uh NAL uh sis): A process by which chemists find out how much of a chemical is present in a substance.

Radial circuit: An arrangement in an electrical substation where several lines run off the substation to serve various side branches.

Radiation (ray dee AY shuhn): The transfer of heat by light.

Radioactive: Unstable elements which break down, shooting off nuclear particles.

Radical: Part of a compound, consisting of several different atoms, which reacts as though it were a metal or nonmetal ion.

Rare earth series: A term used to describe the elements in the next-to-the-bottom row of the periodic table.

Rarefaction (rar uh FAK shuhn) **wave:** A wave of low pressure.

Resistance (rih ZIS tunts): Opposition which prevents or slows down motion.

Reaction: A force which squals an action force but moves in the opposite direction.

Reactor: A nuclear furnace.

Real image: Image formed by rays of light which are focused.

Reclamation: Reclaiming or repairing the land so that it is returned to its original condition.

Reed: A thin piece of wood used as a sound generator in an instrument such as a clarinet or saxophone.

Refraction: The bending of light as it passes from one substance to another.

Resistance (rih ZIS tunts): A device which reduces the flow of electricity.

Resonance: The vibration of two or more waves at the same frequency.

Retro-rockets: Rockets which fire in the opposite direction from which a space vehicle is traveling in order to slow it down.

Rotor: The spinning part of an alternator which contains field electromagnets.

Salt: A compound made up of a positive ion from a base and a negative ion from an acid.

Saturated: (SACH uh rayt uhd): Solution which has dissolved the maximum amount of the solid substance that it can at that temperature.

Secondary circuit: A connection made to an electric machine.

Series circuit: A circuit connected so that the current flows through all parts of it.

Short circuit: Disrupted electrical circuit in which a new circuit is created which has a very low electrical resistance.

Simple machines: Simple devices, such as inclined planes and levers, which can be used to exert large forces.

Single replacement reaction: A chemical reaction in which one element in a compound is exchanged with a different element.

Slip rings: Metal rings mounted on a rotor to lead current away from or into the rotor.

Small calorie: The amount of heat needed to raise the temperature of one gram of water one degree Celsius.

Solubility (sahl yuh Bill uht ee): The physical property of being able to dissolve.

Solute: A substance which is dissolved.

Solution: A liquid mixture.

Solvent: The substance the solute dissolves in.

Sound diaphragm (DI ah fram): A thin disk or cone that vibrates in response to sound waves.

Sounding board: A thin board set below a sound radiator to amplify the sound.

Sound radiator: Any vibrating object that sets up sound waves.

Spectrum (SPEK truhm): The different colors which form when white light is separated.

Stable: Term used to describe a family of elements which do not combine naturally with other elements to form compounds.

Starting distance: Applying a force through a certain distance (which varies) to accelerate an object.

Static electricity: Electricity produced by charged bodies in which the electrons do not move.

Stator: The stationary part of an alternator which contains coils.

Storage cell: Electric cells which can be recharged by passing a current through them.

Sublimation (suhb luh MAY shun): The direct change of a substance from the gaseous to the solid state.

Suction: The result of two unbalanced pressures which result in a push.

Supersaturated: A solution containing more solute than it would usually take to be saturated at a given temperature.

Synchronize (SIN kra nyze): Cause to agree in rhythm.

Tension (TEN shuhn): The force which tends to stretch.

Theory: An explanation of what is known about something.

Thermocouple (THUHR muh kup uhl): A device made of two different metals twisted together which is used to produce an electric current.

Thermodynamics (thur moe dy NAM iks): Science which deals with the changing of heat energy to another form of energy.

Thermostat: A device that regulates heating and cooling systems.

Torque (TORK): Turning action.

Total potential drop: The loss in energy from one level through several levels to the lowest level.

Transition elements: Elements grouped in the center of the periodic table which include metals with greater masses, such as iron (26) and mercury (80).

Trip: A type of circuit breaker in which the switch is held shut by a strip of metal.

Tungsten: A gray-white metal that has a high melting point and doesn't melt at white-hot temperatures.

Turbine: Engine with a central shaft that is connected to large blades.

Two-power lens: Lens which magnifies an object two times.

Uncontrolled chain reaction: An atomic bomb.

Universal motor: A type of motor which can be used with either direct or alternating current; commonly used in vacuum cleaners, mixers, and electric drills.

Valence (VAY lunce): The ability of an element or ion to combine in forming a compound.

Vibrate: The back-and-forth movement of an object in motion.

Visible spectrum: The colors made from light we can see.

Vocal cords: Cords in the larynx which vibrate to produce sound.

Volt: Unit of electromotive force, or potential.

Water turbine: A turbine which spins by the force of water hitting the blades.

Watt: The unit for measuring the rate of electrical energy conversion, equal to one joule of work per second.

Watt-second: The amount of energy converted by a device in one second.

Wet cell: An electric cell which consists of two different metals in a liquid.

ACKNOWLEDGMENTS

Illustrators for this edition:

Lee Ames Robert Dustin Mel Erikson Paul Field
Frank Schwarz Marilyn Dustin Holly Moylan Lewis Johnson
Andre LaBlanc Anthony D'Adamo Ben Palagonia Walter Storozuk

Photographs are credited below. (Photographs credited to A & B have been taken by Talbot D. Lovering, the Allyn and Bacon Staff Photographer.)

UNIT 1—CHAPTER 1

P. 1—A & B. 2–3—Robert E. Kilburn. 4 *top*—Courtesy of the National Society for the Prevention of Blindness; *middle and bottom*—A & B. 5 *top*—Peter S. Howell; *bottom*—A & B. 6 *both*—A & B. 7 *both*—A & B. 8 *all*—A & B. 9 *both*—A & B. 10 *all*—A & B. 11 *top left*—Robert E. Kilburn; *top right*—A & B; *bottom left*—A & B; *bottom right*—Robert E. Kilburn.

CHAPTER 2

P. 13—Dr. E. R. Degginger. 14—A & B. 15 *both*—A & B. 20 *top left*—John H. Gerard; *top right*—A & B; *bottom*—A & B. 21 *all*—A & B (Antique teakettle courtesy of Gene A. Moulton). 23—Terry McKoy. 24 *top*—Cornelius Mead from Editorial Photocolor Archives; *middle*—Roger J. Cheng, Atmospheric Sciences Research Center, SUNY; *bottom*—A & B.

CHAPTER 3

P. 25—A & B. 26—City of Pittsburgh/Department of Water. 32—B. M. Shaub. 33 *both*—A & B. 35—A & B. 36 *top*—Phil Degginger; *middle and bottom*—A & B.

CHAPTER 4

P. 37—BBC, Broadcasting House, London W1A 1AA; 40—A & B. 42 *top*—Chemetron Corporation; *bottom*—A & B. 44—White Motor Corp. 46—A & B. 53—A & B. 56 *all*—A & B.

CHAPTER 5

P. 59—A & B. 60—A & B. 61 *both*—A & B. 62 *top*—Old Sturbridge Village Photo; *middle*—A & B; *bottom*—Courtesy of the A. I. Root Company. 64 *both*—Fabric Research Laboratories, Inc. 66—A & B. 67—Fisher Scientific Company.

CHAPTER 6

P. 69—Peter Schweitzer; *insert*—Central Scientific. 70 *top and bottom*—'Spectrum Chart,' Courtesy of Welch Scientific Company; *next-to-top*—Sargent-Welch Scientific Company. 71 *top*—'Spectrum Chart', Courtesy of Welch Scientific Company; *next-to-top*—A & B. 72 *neon spectrum*—Bausch & Lomb; *mercury, strontium, barium, calcium, and sodium spectra*—'Spectrum Chart', Courtesy of Welch Scientific Company; *bottom*—A & B. 73 *top*—'Spectrum Chart', Courtesy of Welch Scientific Company; *next-to-top*—Bausch & Lomb; *all others*—A & B. 74 *top*—Bausch & Lomb; *bottom*—'Spectrum Chart', Courtesy of Welch Scientific Company. 75 *top*—Bausch & Lomb; *middle*—Courtesy of Beckman Instruments, Inc.; *bottom*—F. B. I. 76 *top and next-to-top*—Bausch & Lomb; *bottom*—Yerkes Observatory Photograph. 77—Yerkes Observatory Photograph. 78 *all*—Mount Wilson & Palomar Observatories. 79 *all*—Mount Wilson & Palomar Observatories. 80 *top two spectra*—'Spectrum Chart', Courtesy of Welch Scientific Company; *third spectrum*—Courtesy of Bausch & Lomb; *middle*—Danish Information Office; *bottom spectrum*—Courtesy of Bausch & Lomb.

CHAPTER 7

P. 81—A & B. 82—Brown Brothers. 88—A & B. 91 *top*—Brookhaven National Laboratory; *bottom*—University of California, Lawrence Berkeley Laboratory. 96—Stanford University.

UNIT 2—CHAPTER 1

P. 97—Jerry Irwin. 98—99—Robert P. Foley. 100—A & B. 102—A & B. 104—A & B. 112 *both*—A & B. 114—Morris Rosenfeld. 116 *both*—A & B.

CHAPTER 2

P. 119—A & B. 120—Peter Travers. 121 *middle*—Jennifer M. Phillips, TNT, Orange, Mass.; *bottom*—Jerry Irwin. 123—Charles G. Kulick. 124 *left*—Dr. Harold E. Edgerton; *right*—A & B. 125 *top*—Central Press Photos—Pictorial Parade; *bottom*—Jacques Jangoux. 126—Dr. Harold E. Edgerton. 127—Dr. Harold E. Edgerton. 128 *top*—Courtesy of Wilson Sporting Goods, Inc.; *middle*—H. Armstrong Roberts; *bottom*—Los Angeles Dodgers. 129—TOM STACK & ASSOCIATES, Julian E. Carabello. 131 *top*—DPI—Jules Zalon; *middle*—Sports Camera West/Joyce R. Wilson; *bottom*—London Daily Express—Pictorial Parade. 132 *top and middle*—H. Armstrong Roberts; *bottom*—Weston Kemp. 134—Harold Lambert. 135 *top*—Sports Camera West/Paul Landfried; *middle*—H. Armstrong Roberts; *bottom*—M. Vanderwall/Leo de Wys, Inc. 137—NASA. 138—A & B.

CHAPTER 3

139 *top left*—Peter S. Howell; *top right*—A & B; *bottom right*—A & B. 140 *both*—Peter S. Howell. 144 *all*—Robert E. Kilburn. 145 *both*—Peter S. Howell. 153—A & B. 155—A & B. 156—A & B. 160 *middle*—A & B; *bottom*—Peter S. Howell.

CHAPTER 4

P. 163 *left*—Walker-Gordon Certified Milk Farm; *right*—Rolls-Royce Aero Engines, Inc. 164—Photo Science Museum, London. 167—A. Devaney, Inc. 168 *top*—Peter S. Howell; *bottom*—A & B. 169—Howard Levy.

CHAPTER 5

P. 171—New York State Department of Commerce.

CHAPTER 6

P. 180—NASA. 181—Dr. Harold E. Edgerton. 182—Sol Libsohn. 183—From PSSC Physics, D. C. Heath and Company, Lexington, MA 1965. 188—NASA. 190—NASA. 192 *both*—NASA.

UNIT 3—CHAPTER 1

P. 193—Keith Murikami/TOM STACK & ASSOCIATES. 194–195—Jones and Laughlin Steel Corporation. 199—Courtesy of the Illinois Department of Transportation. 200—Burlington Northern. 206—Edwin L. Shay. 207 *both*—A & B. 210—Peter S. Howell. 215 *top*—Joseph Miller; *bottom*—Southern Pacific Photo.

CHAPTER 2

P. 217—Richard A. Chase. 218—John A. Rizzo. 223—Whitestone Photo. 224—A & B. 225—A & B. 229—A & B. 231—Boris & Milton. 232 *top*—Betsy Cole; *bottom*—Richard Chase. 234 *both*—Peter S. Howell. 235—Berklee College of Music. 236 *across bottom*—Courtesy of Bell Telephone Laboratories, Inc.

CHAPTER 3

P. **239**—High Altitude Observatory/National Center for Atmospheric Research. **240** top—U. S. Forest Service; bottom —A & B. **246** both—A & B. **247**—A & B. **248**—Weston Kemp. **249** center—General Electric Company; bottom— William G. Gerow, Gerow Sound & Light systems, San Diego, CA. **253**—Eastman Kodak Corporation. **254**—Monkmeyer/Patrick Morin. **256**—top and bottom—Bausch & Lomb. **258** both—A & B. **260** both—A & B.

CHAPTER 5

P. **271** top left and right—B. M. Shaub; bottom left—Acorn Structures, Inc.; bottom right—Donald Johnson & Associates. **272**—Bausch & Lomb. **273**—From PSSC Physics, D. C. Heath and Company, Lexington, MA 1965. **274** top—A & B; bottom—From PSSC Physics, D. C. Heath and Company, Lexington, MA 1965. **275**—From PSSC Physics, D. C. Health and Company, Lexington, MA 1965. **280** top and middle— Laurence Lowry; bottom—U. S. Army Cold Regions Research and Engineering Laboratory.

CHAPTER 6

P. **281** top left—A & B; top right—Worcester Polytechnic Institute; bottom left—The Asphalt Institute; bottom right— Courtesy of E. I. duPont de Nemours & Co. **283** top—A & B; next-to-top and next-to-bottom—Peter S. Howell; bottom— A & B. **284** top—A & B; center left—Northeast Metropolitan Regional Vocational School District; center middle—A & B; center right—Northeast Metropolitan Regional Vocational School District; bottom left—Peter Schweitzer; bottom right —National Education Association Publishing, Joe Di Dio. **285**—Northeast Metropolitan Regional Vocational School District. **286** top—Courtesy of E. I. duPont de Nemours & Co.; bottom—A & B; right—Worcester Polytechnic Institute; far right—Peter Schweitzer. **287** left—Brookhaven National Laboratory; center top—Courtesy of E. I. duPont de Nemours & Co.; center bottom—Hale Observatories; right—Peter Schweitzer.

UNIT 4—CHAPTER 1

P. **289**—Peter H. Dreyer. **290–291**—Robert E. Kilburn. **293** —A & B. **301**—A & B. **302**—A & B.

CHAPTER 2

P. **303**—Peter S. Howell. **305**—A & B. **314**—Courtesy of Salt River Project. **317** top left—Ford Motor Company; top right —Peter Schweitzer; bottom—Central Scientific Company. **318** left—Weston Kemp; right—Peter Schweitzer. **322**— NASA.

CHAPTER 3

P. **323**—A & B. **325** both—Robert E. Kilburn. **327**—General Electric Company. **328**—Courtesy of Bell Telephone Laboratories, Inc. **330**—Westinghouse Electric Corporation. **333** top left—A & B; top right—Peter Schweitzer; next-to-bottom— Black and Decker; bottom—Ametek, Lamb Electric Division. **335**—Ford Motor Company. **339**—Peter Schweitzer.

CHAPTER 4

P. **341**—A & B. **342** both—General Electric Company.

CHAPTER 5

P. **351**—A & B. **358** top and middle—Dave Lawlor; bottom —Ford Motor Company. **360**—Boston Edison Co.

CHAPTER 6

P. **361**—Peter S. Howell. **362** top—David H. Shepard; bottom—Courtesy International Harvester Company. **363**— Allis-Chalmers Corporation. **364**—General Electric Company. **365**—Boston Edison Co. **366**—Boston Edison Co. **367** all—Peter S. Howell. **369**—Peter S. Howell. **370**—Peter S.

Howell. **371**—Peter S. Howell. **373**—General Electric Company.

CHAPTER 7

P. **375**—Peter S. Howell. **377**—Peter S. Howell. **378** all— Encyclopaedia Britannica Films, Inc., 'Elements of Electrical Circuits.' **383**—Terry McKoy. **384** top—Nationwide Insurance Company; middle—Peter S. Howell; bottom—Boston Edison Co.

UNIT 5—CHAPTER 1

P. **385**—EPA-DOCUMERICA–Dick Swanson. **386–387**— Boyer, Gamma-Liaison. **388** top—A & B; bottom—William R. Radomski. **390**—William R. Radomski. **392** top—Robert A. Tyrrell; second row left—Robert E. Kilburn; second row middle—Mary S. Shaub; second row right—A & B; third row left—Photo by Bill Osmun for Air Transport Association; third row middle—U. S. Department of Energy; third row right—Celestron Pacific; bottom left—Electro-Motive Division, General Motors Corporation. **393** top—William R. Radomski; bottom—A & B. **394** top—Ewing Galloway; bottom—Peter S. Howell. **395** top—Vulcan Iron Works, Inc.; bottom—U. S. Department of the Interior, Bureau of Reclamation. **397**—A & B. **399** top—A & B; bottom—State of Indiana, Dept. of Commerce.

CHAPTER 2

P. **402**—Robert E. Kilburn. **404**—A & B. **406**—Reynolds Aluminum. **407**—Atlanta Chamber of Commerce. **409–410** —Manolo Guevara, Jr. **410**—Robert E. Kilburn.

CHAPTER 3

P. **413** both—Peter S. Howell. **414**—A & B. **416** both—Peter S. Howell. **418** both—Robert E. Kilburn. **420**—Solar Power Corporation. **422**—Courtesy of AGA Corporation. **425**— Peter S. Howell.

CHAPTER 4

P. **427**—Peter Gridley, FPG. **429**—A & B. **434**—Robert E. Kilburn. **438** top left—Dennis O'Reilly; top right—A & B; bottom left and right—A & B. **439** all—A & B. **440** top—A & B; bottom—United Nations/UNICEF/D. Mangurian.

CHAPTER 5

P. **441**—A & B. **442**—A & B. **444** left—Courtesy of M. I. T.; right—General Electric Company/Space Division. **445** both— A & B. **448** top—Peter S. Howell; bottom—A & B. **451** both —Robert E. Kilburn. **452**—Robert E. Kilburn.

CHAPTER 6

P. **453** top left—Portland General Electric Company; top middle—Peter Smolens; top right—J. C. Allen & Son; bottom left—Photo Courtesy of Consolidation Coal Company; bottom right—American Iron and Steel Institute. **454** top— The Harry T. Peters Collection/Museum of the City of New York; bottom—Old Sturbridge Village Photo. **464**—B. M. Shaub. **465** top—U. S. Coast Guard Official Photo; bottom— Laurent MAOUS/Gamma-Liaison. **466** top—Edwin L. Shay; bottom—Carnegie-Illinois Steel Corporation.

CHAPTER 7

P. **468** top—Courtesy of Volunteers in Technical Assistance; bottom—Peter S. Howell. **469**—Acorn Structures, Inc. **470** top—The Bettmann Archive, Inc.; bottom—NASA. **471**— Pacific Gas and Electric Company. **472**—French Embassy Press & Information Division; **473** top—Chicago Historical Society; bottom—M. J. Schmidt. **476–477**—Official U. S. Navy Photograph. **478** top—Commonwealth Edison Company; middle—Los Alamos Scientific Laboratory; bottom— Battelle-Northwest. **479** top—Wide World Photo; bottom— Photo Courtesy of Princeton University Plasma Physics Laboratory, Princeton, New Jersey. **481**—A & B.

INDEX

Numerals in boldface (**142**) indicate the page on which the terms are identified or explained.

Fuel cells, 419
Fulcrum, **157**
Fusion, **479**

Galileo, 278
Galvanometer, **310**, 322
Gamma rays, **90**–91
Gases, effect of heat on, 261–263; in forming oxygen, 42–43; in jet engines, 137; leaving solution, 34–35; reaction when formed, 56; in rusting and burning, 41; in tap water, 35
Gasoline, consumption, 408–409; energy in, 408
Gasoline engines, 175–179
Generators, **314,** 315; alternating current, 366–367; direct current, 318
Geothermal energy, 471
Graphs, analyzing, 147
Gravity, 119, 139, 142, 144, 389; around the world, 143; effect on force, 120; effects of, 144; escaping earth's, 190; measuring, 143; and moon's orbit, 182; slope and the effect of, 144
Greenhouse effect, 469

Half life, of element, **91**
Halogens, **93**
Hardwoods, 474
Heat, breaking down copper sulphate, 48; breaking down pyrite, 45; breaking down wood molecules, 45; capacity, 213–216; and changes in volume, 202–207; from chemical energy, 210, 425; of combustion, 210; effect on air molecules, 15, 16; effect on liquids, 202, 203; from electrical energy, 209, 425; electricity from, 321; energy needs for, 403–405; and expansion, 199–201; from friction, 424; and gas, 261–263; and ice, 17; measuring energy of, 424–427; safety rules for, 7; solar energy for, 469; from steam, 212; and temperature, **213**
Heat energy, converting to, 416–417; and evaporation, 212; exchange of, 209; from friction, 208–209; losses at

home, 452; measuring, 208–212
Heat transfer, 196–198
Helium, **76**
Herschel, William, 279
Hot glass, working safely with, 8
Huygen, Christian, 274
Huygen's Wave Theory, 274
Hydrogen, spectrum, 71; test for, 50
Hydrogen ions, replacing with metal ions, 52
Hydrogen peroxide, 46

Ice, changing to steam, 210–211; cooling effect of, 211; heating, 17
Ignition coil, **339**
Image, apparent, **256;** changes in size, 242; explaining, 242; inverted, **247;** locating, 246; mirror, **246;** position of, 245; real, **256;** size and distance of, 257
Incandescent, **326**
Inclined planes, 151
Incomplete knowledge, 95
Independent variable, **352**
Induced charge, **309**
Induction, **309;** attraction by, 309; coils, 339; electromagnetic, 336–337; magnetic, 314; motors, 335
Infrared light, 70, 71
Infrared radiation, **279**–280
Inner transition elements, **88**
Insoluble, **28**
Insulation, 437
Insulators, 299
Intake valve, **177,** 178
Interference, in waves, 272, **275**
Interrupted direct current, **319**
Inverted image, **247**
Investigations and projects, 66–68, 161–170, 261–270, 351–360, 441–452
Ionic theory, 94
Ions, **47;** in electric cells, 312–313; noble gas neighbors as, **94;** and radicals, 54; theory of, 50–51
Iron sulfate, solution of, 52
Isotopes, **89**

Joule, James Prescott, 151, 390
Joule, 151, **390,** 395

Jupiter's moon, Roemer and, 277–278

Kilowatt hour, **349,** 373
Kinetic energy, **393**–395; calculating, 395

Labeling, of containers, 8
Laboratory-made elements, 91
Laboratory work, techniques of, 10; using safety in, 1–12
Lamps, bulb of, 327; incandescent, 326; in model home, 293–294, 298; sockets, 293, 298, 299
Langley, **421**
Larynx, **236**
Law of Conservation of Energy and Mass, 391
Lawrence, E.O., 91
Lenses, camera, 256–257, 258; defects of, 256; field of view, 259; focal length of, 255; liquid, 259; magnifying, 258, 259–260; reducing, 258; shape and behavior, 255; two-power, **259**
Level, **144**
Levers, 152–153; efficiency of, 159; mechanical advantage of, 158; types of, 158, 159
Life styles, in America, 438–439
Light, absorption of, **243;** ancient beliefs about, 272; from atoms, 69–79; behavior of, 239–260; bending of, 250–260, 273; comparing brightness of, 268–269; convergent and divergent beams of, 254–**255;** diffraction of, **274;** direct, 243; distance traveled, 241; effect of distance on brightness, 268; electricity from, 321; flourescent, 73; as form of energy, 271; Huygen's Wave Theory of, 274; infrared, 70; intensities of, 241; locating a beam of, 244; measuring speed of, 277, 278–279; modern theory of, 276; Newton's Corpuscular Theory of, 272–274; outward spread of, 240; and photosynthesis, 417; reflected, 243, 272; refraction of, **251**–253; rising